Maulin P. Shah (Ed.)
Environmental Microbiology

Also of Interest

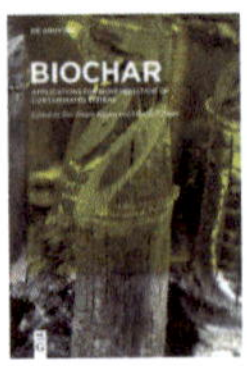

BioChar.
Applications for Bioremediation of Contaminated Systems
Edited by Amit Kumar, Puneet Pathak, Dharm Dutt, 2022
ISBN 978-3-11-073858-2, e-ISBN 978-3-11-073400-3

Industrial Waste.
Characterization, Modification and Applications of Residues
Edited by Herbert Pöllmann, 2021
ISBN 978-3-11-067486-6, e-ISBN 978-3-11-067494-1

Photosynthesis.
Biotechnological Applications with Micro-Algae
Edited by Matthias Rögner, 2021
ISBN 978-3-11-071691-7, e-ISBN 978-3-11-071697-9

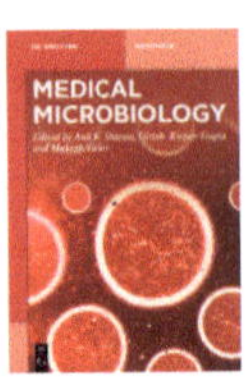

Medical Microbiology.
for Water and Wastewater Treatment Applications
Edited by Anil K. Sharma, Girish Kumar Gupta, Mukesh Yadav, 2022
ISBN 978-3-11-051764-4, e-ISBN 978-3-11-051773-6

Green Chemistry.
Principles and Designing of Green Synthesis
yed Kazim Moosvi, Waseem Gulzar Naqash, Mohd. Hanief Najar, 2021
ISBN 978-3-11-075188-8, e-ISBN 978-3-11-075189-5

Bioelectrochemistry.
Design and Applications of Biomaterials
Edited by Serge Cosnier, 2019
ISBN 978-3-11-056898-1, e-ISBN 978-3-11-057052-6

Environmental Microbiology

Emerging Technologies

Edited by
Maulin P. Shah

DE GRUYTER

Editor
Dr. Maulin P. Shah
Environmental Microbiology Lab
Opp. Champapuri Jain Mandir
A/103 Satsang Park
Ankleshwar-393002
India
maulinshah1979@gmail.com

ISBN 978-3-11-072721-0
e-ISBN (PDF) 978-3-11-072722-7
e-ISBN (EPUB) 978-3-11-072739-5

Library of Congress Control Number: 2022936819

Bibliographic information published by the Deutsche Nationalbibliothek
The Deutsche Nationalbibliothek lists this publication in the Deutsche Nationalbibliografie; detailed bibliographic data are available on the Internet at http://dnb.dnb.de.

Typesetting: Integra Software Services Pvt. Ltd.
Printing and binding: CPI books GmbH, Leck
Cover image: Inkoly/iStock/Getty Images Plus

www.degruyter.com

Contents

List of authors by chapter —— VII

M. Kouselya, Muhilan B. M. and Indranil Chattopadhyay
1 Omics approaches for characterization of environmental microorganisms —— 1

Illathu Kandy Nidhin and Indranil Chattopadhyay
2 Emerging technologies in environmental microbiology —— 33

Nelson Libardi Junior, Nathan Pacheco Amin Vieira da Costa, Rejane Helena Ribeiro da Costa, Carlos Ricardo Soccol and Luciana Porto de Souza Vandenberghe
3 Pollutants removal using aerobic granular sludge technology —— 59

Sreejita Ghosh, Dibyajit Lahiri, Sougata Ghosh, Moupriya Nag and Rina Rani Ray
4 Biofilm-mediated industrial wastewater treatment —— 79

Getu Kassegn Weldegebrieal and Assefu Kassegn Sibhatu
5 Photocatalytic degradation of organic contaminants in wastewater treatment —— 113

Sarangi N. V. and Rajkumar R.
6 Biodegradation of organic pollutants by microbial process —— 137

Satarupa Dey and A. K. Paul
7 Bauxite residue (red mud): microbial resources and their possible exploitation in remediation and rehabilitation —— 161

Oluka N. Ngofa and George Z. Kyzas
8 Advancements in application of dispersants to oil spills —— 185

Arundhati Pal and A. K. Paul
9 Serpentinophytes-associated microbiota: a review of applications in environmental biotechnology —— 221

Komal Agrawal and Pradeep Verma
10 Microbial desalination cell: a versatile insight towards sustainable contribution —— 249

Can Burak Özkal
11 Control of antibiotic resistance by advanced treatment: recent advances, in a nutshell — 265

Prasann Kumar, E. Lokesh Goud, Priyanka Devi
and Bhupendra Koul
12 Metal pollutants in the environment — 291

Atif Aziz Chowdhury, Nilendu Basak, Taniya Roy and Ekramul Islam
13 The language of gram-positive and gram-negative bacteria — 325

Index — 341

List of authors by chapter

M. Kouselya, Muhilan B. M., Indranil Chattopadhyay
Department of Life Sciences,
Central University of Tamil Nadu,
Thiruvarur, Tamil Nadu,
India
indranil@cutn.ac.in

Illathu Kandy Nidhin, Indranil Chattopadhyay
Department of Life Sciences,
Central University of Tamil Nadu,
Thiruvarur, Tamil Nadu,
India
indranil@cutn.ac.in

Nelson Libardi Junior, Nathan Pacheco da Costa, Rejane Helena Ribeiro da Costa, Carlos Ricardo Soccol, Luciana Porto de Souza Vandenberghe
Departamento de Engenharia de Bioprocessos e Biotecnologia, Universidade Federal do Paraná – UFPR, Curitiba-PR,
Brasil, 81531-980
lvandenberghe@ufpr.br

Sreejita Ghosh, Dibyajit Lahiri, Sougata Ghosh, Moupriya Nag, Rina Rani Ray
Department of Biotechnology,
Maulana Abul Kalam Azad University of Technology,
Haringhata, West Bengal,
India
raypumicro@gmail.com

Getu Kassegn Weldegebrieal
Department of Chemistry,
College of Natural and Computational Sciences,
Debre Berhan University,
Ethiopia
gk21@dbu.edu.et

Sarangi N. V, Rajkumar R.
Department of Environmental Sciences,
Bharathiar University,
Coimbatore-641 046,
Tamil Nadu
micro_rajkumar@yahoo.co.in

Satarupa Dey, A. K. Paul
Microbiology Laboratory
Department of Botany University of Calcutta
Kolkata,
India
amalk_paul@yahoo.co.in

Oluka N. Ngofa, George Z. Kyzas
Department of Chemistry,
International Hellenic University,
GR-654 04 Kavala,
Greece
kyzas@chem.ihu.gr

Arundhati Pal, A. K. Paul
Microbiology Laoratory,
Department of Botany,
University of Calcutta, Kolkata
India
amalk_paul@yahoo.co.in

Komal Agrawal, Pradeep Verma
Bioprocess and Bioenergy Laboratory,
Department of Microbiology,
Central University of Rajasthan,
Bandarsindri, Kishangarh,
Ajmer Rajasthan, 305817,
India
pradeepverma@curaj.ac.in

Can Burak ÖZKAL
Department of Environmental Engineering,
Tekirdağ Namik Kemal University,
Çorlu 59860, Tekirdağ,
Turkey
cbozkal@nku.edu.tr

https://doi.org/10.1515/9783110727227-203

Prasann Kumar, E. Lokesh Goud, Priyanka Devi, Bhupendra Koul
Department of Biotechnology,
School of Bioengineering & Biosciences,
Lovely Professional University,
Jalandhar, Punjab,
India, 144411
bhupendra.18673@lpu.co.in

Atif Aziz Chowdhury, Nilendu Basak, Taniya Roy, Ekramul Islam
Department of Microbiology,
University of Kalyani, Kalyani,
Nadia, West Bengal,
India, 741235
ekramul.rs@gmail.com

M. Kouselya, Muhilan B. M. and Indranil Chattopadhyay

1 Omics approaches for characterization of environmental microorganisms

Abstract: Microorganisms constitute two of the three domains of life on the Earth. Microbiota plays a critical role in the ecosystem and human health. They exhibit huge biodiversity and metabolic versatility that allows them to abide and survive even in the harshest environmental conditions, making them all ubiquitous and be efficient and inexhaustible tools for the discovery of novel bioactive molecules. Although they possess ample potential, about 99% of the microbial world remains to be discovered. The primary explanation for this is that in vitro cultivation of these microorganisms remains fractious. The study of microbial heterogeneity and population dynamics has fundamentally been reshaped by the development of culture-independent techniques. The use of omics approaches by genomics, metagenomics, transcriptomics, proteomics, and metabolomic techniques contributes significantly to the discovery and characterization of a real microbial community. These omic techniques can revolutionize biological treatment in environmental engineering through extremely sensitive characterization of already uncultured microorganisms. As an environmentally friendly and economical replacement to minimize the load of pollution, biological approaches using microbes are gaining importance. New omic technologies such as metagenomics, metatranscriptomics, metaproteomics, and metabolomics now have indeed been utilized to develop methods to examine ecology and microorganism heterogeneity and to discover novel microorganisms for use in bioaugmentation, environment surveillance, and bioremediation. To obtain an accurate picture of living microorganisms, convergence of various layers of data, the "multi-omics" strategy, is therefore necessary. In this chapter, we present the evolving approaches and advanced tools for microbial population evaluation that support and extend conventional metagenomic profiling. They comprise measuring omic data types like metabolomics, proteomics, transcriptomics, genomics, and integrating various forms of multi-omic data in a combined framework for characterization of environmental microorganisms.

Keywords: Omics, genomics, proteomics, transcriptomics, metabolomics, microorganisms

M. Kouselya, Muhilan B. M., Dept. of Life Sciences, Central University of Tamil Nadu, Thiruvarur, Tamil Nadu, India
Indranil Chattopadhyay, Dept. of Life Sciences, Central University of Tamil Nadu, Thiruvarur, Tamil Nadu, India, e-mail: indranil@cutn.ac.in, indranil_ch@yahoo.com

https://doi.org/10.1515/9783110727227-001

1.1 Introduction

In the field of biology, the word "'ome" usually refers to the totality or entirety of a set of unique things. For example, a biome is a set of living organisms, and a genome is a collection of genes within a single organism. 'Omics, then, are fields of study which deal with these collections and require the characterization and consideration of numerous molecules concurrently [1]. It refers to the characterization or quantification of a specific molecule that is normally biological in nature to detect the structure, function, molecular mechanisms, and physiology of a set of organisms [2]. The 'omics theory, however, goes far beyond DNA and may include transcripts of RNAs, proteins, and metabolites, also referred to as "omics cascades." And in this cascade:

1) the genome (or metagenome) provides details about what could happen (i.e., the functional potential);
2) the transcriptome (or metatranscriptomic) provides data about what happens to be occurring (i.e., which genes are all being expressed);
3) the proteome (or metaproteome) provides knowledge about molecules that actually make things happen;
4) the metabolome provides details on what has already happened or is occurring currently [1].

Metagenomics is designed to research organisms whose genetic materials could be specifically evaluated from samples retrieved from the field, while transcriptomics and proteomics are indeed a study of the function, components, and structure of various RNA and protein molecules [3]. Metagenomics based on microarray and sequencing is by far the highest saturated method of all metagenomics research, offering consistent and straightforward knowledge on other "omics" technologies [2]. While the 'omics cascade encompasses many of the main "omics" disciplines within review today, several other "omics" have appeared in recent times. Few of them seem to be subdisciplines of the broad "omics" fields mentioned before (e.g., glycomics, intercomics, and lipidomy), whereas others remain new topics but have yet to be accepted as stand-alone fields in modern science. In addition, the area of bioinformatics has been built to include the mathematical and analytical methods required to analyze the increasingly broad and complex datasets that "omics" technologies produce [1]. Recent developments in molecular tools have allowed the cutting-edge evaluation of unculturable species from natural ecosystems to take place [4]. Molecular methods such as fluxomics, metabolomics, proteomics, transcriptomics, and genomics have provided more insight into the microbial populations occupying a specific environment [5, 6]. The advent of next-generation sequencing (NGS) technologies and in-silicon studies has facilitated scientists to resolve issues related to unculturable microorganisms [7]. Such developments also improved the processing of comprehensive biological data on microbes, biomolecules, such as enzymes, and their molecular processes correlated with bioremediation [8, 9]. Due to ground-breaking advances in high-performance DNA sequencing technology, closer to 1,000 microbial genomes from most of the recognized main phylogenetic groups, have also been sequenced completely and several are approaching finalization [10]. Comparative genomic and computational annotation study of DNA sequences supported scientists with knowledge on genome functions, gene regulation, biological mechanisms, the development of microbial genomes, and regulatory and metabolic networks which significantly improved our knowledge on metabolism of microbes [11]. It is important to provide precise quantification and functional characterization of microbial mRNAs, proteins, and metabolites along with their interactions with the environments. Important attempts have been made over the last decade to develop analytical techniques for the measurement of mRNA, proteins, and metabolites. These activities have contributed to the development of many new areas of "omics" studies: transcriptomics, proteomics, metabolomics, inter-comics, and so on [12]. Figure 1.1 shows different types of omics techniques for microbial community analyses. Generally, in comparison to conventional methods, all experimental "omics" techniques can be found to possess three main characteristics. The very first aspect would be that, unlike conventional techniques, "omics" methods are elevated, information-driven, comprehensive, and top-down techniques; the second attribute is that these high-performance "omics" methods produce vast volumes of data, and the study of these data also needs considerable mathematical and analytical effort. Till date, a significant amount of knowledge on cellular metabolism was being gained by the use of human "omics" methods [13]. The final aspect is an effort to study cell metabolism as a single "integrated mechanism" instead of a simple set of separate parts using knowledge on the relationship among multiple determined molecular organisms [12]. Although high-throughput omics methods to study substances at various cellular stages are increasingly being available, it has also become apparent that any particular "omics" method might not be adequate to describe the complexities of metabolism [14]. Embedded multi-"omics" methods have currently been introduced, and studies have helped scientists to uncover the intricate metabolic networks and universal regulatory mechanisms in different eukaryotic species [15]. Although further work also needs to done to develop testing protocols and computational methodology such that the interdisciplinary study of several huge "omics" databases could be used to produce new information that cannot be reached by an individual data from analysis alone, such initial research has shown that comprehensive omics assessment could be the main solution to decipher intricate biological datasets [1]. In this chapter, we discuss the main 'omics-based approaches commonly used to characterize environmental microorganisms, and also the methods for analyzing and interpreting the "bioinformation" which these studies produce.

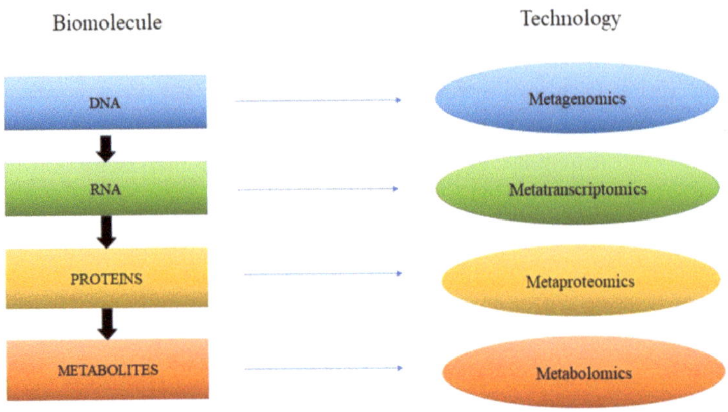

Figure 1.1: Different types of omics technique for microbial community analyses.

1.2 Metagenomics

The scientific community has extended the concept of metagenomics to generally include any technique centered on analyzing DNA derived from environmental samples [1]. Metagenomic-based assays have become the standard for characterizing microbial communities over the past two decades and have been used in numerous studies to investigate the structure, function, and metabolic capacity of microbial communities in a wide range of environments [1]. Due to the heterogeneity of microbial communities in these habitats, and the lack of a good collection of reference sequences from a broad microbial community to function as a scaffold for assembling the sequences, the integration of metagenomics data from environmental samples is extremely challenging currently [17]. Metagenomics data collection is typically restricted to only highly dominant members of simple groups, such as those in the drainage of acid mines or polluted groundwater. The relatively short read length (500 bp) of many commonly used sequencing methods is yet another difficulty for assembly [1]. The advancement of newer sequencing technologies, such as Pacific Biosciences, however, promises the potential to provide longer reads (3,000 bp) covering entire genes and probably even operons, thus allowing improved taxonomic classification and/or functional prediction [1]. An overview of metagenomic approaches is displayed in Figure 1.2.

Few metagenomic techniques are discussed below.

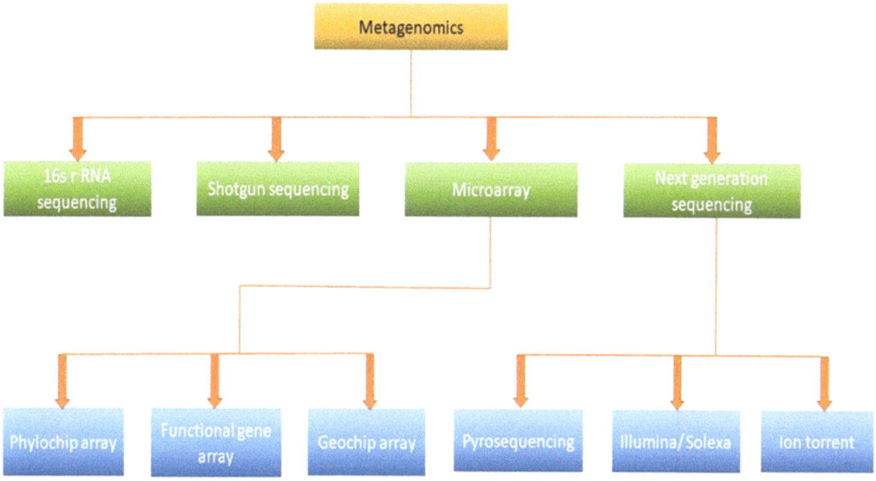

Figure 1.2: Overview of metagenomic approaches.

1.2.1 16S rRNA

In all microorganisms, the identification of a highly conserved and variable 16S rRNA gene sequence is considered the standard for defining phylogenetic similarity between individuals in microbial communities [18]. For a thorough evaluation of microbial diversity, 16S rRNA gene sequence analysis can be employed by specifically amplifying and sequencing the 16S rRNA gene's hypervariable regions. This technique is a hugely versatile and also cost-efficient technology that numerous bioinformatic tools can easily access and has emerged as a widely employed method for characterizing complex microbial populations [19]. This can be employed to classify microbes that are novel, unculturable, and morphologically unrecognizable [20]. A major breakthrough for microbial ecology studies was the use of the 16S ribosomal RNA gene as a phylogenetic marker, with many culture-independent techniques being introduced after the direct cloning of environmental DNA [21]. In bacteria, the 16S rRNA gene consists of interspersed variable and conserved sequences, namely eight hypervariable regions (V1–V8). These regions contained nucleotides 69–99, 137–242, 433–497, 576–682, 822–879, 986–1,043, 1,117–1,173, and 1,243–1,294, respectively, for V1–V8 [22]. The size of these hypervariable regions ranges from about 50 to 100 bases in length, whereas the sequences vary in terms of variation and corresponding usefulness for universal microbial identification. Conserved sequences flank hypervariable regions of the 16S rRNA gene, and this allows the device of "universal" PCR primers which can be used to amplify 16S rRNA hypervariable regions from a wide range of distinct species of bacteria [23].

The selection of primers is an important aspect for the precise characterization of archaeal and bacterial species utilizing rDNA amplicon analysis. The use of suboptimal pairs of primers can lead to underrepresentation or selection against whole groups or single species, which may bring about uncertain biological inferences [24]. Various hypervariable regions develop at different rates, and in some hypervariable regions, different species of the same genus might be identical and in others more distinct [25]. For phylogenetic differentiation at species levels, sequences of 500–700 bp have shown to be necessary [20]. Segments of up to 700 bp are routinely sequenced with NGS technologies, with studies promoting the utilization of V1, V2, and V3 regions for archaeal and bacterial sequence characterization and deep sequencing (Wang *et al.,* [25]). Some also recommend that regions created using primer sets 967F-1046R and 8F-338R for V6 exaggerate the abundance of species and encourage the V4–V6 produced employing primer sets 967F-1220R, 805F-1046R, and 530F-805R, as the most applicable [25, 26]. Fragments that include the hypervariable regions V3, V7, and V7 + V8 (produced utilizing primer sets 1046F-1392R, 1046F-1220R, and 338F-530R) also underrated the abundance of species [25]. Recent studies have shown that a diverse environment achieves greater microbial population coverage for the V7–V8 fragments [26]. When attempting to characterize highly heterogeneous microbial communities, it is strongly advisable to utilize primers that target two regions of the 16S rRNA gene in all high-throughput sequencing attempts, even though good recognition of a microbial population has been attained by selecting single hypervariable regions [26, 27].

1.2.2 Shotgun sequencing

In microbial ecology, shotgun metagenomic sequencing is an effective method since it offers a robust and accurate assessment of microbial heterogeneity [28]. With no markers like viruses, it is the only way to investigate the microbial community [29, 30]. In the pathway hypotheses and taxonomic evaluation for the practical interpretation of the microbe in analysis, it enables strain-level remodeling [19]. This is an evolving molecular technique for filling the space between the functional and structural capability of the population. It also helps to consider the mechanisms that microorganisms follow to survive in unfavorable surroundings [31]. To construct a taxonomy profile, this protocol for taxonomy study comprises efficient trimming and assessment of a reference database comprising explicitly engineered marker genes or entire genomes. The results could be used for other studies such as antibiotic resistance gene profiling, metagenomic assembly, metabolic function profiling, and metagenomic binning since it includes all genetic information in a sample [32]. It is the nontargeted sequencing ("shotgun") of all ("meta-") genomics of genomes of microbes present in a sample. Shotgun sequencing could be used for profiling microbial communities' taxonomic configuration and functional potential and for retrieving whole genome sequences [29].

1.2.3 Microarray

One of the most promising techniques in functional genomics is DNA microarray, which is commonly recognized as a DNA chip or a biochip. It is a collection of microscopic DNA spots that are deposited or synthesized by covalent or noncovalent interactions in a 2-D or 3-D array on a solid stationary surface such as silicon chips, glass, or nylon membrane. It enables the study, at once, of multiple genes without amplification of the individual genes by PCR [32]. To evaluate the expression pattern of genes involved in catabolism and physiology of microbes collected from environmental specimens, DNA microarrays have been documented to be used [33]. An extensive 50-mer-based oligonucleotide microarray was developed by Rhee et al. [34] to classify genes involved in biodegradation and biotransformation in microbial communities. DNA microarrays have been used in gene quantification in biological samples, transcriptional profiling, microbial genome stress gene analysis, and so on [35, 36]. Below are different microarrays developed to analyze microbial diversity [32]. The basic concept of microarrays constitutes the ability of a single-stranded DNA or RNA molecule to hybridize with complementary probe molecules bound to a solid support [37]. Readouts are marked by the fluorescent dyes inserted in the sample as signals given off [38]. By spotting a synthesized ex situ probe or even by directly arranging the probe in situ, the microarray slides/chips are designed [37]. Microarrays have a rapid and sensitive detection device capable of identifying up to a single mismatch, compared to conventional nucleic acid hybridization approaches [38, 39].

1.2.3.1 Phylochip array

It is a microarray of convenience and quality chips developed by the company Affymetrix to recognize various archaeal and bacterial organisms from intricate microbial populations. Without the need for any culture techniques, it provides a steadily fast, complete, and accurate testing method for samples taken from the environment. These chips contain extensive gene information and are broadly used in the recognition and study of mutations and polymorphisms depending on hybridization, such as single nucleotide polymorphisms or disease-relevant mutation evaluation. In severe biological systems such as effluents from industries, salters powered by the sun, coral reefs, and olive mill squanders, they have been used to examine microbial profiles [40]. Phylochip arrays have supposedly been used in industrial wastewater contaminated by chromium and Australian sugarcane processing plants to investigate bacterial diversity and functional traits [23, 41]. It is updated frequently, and two variants of the microarray are presently accessible. Generation of second-generation PhyloChip (G2) began in 2002 and became accessible in 2006. Presently, PhyloChip G3 is accessible via Second Genome Inc. For the identification of 842 subfamilies or 8,741 taxa, PhyloChip G2 comprises 297,851 mismatch (MM) and perfect match (PM) 16S rRNA gene probes, including 121 archaeal and bacterial orders [42]. In March 2002, the model of PhyloChip G2 was centered on more than 30,000 16S rRNA gene sequences obtained from the database, Greengenes [43]. The huge sensitivity it produces is one of the major benefits that the PhyloChip offers. More than 600 billion sequences are produced by a standard 16S PCR reaction with an output of 500 ng amplicons approximately, which enables even the scarce microbial communities to be detected as well as the dominant ones. And, when compared to rRNA Sanger sequencing clone libraries, the PhyloChip has been shown to show greater variability within a population because of the positioning of the whole gene product on the microarray in contrast to the study by conventional sequencing methods of up to thousands of individual molecules [23]. It has been shown that the PhyloChips G2 and G3 offer family recognition resolution at subfamily stages, and they have been used in about 80 publications [24, 42]. In a wide range of complex habitats such as agricultural waste, solar salterns, olive-mill waste, coral reefs, marine environments, aircraft particulate air, plant tissues, soil, and numerous human microflora, this technique has been used to successfully define the microbial profile [22, 23, 44–50].

1.2.3.2 Functional gene arrays (FGAs)

Functional gene arrays (FGAs) are the specific type of microarrays of DNA integrating probes correlated with the functional advancement of microbes for crucial genes [51]. To determine the network and functional composition of microbial populations, FGAs are beneficial [52]. It is widely used for coordinated examination of several functional genes [36, 51]. Since most microbes in such samples are extremely diverse, metabolically versatile, and unculturable, this technique is commonly utilized to analyze microbial populations from environmental specimens [53]. A unique kind of DNA microarray consisting of probes for critical genes associated with the functional processes of microbes is FGAs. FGAs consist of probes for essential genes of microbes that are associated with functional processes of concern [51]. Rather than PCR-based methods that restrict the number of genes that can be studied at once, this type of array enables the simultaneous analysis of several functional genes [36, 51]. FGAs are particularly useful for the analysis of environmental samples because, due to the shortfall of cultured microorganisms and a greater level of metabolic and diversity versatility that occurs in microbial populations, the specific roles are not established [53].

1.2.3.3 Geochip array

Geochip array is a highly comprehensive DNA microarray available for pondering microbial populations associated with management of the ecosystem, biogeochemical cycling, bioremediation, and preservation of the environment. This technique can be employed in soil, water, air, human, and animal sources to examine microbial associations. It does not require any previous microbial population information and can identify less prevalent microorganisms to avoid annotation bias. This method is relatively fast and utilizes either RNA or DNA and employs 56,990 gene sequences from 292 functional genes including 27,812 probes [32]. GeoChip uses core enzymes or genes to recognize different microbe-mediated processes of resistance mechanisms for heavy metals such as cadmium, silver, copper, nickel, mercury, zinc, and lead; for biogeochemical cycles of nitrogen, sulfur, phosphorus, and so on; for degradation mechanisms to recognize and monitor the breakdown of xenobiotics commonly present in the environment [53, 54]. It has been described as a quantitative, sensitive, insightful, and efficient method for analyzing the functional layout of microbial populations from varied environments like polluted habitats, soils, extreme environments, and aquatic systems [52, 53].

1.2.4 Next-generation sequencing (NGS)

The primary concept in NGS includes DNA molecules that are sequenced in a flow cell in a hugely simultaneous fashion or stepwise recursive manner [55]. Second-generation sequencing technology accelerated the sequencing procedure by producing sequences with quite a high throughput cost-effectively. From several specimens where tiny parts of DNA are connected with adaptors while amplification of DNA for arbitrary reads, it generates a massively parallel analysis that delivers huge data in a short period [56]. This approach includes (i) preparation of library, (ii) sequencing, (iii) marking of base, (iv) alignment to the existing genome, (v) and diverse description. The construction of the library starts with DNA fragmentation into several segments by transposase or enzymatic digestion, and sonication, accompanied by adaptor sequence ligation [32]. To produce DNA replicas, the library which was prepared is then clonally amplified by employing PCR techniques. Using various techniques, sequencing of DNA replicas is done [57]. Illumina, Pyrosequencing (Roche 454 sequencing), SOLiD, PacBio RS, Ion Torrent, and so on are the main platforms used for microbiome studies in NGS [58].

1.2.4.1 Pyrosequencing technique

The GS FLX Pyrosequencer employs pyrosequencing, an NGS technology [59]. The pyrosequencing method is a sequencing by synthesis method centered on the identification of pyrophosphate liberation after the introduction of a nucleotide to the lately synthesized DNA strand [60]. DNA strand is incised and linked with oligonucleotide adapters, attached to a 28-μm bead, and amplified with PCR and pyrosequenced in an emulsion of oil-in-water [61]. These amplicon-comprising beads are then captured in a picoliter-scaled well and pyrosequenced. The presence of nonlabeled nucleotides creates the computationally observable release of inorganic pyrophosphate through enzyme-mediated release, thereby a DNA sequence of near to 1,000 base pair read length is produced by repetitive pyrogenic cycles and can produce 1 million reads/run approximately [62]. It was later used to study microbial communities from different sources of the environment, such as underground mine water, freshwater, marine water, corals, fish, mosquitoes, and terrestrial animals [63]. To remove mixed and low-quality sequences which consist of greater than one original DNA fragment per bead, analysis software analyzes the raw reads employing different quality filters. Via the quality control tool, sequences that do not encompass the initiating TCGA sequence are excluded. The filtered reads yield about 100 Mb of quality information. FLX reads are recognized to be of sufficient length to construct tiny genomes with high quality and contiguity, such as bacterial and viral genomes [64].

1.2.4.2 Illumina/Solexa sequencing

The second commercial NGS platform was developed by Solexa. Eventually, Solexa was obtained by Illumina company and is now known as Illumina [59]. The preparation of the DNA library begins with random DNA fragmentation into sequences proceeded by adapter ligation to either end of every sequence [65]. With complementarity to the adaptor on a glass surface, linkers are used to connect forward and reverse primers. Bridge PCR amplifies the DNA fragment, resulting in several copies of a single DNA template in a "cluster" [32]. Around a million replicas of the beginning sequences are included in each cluster. In the sequences, nucleotides are identified via sequencing by synthesis technique, utilizing reversible terminators in which primers for sequencing, DNA polymerases, and four modified nucleotides are integrated such that the sequences are hybridized by the primers [66]. This technique utilizes altered dNTPs comprising a terminator at the 3′-OH end with a fluorescent tag that inhibits polymerization and enables each growing DNA strand to have only a single base addition [32]. In each step, sequencing continues in loops, and single bases are added to all templates identically, accompanied by the fluorescent terminator's chemical cleavage that identifies the nucleotide inserted [67]. To activate the fluorescent label integrated into the nucleotide base, lasers are transmitted onto the DNA clusters that are subsequently identified and captured by a camera. Numerous software programs are used to translate these signals into different sequences of nucleotides. Until billions of clusters are being sequenced, the process continues with the removal of the fluorescently labeled terminator group by adding a new fluorescently labeled terminator base [68]. Since the DNA sequence is examined base by base, Illumina sequencing is remarkably accurate [32]. The Illumina sequencing technology when compared to 454 sequencing produces (i) good precision with error margins below 1% and (ii) much higher throughput (about 1.5 Gbp per run) at the expense of slightly shorter read lengths [59].

1.2.4.3 Ion torrent sequencing

This technique is analogous to pyrosequencing. This method is based on the liberation of a proton when a dNTP is introduced to the DNA polymer, rather than using fluorescent-labeled nucleotides. To the fragmented RNA or DNA, adaptors are attached, and a single molecule is mounted on beads and using emulsion PCR, amplified. Each bead is placed into a single slide well comprising single dNTPs, polymerase, and buffer species. Polymerase's introduction of nucleotides into the DNA strand releases hydrogen ions as a by-product that reduces pH. A pH sensor at the bottom of the microwell detects the decrease in pH and converts it into a voltage that is corresponding to the number of nucleotides integrated. This technique is capable of generating 600, 400, and 200 bp of reads in length [32].

1.3 Single-cell genomics (SCG)

Single-cell genomics (SCG) is evolving by offering genomic information from individual, uncultured cells as a solid complement to cultivation-based and microbial community-focused research methods [69]. Although metagenomic-based bioprospecting offers an alternative for exploiting bacterial diversity [70], the primary benefit of the SCG method is that the entire genome is sequenced instead of individual genes of the uncultured microorganisms. SCG comprises a series of combined procedures, beginning with the collection and preservation of environmental samples, proceeded by physical separation of individual cells, lysis, and whole genome amplification, and then advancing through either targeted loci or whole genome sequencing and interpretation of sequences [69]. In the presence of betaine or glycerol, almost every environmental specimen may be preserved or treated right away so that the integrity of the cells is preserved [71]. Cell separation is the next step, and this is presently accomplished by the use of fluorescence-activated cell sorting [71]. A straight connection between metabolic and phylogenetic markers of archaea and bacteria that are uncultured is feasible for the first time with the SCG method. SCG is indeed able to generate reference genomes of uncultivated microorganisms, allowing complex ecosystems to be studied. SCG has been instrumental in defining the metabolic properties, evolutionary histories, and inter-organismal interactions of many ecosystems and biogeochemical cycles dominated by uncultured microbes. The SCG method also offers significant potential in microbial microevolution studies and industrial bioprospecting. The performance of SCG recovery varies widely, from 0% to a completed genome, and relies on the intrinsic characteristics of the cell (DNA packaging and the structure of the cell wall) and all elements of the SCG pipeline [69]. Initial attempts of SCG on environment samples were hindered by severe DNA contamination [72], which may occur from the handling, reagents, equipment, and sample itself. Lately, the problem of contamination has been nearly solved by incorporating clean techniques in flow cytometry [71], using laser tweezers in microfluidics [73], and decontaminating MDA reagents before usage [71, 73].

1.4 Metatranscriptomics

The analysis of variations in the gene expression of microbes and their control in natural habitats at a particular time and location by mRNA transcript sequencing that is arbitrarily collected from microbial populations is environmental transcriptomics or metatranscriptomics [74]. The absence of poly-A tails in prokaryotic mRNA transcripts is a significant setback to this strategy, contributing to the difficulties of obtaining cDNA. Latest developments in metatranscriptomics have contributed to the establishment of a new RNA technique that uses exceedingly ample RNA molecules (both rRNA and mRNA), to evaluate the complete phylogenetic array of the microbial diversity that is taxonomically and functionally important [3]. Both the methods to partial and whole population study have their own merits and disadvantages. A clearer analysis of the taxonomic diversity correlated among various groups in the microscopic ecology can be achieved by applying a combination of the appropriate approaches. To provide an in-depth view of environmental microbiology, an interdisciplinary approach should therefore be pursued between different omics tools [3]. In the soil, eukaryotic organisms can express unique functions that could be analyzed by examining the collection of particular eukaryotic polyadenylated mRNA which are precisely extracted from the environmental specimen. It is possible to obtain the required quality of RNA from soil specimens using other extraction techniques. To turn the entire amount of soil mRNA into cDNA, high-throughput sequencing might be used.

Several common strategies for high-throughput transcriptomics include:

(i) first, it recognizes mRNAs that vary under various experimental conditions in their expression level and then determines the respective identity of genes, such as serial evaluation or differential demonstration of gene expression [75];
(ii) conversely, to test the expression patterns of previously identified genes, for example, by cDNA or oligonucleotide microarray, or nanoliter-volume RT-PCR (reverse transcript-PCR) that is chip-based, which concurrently tests gene expression at greater precision and susceptibility for several thousand genes than microarrays [76];
(iii) a current RNA sequencing technology that directly enables RNA molecules to be massively sequenced parallelly without preceding cDNA synthesis or requirement for ligation or amplification steps as in Helicos Genetic Analysis scheme [77];
(iv) a recent strategy for direct sequencing of cDNA converted from whole transcriptomes using next-generation sequencers (i.e., Illumina's Genome Analyzer, Applied Biosystems SOLiD System, and Roche 454 Sequencer) [78].

It is essential to enhance microbial mRNAs by diminishing rRNA previously to metatranscriptomic sequencing since mRNAs in the total microbial RNA pool are overshadowed in abundance by bacterial rRNAs. Microbial mRNA would then be converted by normal methods to cDNA and sequenced. Metagenomic and metatranscriptomic (meta'omic) sequencing can be conducted in parallel with sufficient barcoding of DNA and cDNA samples, rendering RNA sequencing a natural extension of microbial community surveys, and further concern during the design of microbial community sequencing-based surveys [79]. For freshwater and marine microbial populations, metatranscriptomic techniques were first implemented [78]. These findings have shown that microbial total RNA can be used, like DNA, to assess the structure, role, and diversity of the population. Besides, these findings demonstrated that RNA sequencing generated large quantities of novel sequence data, possibly by capturing low copy number genes or organisms that are under-sampled by DNA sequencing alone. Also, metatranscriptomic sequencing provides a way of identifying and quantifying RNA viruses not otherwise included in metagenomic assessments based on DNA (excluding integrated retroviruses) [80]. Integrating metatranscriptomics with taxonomic and functional profiling based on DNA shows prominent over- or under-expression of specific functions and, in some instances, the behaviors of whole species, all according to their metagenomic abundances [81]. From an environmental perspective, meta'omic sequencing of waters polluted by the Deepwater Horizon oil spill showed abundance for organisms and pathways associated with complex hydrocarbon degradation [82]. This indicates that integrated meta'omic sequencing may play a vital role in the formulation of approaches for bioremediation, where it would be appropriate to ensure that degradation pathways in a microbial community are both present and active [81]. For no more than a few minutes after collection, mRNA will usually remain in an environmental sample. In addition, for various environments and microorganisms, and by gene function, the mRNA half-life can differ, with household genes producing more stable mRNA products [83]. There is indeed a range of collection protocols, like commercial (easily standardized) kits and conventional "homemade" methods, which often yield larger amounts and better-quality RNA, but when performing latitudinal studies, standardization can be more difficult. Also, to avoid enzymatic RNA degradation, sample collection requires immediate freezing in liquid nitrogen. Such scenarios can involve the use of buffers that stabilize RNA, like the MO BIO LifeGuard Soil Preservation Solution. These buffers enable the collection, while in the greenhouse or field, of mRNAs from environmental samples with prompt conservation. Most RNA extraction from commercial kits, as with DNA extraction procedures, includes bead-beating technique and RNA trapping in a stable buffer that can be frozen and analyzed further [1]. Once mRNA is obtained and maintained safely, it has to be converted to cDNA (complementary DNA). However, mRNA is often present as a small fraction of the total RNA (mostly rRNA and tRNA). Thus, mRNA is always selectively isolated from total RNA. There is a range of commercial methods available for sample collection and RNA extraction, including the subtractive hybridization utilizing magnetic beads paired with oligos unique to rRNA and tRNA and use of exonuclease treatment (targeting rRNA), which are consequently eliminated from the solution. However, eukaryotic mRNA can be present in high concentrations in clinical and environmental samples, where eukaryotic mRNA can be extracted by aiming mRNA possessing 30 poly-A tails [84]. For most downstream applications, cDNA is most often the template of choice following mRNA enrichment. In these cases, as in many other techniques that involve cDNA synthesis, reverse transcriptase and either specific primers or random oligos are added. Most metatranscriptomics study is currently performed utilizing 454 or Illumina systems, the former generating larger sequence products (500 bp), while the latter accounts for smaller sequences (150 bp), but a greater number of products [1]. Bioinformatic analysis, after sequencing, eliminates low-quality and short-read sequences. Sequence ends are also clipped and data analyzed for the presence of rRNA (which can still be found, despite mRNA enrichment) sequences that are subsequently excluded from the library. Sequences are typically compared with available databases that attribute the function and identification of genes [1]. It is also termed "gene expression profiling" since it offers an overview of down- or upregulation of a gene in microbial communities in different environments. Analysis of mRNAs provides a clear view of the tissue and cell-specific particular gene manifestation such as (a) evaluation of alternative splicing to predict isoforms of protein; (b) transcript evaluation, inexistence, and existence; and (c) assessment of the effect of genotype on expression of gene by expression quantitatively evaluable particular expression of allele or trait loci analysis [32].

For transcriptome, sequencing techniques and microarrays are used [85]. RNA-sequencing employs NGS to determine the amount of RNA in a given sample, while microarrays aid to evaluate gene expression. RNA-microarray approach utilizes previously designed probes which makes this technique affordable, more efficient, and also an improved method for the analysis of expression of a protein. RNA sequencing is detailed because at a much-advanced coverage and wide discovery studies it enables various types of RNA [86, 87]. Primary transcriptome data generation includes RNA of interest's purification and proceeded by conversion of RNA to cDNA, fragmentation of cDNA to create a library via RNA sequencing, operation of the sequence or microarray via an advanced database software and if required QC [32]. In transcriptomics, DNA microarray is an important tool that assists in the analysis and evaluation of mRNA expression of every single gene in an organism. With a dynamic environment, the entire DNA microarray genome technology helps to identify and examine gene expressions exposed to modification [88]. This technique has also been used to analyze variation in catabolic and metabolic gene expressions, to study the physiology of the microbial population from different environments, to characterize novel species of bacteria, and so on [36]. Expression quantitative trait loci mapping is employed to analyze and integrate gene expression (RNA data) with genetic variants (DNA data) as a convincing strategy [89]. Gene expression regulation can also be analyzed through the combined study of the levels of mRNA and micro-RNA [32]. Together with the expression of miRNA, mRNA profiling could also be utilized to examine differences in the transcriptome collection, especially to recognize the miRNA transcripts which are regulated, highlighting the probable molecular pathways that support a specific trait or condition [90]. In addition, comparing metatranscriptomes obtained from samples subjected to different circumstances of cultivation or environment can provide associations between the expression of genes and environmental variables [91]. Metatranscriptomics, as well as metaproteomics, could therefore enable us to understand the distribution and role of expressed genes and corresponding proteins in microbial environment of interest [92].

1.5 Proteomics

One of the important components of living organisms is proteins since proteins are the major constituent of cell structure construction and also act as signal transduction proteins and catalytic enzymes in the metabolic pathways throughout the cells' regulatory pathways [93]. To make an analogy with transcriptomics and genomics, the term "proteomics" was coined as a technique for the study of massive proteins, specifically their structures and functions [94]. The collection of proteins produced within an organism, organ, tissue, or cell is a proteome. Proteomics is the area of science that studies and examines proteomes. Proteomic research leads to the decoding, post-translational modifications, metabolic pathways, molecular mechanisms, and so on in the cell. The ubiquitous protein expression in microbes living in polluted sites attributed to human activities has been tracked and analyzed [95]. The analysis of the complete protein profile of microbial groups living in a particular environment is metaproteomics or community proteomics [96]. The efficacy of protein extraction, methods employed for protein filtration from a complicated mixture and the clear recognition of fractionated proteins or peptides, database search, and interpretation of data depend on metaproteomic analysis [97]. It is also specified as a functional method to genomics as it aids to explore one organism's protein expression pattern and helps to generate a protein graph of all expressed proteins in one organism that flourishes in a specified habitat.

The following measures have been developed to examine the proteome of microbial communities across random ecosystems. Complex environmental samples require unequivocal strategies for sample collection or extraction of proteins. For segregation and identification of proteins, two approaches have been devised. The first approach is in accordance with 1-D or 2-D polyacrylamide gel electrophoresis (PAGE). Proteolytic enzymes such as trypsin are utilized to digest the abscised targeted bands which are protein to peptides that are carefully examined by tandem MS analysis or mass spectrometry, database testing, and bioinformatic assessment [98]. Another approach is centered on tandem MS/MS-coupled liquid chromatography (LC-MS/MS), where proteases are employed to digest the whole proteome into a peptide mixture that is complex and are purified by reverse-phase microcapillary or cation exchange chromatography. It enables proteins' short-term high-throughput detection [97], enables insoluble membrane proteins to be detected [99], and improves proteome distribution. Data with high precision and reproducibility are beneficial for quantitative and comparative proteomics. Under biological circumstances such as treatment versus control, disease versus healthy, wild type versus specific genotype, as well as different gradients in an environment like salinity gradients and nutrient, it seeks to analyze the divergence in protein expression. For proteomics, various labeling techniques such as stable isotope labeling using peptides and amino acids [100], isotope dilution [101], tandem mass tags [102], absolute and relative quantification isobaric tags [103], affinity tags that are isotope-coded [104] were developed. An organism's metabolism is conditioned to the environment. Altering external stimuli induces alteration in the expression of protein and could be useful in bioremediation techniques to analyze such changes [105]. Quantification and assessment of proteins that are expressed by microbes present in various environments such as soil, aquatic, and freshwater habitats, activated sludge, sediments, soil, animal or human microbiome, biofilms of acid mine drainage, and associated plant microorganisms are reported in several studies [106]. Even though DNA-based and RNA-based approaches can offer a thorough insight further into the environmental functions of microorganisms, proteins (not genes) are mainly accountable for most microbial processes. The analysis of such proteins (i.e., enzymes) will also provide a more accurate description of microbial activity. Proteins formed by a provided microorganism within a series of conditions are generally

known to as proteome. In comparison to the genome, the proteome is far more complex (like the transcriptome) with various proteins formed based on the level of cell metabolism as well as the environmental triggers present. Researching the proteome does have the ability to offer unique knowledge on cell structure and the pathways underlying cell responses to various stimuli. Especially, proteomics-based strategies enable the identification of proteins that are expressed differently and are therefore probably to be essential in the microbial responses to environmental conditions. Proteomics-based analyses of environmental impacts on microorganisms usually include the following:
1) The microorganism's exposure to a state of interest
2) Protein isolation from each population
3) Protein separation
4) Identification of protein

The very first two stages in proteomics-based research are reasonably easy to complete. There is a range of efficient strategies possible to purify and isolate heterogeneous protein mixtures formed by microorganisms. Splitting the proteins found in these diverse mixtures, though, is one of the most difficult facets of proteomics. Two different protein methods are widely used: LC-MS and 2D-PAGE . In the latter technique, the proteins are first separated by the pH at which the protein does not have a net charge, the isoelectric point (pI). The second dimension of 2D-PAGE distinguishes proteins depending on the mass using a polyacrylamide gel. The resultant gel includes several spots, each preferably comprising a single protein which can be classified using MS techniques listed further. Conversely, LC may be used to distinguish proteins. In this strategy, proteins from a population group are enzymatically combined and digested into the constituent peptides. Such peptides are differentiated by an LC which enables the segregation of molecules based on hydrophobicity or charge. Mostly on the basis of such peptides, proteins throughout the initial cell population are classified as mentioned. LC-based protein isolation could be more automated and reproducible than 2D-PAGE [1].

1.6 Metabolomics

Metabolomics is a study of metabolites with low molecular weight. Environmental metabolomics contains metabolites formed by associations between microorganisms, small eukaryotes, plants, livestock, predators, and the involvement of abiotic pressures and stimulants. Common metabolites (#1,500 Da) contain organic acids (e.g., glycolytic intermediates), amino acids (e.g., protein intermediates), and different saccharides (e.g., monosaccharides and cleaved sugars) [1]. As with genomics, transcriptomics, and proteomic studies, the aim of metabolomics is always to explicate the role of the microorganism or microbial community; nevertheless, proteomics and metabolomics disclose knowledge about the "final" genome material. Metabolomics often characterizes relationships among microbial components and the environment, or between microbial and the other higher environmental species, such as plants and animals. Metabolomics was utilized as an experimental technique [107] to discover the cognitive function of microbial species and specific single cells in their environment, exposing community and ecological framework. Directed metabolomics allows the consumer to concentrate on a single metabolite, for example, while medication can determine the upregulation or downregulation of the substance, whereas global metabolomics considers the biological system and their metabolites as overall. Metabolites are classified into two groups: exometabolome and endometabolome, which are intracellular and extracellular metabolites. Just like transcriptomics, the analysis of intracellular metabolites may be more challenging because these molecules are much more transient and in a continuous state of flux. The metabolic complexities and research goals concerning intracellular or extracellular metabolites decide the form of extraction and processing. Once the metabolites are collected, they are detected with a variety of instruments like liquid and gas chromatography-MS, Raman spectroscopy, and nuclear magnetic resonance (NMR). In certain situations, relying on the scope of the biological environment, the analysis will involve a mixture of two or three of such methods [107]. Metabolomics characterizes the microbial activity of soil and its relationship with its surroundings and the abiotic stresses and often analyzes the reaction of a single individual to certain other biological influences. Population-related or human experiments are performed to assess the resilience of various populations of distinct microorganisms phylogenetically [3]. The cell creates a variety of secondary and primary metabolites under extrinsic stress, which has allowed us to comprehend and examine the impact of environmental factors on the metabolism of species [7].

Methods based on metabolome have encouraged the creation of systems that can forecast microbial behaviors within bioremediation strategies. It helps one to consider the effective behaviors of the microbial populations and in their practical contributions to the ecosystems within which they survive. Numerous spectroscopic techniques may be utilized to classify metabolites generated by species [108]. In accordance with varying environmental factors, metabolomics analyzes the metabolites developed in the cell that, sequentially, offer data on cells' regulatory activities [109]. Such metabolites may be employed as a bioindicator to track the biological impact of waste pollutants for improved environmental understanding. Significant advancements in analytical methods, computational tools, and bioinformatics have made it possible for diverse metabolites to be analyzed, isolated, and interpreted and their pathways explained [110]. Major global metabolome analysis techniques include metabolite profiling, metabolic fingerprinting, and targeted analysis [111]. The metabolomics research workflow begins with the extraction process and processing of the sample, followed immediately by the isolation and identification of the analytes. Measurement and quantitative analysis of metabolites are typically done by merging chromatography techniques (gas chromatography and LC) and by tracking devices such as NMR and MS [112]. MS and NMR spectrometers are the most commonly used scientific tools in the metabolic study [113]. Among the MS techniques, LC-MS is commonly employed, owing to higher precision, nonnecessity of chemical derivatization measures, and lower ion suppression. Two approaches are widely used for analyzing metabolic results. The first approach uses statistics and clustering algorithms; meanwhile, the second approach uses networks to visualize the temporal and spatial properties of the results. Numerous software instruments such as XCMS, MetaboAnalyst, Matlab, and MetATT are employed for the study of metabolic results. In addition to data analysis, multiple network visualization software applications such as MetScape Plugin, MetaMapp, Pathomx, and MAVEN are being created to simulate a graphical representation of metabolic processes as heat maps or graphs [110]. Metabolomics, together with bioinformatics methods and databases, has made it possible to understand better the microbial population, and the genes responsible for encoding catabolic enzymes and their catabolic pathways. It is therefore an effective technique for characterizing metabolic networks and exploring new metabolic pathways [114]. Also, several pathway databases and libraries are available, like Reactome, Kyoto Encyclopedia of Genes and Genomes (KEGG), and PANTHER pathway, which produce data on particular metabolic pathways, enzymes and reactions, chemical structures, names, pathway, metabolite occurrence, molecular weight, and so on [110]. Metabolomics is now increasingly evolving omics science and can offer a comprehensive view into different metabolites as well as metabolic pathways. The rate of enzyme reactions in the cells is often controlled by the concentration of products and substrates [115]. In comparison, for most cells, there is no causal association among genes and cell metabolites in the same manner as for proteins and mRNA. For example, there are cellular intermediates that are less than 600 low molecular mass in *Saccharomyces cerevisiae* [116], while ~6,200 protein encoding genes present in its genome [117]. As a tool for identifying small molecular diversity in cells and revealing variations in small molecular abundance, metabolomics has many benefits in the form of metabolic analyses, since metabolites are indeed functional agents within cells, and their concentration levels differ significantly as a result of physiological or genetic changes [117]. Techniques like high-performance LC-MS, gas chromatography time-of-flight MS (GC-TOF), NMR spectroscopy, and capillary electrophoresis MS (CE-MS) are usually conducted for metabolomics analysis, and lately, vibrational spectroscopy (in which sensitivity and resolution are regarded to be lesser than MS). Metabolomics research may also be carried out by the simultaneous use of many new technologies in an attempt to obtain high exposure and improved detection [118]. Compared to proteomics and transcriptomics, methods that are employed to characterize gene expression's end-products are lesser advanced, and many of the metabolomics research performed so far have not yet been adequately detailed (on average just a few hundred metabolites reported in the majority of these research) and more progress in their measuring precision is still needed. That being said, such studies have shown that microbial metabolomics may be an effective tool for decoding the metabolism of microbes and integrating the gap between phenotype and genotype as it enhances improvements in the proteome and offers a clearer description of the phenotype of the species than any other process.

1.7 Fluxomics

Microbial metabolomic analyses utilizing isotope-labeled intermediate metabolites in combination with different forms of active metabolic flux design have been proposed as an effective technique in research into diverse and vast metabolic systems in some recent applications, and the word "fluxomics" was coined to characterize this field of study [12]. Fluxomics is a quantitative method that examines the rates of metabolic processes within a biological individual and changes in their metabolic rates. Fluxome is the complete collection of cell metabolic fluxes that provides details about many cellular processes; hence, it is a distinctive cell phenotypic feature. Flux analysis provides important phenotype details as it evaluates the metabolome in its functional interaction toward the genome and surroundings [119]. The in vivo enzymatic functions which cannot be derived from metabolomic, proteomic, or transcriptomic data solely are assessed by metabolic reaction rates. A good fluxomic analysis depends on the accuracy of isotopic labeling measurements, metabolic network reorganization, and quantitative metabolomics data, explaining metabolic reaction regulation and stoichiometry [32]. Tracer-based metabolomics and stoichiometric metabolic flux analysis or flux balance analysis are two methods widely used for flux quantitative analysis, using a firm carbon ^{13}C isotope to track the distribution of carbon across different pathways pursued for identification and characterization by analytical techniques such as mass spectroscopy (MS) analysis or NMR. In order to generate a regulated design of metabolic flux, flux balance study employs the metabolic reactions' stoichiometry in accordance with chemical, biological, and thermodynamic parameters. Cells are treated with ^{13}C-labeled substrates in ^{13}C isotope labeling that can be metabolized such as glucose, leading to ^{13}C atom-containing metabolites. ^{13}C atoms are incorporated into the newly designed metabolites in different positions and amounts, based on the metabolic pathway. Thus, there might be different isotopomers for every specific metabolite (numerous molecules of the same metabolite with different labeling states). To control the concentration and isotopomeric distribution of these metabolites, metabolomic analytical frameworks are being used [120]. In order to estimate the flux distribution, the integrative methods utilize the incorporation of metabolomic data that is tracer-based into a collection of software by incorporating the experimental data into the more comprehensive metabolic model [121].

Examination of the stoichiometry of metabolic flux based on ^{13}C was primarily utilized to measure the intracellular fluxes of a wide variety of microbes [122]. ^{13}C FLUX is an example of software packages that measure flux sets [121]. The existence of cometabolic pathways for the biodegradation of halogenated organic compounds, radionuclides, and harmful metals was revealed by the fluxome profile assessment of *Shewanella* marine bacteria employing GC-MS tests, biochemical, genetic, and statistical algorithms. When exposed to various carbon sources, the bacteria showed a receptive metabolic flux [123]. In an in situ dechlorinating bioreactor, integrated molecular diagnostic and kinetic modeling strengthened the knowledge of organohalide-respiring bacteria and their metabolite dynamics [124]. Important indications of a strain arising from *Pseudomonas putida* S12 that was phenol overproducing amid metabolic engineering were discovered by metabolic flux study [125]. For the investigation of the fundamental metabolic network for the co-utilization of aromatic carbons and sugar in *Pseudomonas putida*, metabolic flux analysis was used [126].

1.8 Integration of omics

Reference databases such as KEGG and MetaCyc may serve as a forum to facilitate the discovery of 'omics knowledge, and new software techniques are evolving to enable these databases to function as an "omics integration" platform. The KEGG database can be accessed instantly through a web-based platform (http://www.genome.jp/kegg/), enabling users to explore genes, compounds, and pathways, or to provide information about genes and compounds that are differentially identified to decide which functional pathways may be influenced [1]. In the potential of one data form to affirm or contradict the outcomes of another, the importance of multiple-'omics datasets might be appreciated. For instance, Mason et al. [82] used a combination of metagenomic, metatranscriptomic, and SCG sequencing to describe microbial community behaviors after the Deepwater Horizon oil spill in the Gulf of Mexico in 2010. Interestingly, however, the transcriptomic analysis showed that there was no alteration in the expression of genes related to the degradation of more recalcitrant compounds, suggesting that while variations in patterns of gene abundance can provide clear clues about how a system functions, it may be necessary to add transcriptomic and/or metabolomic data to determine which genes, compounds, and metabolites are actually being used [1]. In comparison, correlation and network-based methods can be used within and between 'omics datasets to detect co-occurrence and/or co-abundance trends. Correlation and network-based methods are relatively easy, and the errors found in 'omics measurements and the biology they are used to portray are often naïve [1]. Correlation-based analyses have been used with effectiveness in several studies, despite the possible drawbacks they could expose new insights into the biology of a system of interest. For example, associations within 'omics (i.e., analyses using a single-'omic technology) have been used to allocate a functional value to unknown genes [133], to identify bacteria, metabolites, or genes that are correlated with, or distinctive of, environmental or ecological subtypes [127], and to provide perception into the culture of formally uncultivable organisms [128]. The incorporation of mixed 'omics datasets is regarded to be at the frontline of science and has enormous potential for environmental microorganism characterization; however, it is a technical challenge that appears to be fully addressed, particularly for the analysis of various microbial communities [1].

Metagenomic libraries have become strong sources for bioprospecting novel biocatalysts and biomolecules, even from species that have not yet been cultivated. The fusion of metagenomic methods and heterologous forms of expression system help to allow significant usage of microbial biodiversity for human welfare [37]. Global protein and metabolite profiling are indeed crucial in addition to gene profiling, especially when studying mechanisms of complex pathways like bioremediation [129, 130]. This makes it a daunting task to comprehensively classify the physiological functions and elucidate the ecological roles of novel species. In the field of environmental microbiology, an effective combination of different high-throughput approaches is especially fruitful. They offer the benefit of miniaturization, automation, and huge time-consuming step "parallelization," along with an evaluation of "real time" [63]. Detailed characterization of the biological and ecological role of microbial communities in an ecosystem can be given to achieve synergies between emerging-omic approaches in the long term. To provide deeper insight into the genome structure, microheterogeneity, lateral gene transfer, and nutrient cycling between individuals of a region's microbial population, integration of complementary "omics" techniques is proposed. In a way, this "systems biology" perspective promises to provide the clear view needed to understand microbial ecosystems and solve some important microbial ecology and evolution concerns [37]. Finally, for finding correlations between various types of data and for covering mechanistic gaps, integrative techniques are important. No particular assay can characterize a microbial community in complete mechanical detail, and a deeper community classification can only be created by simultaneously considering different data types [81].

1.9 Conclusion

In this decade, the dramatic rise in sequence data from metagenomes, metatranscriptomes, and metaproteomes continues to reveal the immense variety of biological potential still unexplored in nature. For the study of individual bacteria, microbial populations, their metabolism, and associations, the multi-omics age offers a wide range of opportunities. The metabolic capacity of microorganisms is described by metagenomics; new microbes can be discovered in a population, and novel pathways and genes can be identified. By illustrating active pathways in a metagenome during given conditions, metatranscriptomics can further broaden our knowledge. The function of less prevalent members of the microbial population can also be elucidated, and the significance of metabolic networks between members of the community is underlined. Metaproteomic studies have identical findings, although metabolomics can confirm the chemical reactions that occur. Alone or in combination, omics techniques are an effective tool to characterize microbial communities [131]. High-end molecular tools' knowledge offers highly functional benefits that can result in massive breakthroughs. In cultivation studies, the integration of multi-omics knowledge improves the likelihood of succeeding and reduces the search space in search for new microbial isolates. Although more efforts are still required to improve the coverage, sensitivity, and specificity of "omic" technologies in the context of their identification, the findings of these experiments based on microbes so far highly indicate that the combination of information at various levels in sequence from genes to proteins and then to fluxes in metabolites is an efficient technique and can be crucial for comprehension of microbes [75]. Omics techniques can predict the metabolism of microbes in contaminated areas. These studies will help monitor novel bioremediation organisms and provide incredible and novel perspectives into their major molecular biodegradation pathways [32]. The decadal challenge is to further explore and use such sets of data to address our ecological and investigatory questions regarding microbial members and their function in the natural ecosystems we are examining. But there is far more scope for multi-omics approaches than only functioning as informational resources. They include hypotheses that require testing to introduce a variety of earlier obstinately existing microbes into cultivation employing improved cultivation methods [132].

References

[1] Hollister, E.B., Brooks, J.P., Terry, J.G. "Bioinformation and 'Omic Approaches for Characterization of Environmental Microorganisms". In: eds. Pepper, I.L., Gerba, C.P., Gentry, T.J. Environmental Microbiology (Third Edition), Academic Press, 2015, 483–505. ISBN 9780123946263. https://doi.org/10.1016/B978-0-12-394626-3.00021-1. https://www.sciencedirect.com/science/article/pii/B9780123946263000211.

[2] Feng, Y.Y. Omics breakthroughs for environmental microbiology. Omics Environ Microbiol. 2013, 40, 18–33.

[3] Biswas, R., Sarkar, A. (2018). 'Omics' Tools in Soil Microbiology: The State of the Art. In: Adhya, T., Lal, B., Mohapatra, B., Paul, D., Das, S. (eds). Advances in Soil Microbiology: Recent Trends and Future Prospects. Microorganisms for Sustainability, vol 3. Springer, Singapore. https://doi.org/10.1007/978-981-10-6178-3_3

[4] Gilbert, J.A., Dupont, C.L. Microbial metagenomics: Beyond the genome. Annu Rev Mar Sci. 2011, 3, 347–371. doi: 10.1146/annurev-marine-120709-142811.

[5] Gupta, K., Biswas, R., Sarkar, A. “Advancement of omics: Prospects for bioremediation of contaminated soils,”. In: ed Shah, M. Microbial Bioremediation & Biodegradation. Singapore, Springer, 2020, 113–142.
[6] Rodríguez, A., Castrejón-Godínez, M.L., Salazar-Bustamante, E., Gama-Martínez, Y., Sánchez-Salinas, E., Mussali-Galante, P. et al., Omics approaches to pesticide biodegradation. Curr Microbiol. 2020, 77, 545–563. doi: 10.1007/s00284-020-01916-5.
[7] Malla, M.A., Dubey, A., Yadav, S., Kumar, A., Hashem, A., Abd Allah, E.F. Understanding and designing the strategies for the microbe-mediated remediation of environmental contaminants using omics approaches. Front Microbiol. 2018, 9, 1132. doi: 10.3389/fmicb.2018.01132.
[8] Misra, B.B., Langefeld, C.D., Olivier, M., Cox, L.A. Integrated omics: Tools, advances, and future approaches. J Mol Endocrinol. 2018, 62, R21–R45. doi: 10.1530/JME-18-0055.
[9] Pandey, A., Tripathi, P.H., Tripathi, A.H., Pandey, S.C., Gangola, S. “Omics technology to study bioremediation and respective enzymes,”. In: ed Bhatt, P. In Smart Bioremediation Technologies: Microbial Enzymes. New Delhi, Academic Press, 2019, 23–43. doi: 10.1016/B978-0-12-818307-6.00002-0.
[10] Kyrpides, N.C. Fifteen years of microbial genomics: Meeting the challenges and fulfilling the dream. Nat Biotechnol. 2009, 27, 627–632.
[11] Cardenas, E., Tiedje, J.M. New tools for discovering and characterizing microbial diversity. Curr Opin Biotechnol. 2008, 19, 544–549.
[12] Ishii, N., Tomita, M. Multi-omics data-driven systems biology of E. coli. In: Edited by Lee, S.Y., In Systems Biology and Biotechnology of Escherichia Coli. 41–57. Dordrecht, Netherlands, Springer, 2009.
[13] Park, S.J., Lee, S.Y., Cho, J., Kim, T.Y., Lee, J.W., Park, J.H., Han, M.J. Global physiological understanding and metabolic engineering of microorganisms based on omics studies. Appl Microbiol Biotechnol. 2005, 68, 567–579.
[14] Gygi, S.P., Rochon, Y., Franza, B.R., Aebersold, R. Correlation between protein and mRNA abundance in yeast. Mol Cell Biol. 1999, 19, 1720–1730.
[15] Alter, O., Golub, G.H. Integrative analysis of genome scale data by using pseudoinverse projection predicts novel correlation between DNA replication and RNA transcription. Proc Natl Acad Sci U S A. 2004, 101, 16577–16582.
[17] Thomas, T., Gilbert, J., Meyer, F. Metagenomics-a guide from sampling to data analysis. Microb Inform Exp. 2012, 2(1), 1–12.
[18] Lovley, D.R. Cleaning up with genomics: Applying molecular biology to bioremediation. Nat Rev Microbiol. 2003, 1, 35–44. doi: 10.1038/nrmicro731.
[19] Han, D., Gao, P., Li, R., Tan, P., Xie, J., Zhang, R. et al., Multicenter assessment of microbial community profiling using 16S rRNA gene sequencing and shotgun metagenomic sequencing. J Adv Res. 2020, doi: 10.1016/j.jare.2020.07.010.
[20] Clarridge, J.E. Impact of 16S rRNA gene sequence analysis for identification of bacteria on clinical microbiology and infectious diseases. Clin Microbiol Rev. 2004, 17, 840–862. doi: 10.1128/CMR.17.4.840-862.2004.
[21] Pace, N.R., Stahl, D.A., Lane, D.J., Olsen, G.J. The analysis of natural microbial populations by ribosomal RNA sequences. ASM News. 1985, 51(4), 4–12.
[22] Hazen, T.C., Dubinsky, E.A., DeSantis, T.Z. et al., Deep-sea oil plume enriches indigenous oil-degrading bacteria. Science. 2010, 330(6001), 204–208.
[23] Katsaveli, K., Vayenas, D., Tsiamis, G., Bourtzis, K. Bacterial diversity in Cr(VI) and Cr(III)-contaminated industrial wastewaters. Extremophiles. 2012, 16, 285–296. doi: 10.1007/s00792-012-0429-0.

[24] DeSantis, T.Z., Brodie, E.L., Moberg, J.P., Zubieta, I.X., Piceno, Y.M., Andersen, G.L. High-density universal 16S rRNA microarray analysis reveals broader diversity than typical clone library when sampling the environment. Microb Ecol. 2007, 53(3), 371–383.
[25] Youssef, N., Sheik, C.S., Krumholz, L.R., Najar, F.Z., Roe, B.A., Elshahed, M.S. Comparison of species richness estimates obtained using nearly complete fragments and simulated pyrosequencing-generated fragments in 16S rRNA gene-based environmental surveys. Appl Environ Microbiol. 2009, 75(16), 5227–5236.
[26] Kumar, P.S., Brooker, M.R., Dowd, S.E., Camerlengo, T. Target region selection is a critical determinant of community fingerprints generated by 16S Pyrosequencing. PLoS One. 2011, 6 (6), Article ID e20956.
[27] Vasileiadis, S., Puglisi, E., Arena, M., Cappa, F., Cocconcelli, P.S., Trevisan, M. Soil bacterial diversity screening using single 16S rRNA gene V regions coupled with multi-million read generating sequencing technologies. PLoS One. 2012, 7(8), e42671. doi: 10.1371/journal.pone.0042671
[28] Hillmann, B., Al-Ghalith, G.A., Shields-Cutler, R.R., Zhu, Q., Gohl, D.M., Beckman, K.B. et al., Evaluating the information content of shallow shotgun metagenomics. mSystems. 2018, 3, e00069–18. doi: 10.1128/mSystems.00069-18.
[29] Quince, C., Walker, A.W., Simpson, J.T., Loman, N.J., Segata, N. Shotgun metagenomics, from sampling to analysis. Nat Biotechnol. 2017, 35, 833–844. doi: 10.1038/nbt.3935.
[30] Vermote, L., Verce, M., de Vuyst, L., Weckx, S. Amplicon and shotgun metagenomic sequencing indicates that microbial ecosystems present in cheese brines reflect environmental inoculation during the cheese production process. Int Dairy J. 2018, 87, 44–53. doi: 10.1016/j.idairyj.2018. 07.010.
[31] Ranjan, R., Rani, A., Metwally, A., McGee, H.S., Perkins, D.L. Analysis of the microbiome: Advantages of whole genome shotgun versus 16S amplicon sequencing. Biochem Biophys Res Commun. 2016, 469, 967–977. doi: 10.1016/j.bbrc.2015.12.083.
[32] Chandran, H., Meena, M., Sharma, K. Microbial Biodiversity and Bioremediation Assessment through Omics Approaches. Frontiers in Environmental Chemistry. 25 September 2020, doi: 10.3389/fenvc.2020.570326.
[33] Chandran, H., Meena, M., Barupal, T., Sharma, K. Plant tissue culture as a perpetual source for production of industrially important bioactive compounds. Biotechnol Rep. 2020, 26, e00450. doi: 10.1016/j.btre.2020.e00450.
[34] Rhee, S.K., Liu, X., Wu, L., Chong, S.C., Wan, X., Zhou, J. Detection of genes involved in biodegradation and biotransformation in microbial communities by using 50-mer oligonucleotide microarrays. Appl Environ Microbiol. 2004, 70, 4303–4317. doi: 10.1128/AEM.70.7.4303-4317.2004.
[35] Chittora, D., Meena, M., Barupal, T., Swapnil, P., Sharma, K. Cyanobacteria as a source of biofertilizers for sustainable agriculture. Biochem Biophys Rep. 2020, 22, 100737. doi: 10.1016/j.bbrep.2020.100737.
[36] Greene, E.A., Voordouw, G. Analysis of environmental microbial communities by reverse sample genome probing. J Microbiol Methods. 2003, 53, 211–219. doi: 10.1016/S0167-7012(03)00024-1.
[37] Akondi, K.B., Lakshmi, V.V. Emerging Trends in Genomic Approaches for Microbial Bioprospecting. OMICS. 2013 Feb;17(2):61–70. doi: 10.1089/omi.2012.0082.
[38] Zhou, J. Microarrays for bacterial detection and microbial community analysis. Curr Opin Microbiol. 2003, 6(3), 288–294.
[39] Gao, H., Yang, Z.K., Gentry, T.J., Wu, L., Schadt, C.W., Zhou, J. Microarray-based analysis of microbial community RNAs by whole-community RNA amplification. Appl Environ Microbiol. 2007, 73, 563–571.

[40] Nikolaki, S., Tsiamis, G. Microbial diversity in the era of omic technologies. Biomed Res Int. 2013, 2013, 958719. doi: 10.1155/2013/958719.
[41] Sharmin, F., Wakelin, S., Huygens, F., Hargreaves, M. Firmicutes dominate the bacterial taxa within sugar-cane processing plants. Sci Rep. 2013, 3, 3107. doi: 10.1038/srep03107.
[42] Brodie, E.L., DeSantis, T.Z., Joyner, D.C. et al., Application of a high-density oligonucleotide microarray approach to study bacterial population dynamics during uranium reduction and reoxidation. Appl Environ Microbiol. 2006, 72(9), 6288–6298.
[43] DeSantis, T.Z., Dubosarskiy, I., Murray, S.R., Andersen, G.L. Comprehensive aligned sequence construction for automated design of effective probes (CASCADE-P) using 16S rDNA. Bioinformatics. 2003, 19(12), 1461–1468.
[44] Cox, M.J., Huang, Y.J., Fujimura, K.E. et al., Lactobacillus casei abundance is associated with profound shifts in the infant gut microbiome. PLoS One. 2010, 5(1) Article ID, e8745.
[45] Kellogg, C.A., Piceno, Y.M., Tom, L.M., DeSantis, T.Z., Zawada, D.G., Andersen, G.L. PhyloChip microarray comparison of sampling methods used for coral microbial ecology. J Microbiol Methods. 2012, 88(1), 103–109.
[46] Korves, T.M., Piceno, Y.M., Tom, L.M. et al., Bacterial communities in commercial aircraft high-efficiency particulate air (HEPA) filters assessed by PhyloChip analysis. Indoor Air. 2013, 23(1), 50–61.
[47] Reith, F., Brugger, J., Zammit, C.M. et al., Influence of geogenic factors on microbial communities in metallogenic Australian soils. ISME J. 2012, 6, 2107–2118.
[48] Sagaram, U.S., Deangelis, K.M., Trivedi, P., Andersen, G.L., Lu, S.-E., Wang, N. Bacterial diversity analysis of huanglongbing pathogen-infected citrus, using phyloChip arrays and 16S rRNA gene clone library sequencing. Appl Environ Microbiol. 2009, 75(6), 1566–1574.
[49] Tsiamis, G., Katsaveli, K., Ntougias, S. et al., Prokaryotic community profiles at different operational stages of a Greek solar saltern. Research in Microbiology. 2008, 159(9-10), 609–627.
[50] Tsiamis, G., Tzagkaraki, G., Chamalaki, A. et al., Olive-mill wastewater bacterial communities display a cultivar specific profile. Curr Microbiol. 2012, 64(2), 197–203.
[51] Palmer, C., Bik, E.M., Eisen, M.B., Eckburg, P.B., Sana, T.R., Wolber, P.K. et al., Rapid quantitative profiling of complex microbial populations. Nucleic Acids Res. 2006, 34, e5. doi: 10.1093/nar/gnj007.
[52] Shi, Z.J., Yin, H., Van Nostrand, J.D., Voordeckers, J.W., Tu, Q., Deng, Y. et al., Functional gene array-based ultrasensitive and quantitative detection of microbial populations in complex communities. mSystems. 2019, 4, e00296–e00219. doi: 10.1128/mSystems.00296-19.
[53] van Nostrand, J.D., He, Z., Zhou, J. "Dynamics of microbes in the natural setting: Development of the Geochip,". In: eds Sen, K., Ashbolt, N.J. In Environmental Microbiology: Current Technology and Water Applications. Norfolk, Caister Academic Press, 2011, 316.
[54] Zehra, A., Meena, M., Swapnil, P., Raytekar, N.A., Upadhyay, R.S. Sustainable approaches to remove heavy metals from water. In: eds Singh, J., Vyas, A., Wang, S., Prasad, R. Microbial Biotechnology: Basic Research and Applications. Singapore, Springer, 2020, 127–146. doi: 10.1007/978-981-15-2817-0_6.
[55] Pareek, C.S., Smoczynski, R., Tretyn, A. Sequencing technologies and genome sequencing. J Appl Genet. 2011, 52(4), 413–435.
[56] Zhang, J., Chiodini, R., Badr, A., Zhang, G. The impact of next-generation sequencing on genomics. J Genet Genom. 2011, 38, 95–109. doi: 10.1016/j.jgg.2011.02.003.
[57] Samorodnitsky, E., Jewell, B.M., Hagopian, R., Miya, J., Wing, M.R., Lyon, E. et al., Evaluation of hybridization capture versus amplicon-based methods for whole-exome sequencing. Hum Mutat. 2015, 36, 903–914. doi: 10.1002/humu.22825.

[58] Hodkinson, B.P., Grice, E.A. Next-generation sequencing: A review of technologies and tools for wound microbiome research. Adv Wound Care. 2015, 4, 50–58. doi: 10.1089/wound.2014.0542.

[59] Nikolaki, S., Tsiamis, G. Microbial Diversity in the Era of Omic Technologies. Biomed Res Int. 2013, 2013, 958719. doi: 10.1155/2013/958719.

[60] Kchouk, M., Gibrat, J.F., Elloumi, M. Generations of sequencing technologies: From first to next generation. Biol Med. 2017, 9, 395. doi: 10.4172/0974-8369.1000395.

[61] Ronaghi, M., Karamohamed, S., Pettersson, B., Uhlén, M., Nyrén, P. Real-time DNA sequencing using detection of pyrophosphate release. Anal Biochem. 1996, 242, 84–89. doi: 10.1006/abio.1996.0432.

[62] El-Metwally, S., Ouda, O.M., Helmy, M. Next Generation Sequencing Technologies and Challenges in Sequence Assembly. New York, NY, Springer Verlag, 2014.

[63] Dinsdale, E.A., Edwards, R.A., Hall, D., Angly, F., Breitbart, M., Brulc, J.M. et al., Functional metagenomic profiling of nine biomes. Nature. 2008, 452, 629–632. doi: 10.1038/nature06810.

[64] Morozova, O., Hirst, M., Marra, M.A. Applications of new sequencing technologies for transcriptome analysis. Annu Rev Genomics Hum Genet. 2009, 10, 135–151.

[65] Turcatti, G., Romieu, A., Fedurco, M., Tairi, A.P. A new class of cleavable fluorescent nucleotides: Synthesis and optimization as reversible terminators for DNA sequencing by synthesis. Nucleic Acids Res. 2008, 36, e25. doi: 10.1093/nar/gkn021.

[66] Meena, M., Samal, S. Alternaria host-specific (HSTs) toxins: An overview of chemical characterization, target sites, regulation and their toxic effects. Toxicol Rep. 2019, 6, 745–758. doi: 10.1016/j.toxrep.2019.06.021.

[67] Arazoe, T., Kondo, A., Nishida, K. Targeted nucleotide editing technologies for microbial metabolic engineering. Biotechnol J. 2018, 13, e1700596. doi: 10.1002/biot.201700596.

[68] Reuter, J.A., Spacek, D.V., Snyder, M.P. High-throughput sequencing technologies. Mol Cell. 2015, 58, 586–597. doi: 10.1016/j.molcel.2015. 05.004.

[69] Stepanauskas, R. Single cell genomics: An individual look at microbes. Curr Opin Microbiol. 2012, 15(5), 613–620.

[70] Hess, M., Sczyrba, A., Egan, R. et al., Metagenomic discovery of biomass-degrading genes and genomes from cow rumen. Science. 2011, 331(6016), 463–467.

[71] Swan, B.K., Martinez-Garcia, M., Preston, C.M. et al., Potential for chemolithoautotrophy among ubiquitous bacteria lineages in the dark ocean. Science. 2011, 333(6047), 1296–1300.

[72] Marcy, Y., Ouverney, C., Bik, E.M., Losekann, T., Ivanova, N., Martin, H.G., Szeto, E., Platt, D., Hugenholtz, P., Relman, D.A. et al., Dissecting biological "dark matter" with single-cell genetic analysis of rare and uncultivated TM7 microbes from the human mouth. Proc Natl Acad Sci USA. 2007, 104, 11889–11894.

[73] Blainey, P.C., Mosier, A.C., Potanina, A., Francis, C.A. Quake SR: Genome of a low-salinity ammonia oxidizing archaeon determined by single-cell and metagenomic analysis. PLoS ONE. 2011, 6, e16626.

[74] Moran, M.A. Metatranscriptomics: Eavesdropping on complex microbial communities. Microbe. 2009, 4(7), 7.

[75] Zhang, W., Feng, L., Nie, L. Integrating Multiple 'Omics' Analysis for Microbial Biology: Application and Methodologies. Microbiology (Reading). 2010 Feb, 156(Pt 2), 287–301. doi: 10.1099/mic.0.034793-0.

[76] Stedtfeld, R.D., Baushke, S.W., Tourlousse, D.M., Miller, S.M., Stedtfeld, T.M., Gulari, E., Tiedje, J.M., Hashsham, S.A. Development and experimental validation of a predictive threshold cycle equation for quantification of virulence and marker genes by high-throughput nanoliter-volume PCR on the OpenArray platform. Appl Environ Microbiol. 2008, 74, 3831–3838.

[77] Ozsolak, F., Platt, A.R., Jones, D.R., Reifenberger, J.G., Sass, L.E., Mclnerney, P., Thompson, J.F., Bowers, J., Jarosz, M., Milos, P.M. Direct RNA sequencing. Nature. 2009, 461, 814–818.
[78] Frias-Lopez, J., Shi, Y., Tyson, G.W., Coleman, M.L., Schuster, S.C., Chisholm, S.W., Delong, E.F. Microbial community gene expression in ocean surface waters. Proc Natl Acad Sci U S A. 2008, 105, 3805–3810.
[79] Giannoukos, G. et al., Efficient and robust RNA-seq process for cultured bacteria and complex community transcriptomes. Genome Biol. 2012, 13(3), R23. doi: 10.1186/gb-2012-13-3-r23.
[80] Willner, D. et al., Metagenomic analysis of respiratory tract DNA viral communities in cystic fibrosis and non-cystic fibrosis individuals. PLoS ONE. 2009, 4, e7370. [PubMed: 19816605].
[81] Franzosa, E.A., Tiffany, H., Sirota-Madi, A., Shafquat, A., Galeb Abu-Ali, X.C. Morgan, and Curtis Huttenhower. Sequencing and Beyond: Integrating Molecular 'Omics for Microbial Community Profiling. 2015 Jun, 13(6), 360–72. doi: 10.1038/nrmicro3451.
[82] Mason, O.U., Hazen, T.C., Borglin, S., Chain, P.S.G., Dubinsky, E.A., Fortney, J.L. et al., Metagenome, metatranscriptome and single-cell sequencing reveal microbial response to Deepwater Horizon oil spill. ISME J. 2012, 6, 1715–1727.
[83] Selinger, D.W., Saxena, R.M., Cheung, K.J., Church, G.M., Rosenow, C. Global RNA half-life analysis in Escherichia coli reveals positional patterns of transcript degradation. Genome Res. 2003, 13, 216_223.
[84] Bailly, J., Fraissinet- Tachet, L., Verner, M.C., Debaud, J.C., Lemaire, M., Wesolowski Louvel, M. et al., Soil eukaryotic functional diversity, a metatranscriptomic approach. ISME J. 2007, 1, 632–642.
[85] Maroli, A.,.S., Gaines, T.A., Foley, M.E., Duke, S.O., Do˘gramacı, M., Anderson, J.V. et al., Omics in weed science: A perspective from genomics, transcriptomics, and metabolomics approaches. Weed Sci. 2018, 66, 681–695. doi: 10.1017/wsc.2018.33.
[86] Nagalakshmi, U., Waern, K., Snyder, M. RNA-Seq: A method for comprehensive transcriptome analysis. Curr Protoc Mol Biol. 2010, 4, 1–13. doi: 10.1002/0471142727.mb0411s89.
[87] Shendure, J. The beginning of the end for microarrays? Nat. Methods. 2008, 5, 585–587. doi: 10.1038/nmeth0708-585.
[88] Gao, H., Wang, Y., Liu, X., Yan, T., Wu, L., Alm, E. et al., Global transcriptome analysis of the heat shock response of Shewanella oneidensis. J Bacteriol. 2004, 186, 7796–7803. doi: 10.1128/JB.186.22.7796-7803. 2004.
[89] Yao, C., Chen, B.H., Joehanes, R., Otlu, B., Zhang, X., Liu, C. et al., Integromic analysis of genetic variation and gene expression identifies networks for cardiovascular disease phenotypes. Circulation. 2015, 131, 536–549. doi: 10.1161/CIRCULATIONAHA.114.010696.
[90] Pantaleo, M.A., Ravegnini, G., Astolfi, A., Simeon, V., Nannini, M., Saponara, M. et al., Integrating miRNA and gene expression profiling analysis revealed regulatory networks in gastrointestinal stromal tumors. Epigenomics. 2016, 8, 1347–1366. doi: 10.2217/epi-2016-0030.
[91] Bomar, L., Maltz, M., Colston, S., Graf, J. Directed culturing of microorganisms using metatranscriptomics. MBio. 2011, 2, 1–8.
[92] Keller, M., Hettich, R. Environmental proteomics: A paradigm shift in characterizing microbial activities at the molecular level. Microbiol Mol Biol Rev. 2009, 73, 62–70, 1.
[93] Graham, R.L.J., Graham, C., McMullan, G. Microbial proteomics: A mass spectrometry primer for biologists. Microb Cell Fact. 2007, 6, 26.
[94] James, P. Protein identification in the post-genome era: The rapid rise of proteomics. Q Rev Biophys. 1997, 30, 279–333.

[95] Kim, H.J., Ishidou, E., Kitagawa, E., Momose, Y., Iwahashi, H. A yeast DNA microarray for the evaluation of toxicity in environmental water containing burned ash. Environ Monit Assess. 2004, 92, 253–272. doi: 10.1023/B:EMAS.0000014504.03500.41.
[96] Hart, E.H., Creevey, C.J., Hitch, T., Kingston-Smith, A.H. Metaproteomics of rumen microbiota indicates niche compartmentalisation and functional dominance in a limited number of metabolic pathways between abundant bacteria. Sci Rep. 2018, 8, 10504. doi: 10.1038/s41598-018-28827-7.
[97] Verberkmoes, N.C., Denef, V.J., Hettich, R.L., Banfield, J.F. Systems biology: Functional analysis of natural microbial consortia using community proteomics. Nat Rev Microbiol. 2009, 7, 196–205. doi: 10.1038/nrmicro 2080.
[98] Wang, D.Z., Xie, Z.X., Zhang, S.F. Marine meta proteomics: Current status and future directions. J Proteom. 2014, 97, 27–35. doi: 10.1016/j.jprot.2013.08.024.
[99] Wu, C.C., Yates, J.R. The application of mass spectrometry to membrane proteomics. Nat Biotechnol. 2003, 21, 262–267. doi: 10.1038/nbt0303-262.
[100] Lau, H.T., Suh, H.W., Golkowski, M., Ong, S.E. Comparing SILACand stable isotope dimethyl-labeling approaches for quantitative proteomics. J Proteome Res. 2014, 13, 4164–4174. doi: 10.1021/pr500630a.
[101] Bettmer, J. Application of isotope dilution ICP-MS techniques to quantitative proteomics. Anal Bioanal Chem. 2010, 397, 3495–3502. doi: 10.1007/s00216-010-3861-y.
[102] Ward, M.A., Hyman, B., Scherzer, C., Trisini-Lipsanopoulos, A., Ivinson, A.J., Lößner, C. et al., Exploration of plasma biomarkers for Alzheimer's disease using isotopic tandem mass tags and a combined directed/data-dependent acquisition NLC-MS/MS method. Alzheimer's Dement J Alzheimer's Assoc. 2014, 10, P674. doi: 10.1016/j.jalz.2014.05.1215.
[103] Martínez-Esteso, M.J., Casado-Vela, J., Sellés-Marchart, S., Pedreño, M.A., Bru-Martínez, R. Differential plant proteome analysis by isobaric tags for relative and absolute quantitation (iTRAQ). Methods Mol Biol. 2014, 1072, 155–169. doi: 10.1007/978-1-62703-631-3_12.
[104] Hägglund, P., Bunkenborg, J., Maeda, K., Svensson, B. Identification of thioredoxin disulfide targets using a quantitative proteomics approach based on isotope-coded affinity tags. J Proteome Res. 2008, 7, 5270–5276. doi: 10.1021/pr800633y.
[105] Mattarozzi, M., Manfredi, M., Montanini, B., Gosetti, F., Sanangelantoni, A.M., Marengo, E. et al., A metaproteomic approach dissecting major bacterial functions in the rhizosphere of plants living in serpentine soil. Anal Bioanal Chem. 2017, 409, 2327–2339. doi: 10.1007/s00216-016-0175-8.
[106] Wang, D.Z., Kong, L.F., Li, Y.Y., Xie, Z.X. Environmental microbial community proteomics: Status, challenges and perspectives. Int J Mol Sci. 2016, 17, 1275. doi: 10.3390/ijms17081275.
[107] Dunn, W.B. Current trends and future requirements for the mass spectrometric investigation of microbial, mammalian, and plant metabolomes. Phys Biol. 2008, 5, 1–24.
[108] Bargiela, R., Herbst, F.A., Martínez-Martínez, M., Seifert, J., Rojo, D., Cappello, S. et al., Metaproteomics and metabolomics analyses of chronically petroleum-polluted sites reveal the importance of general anaerobic processes uncoupled with degradation. Proteomics. 2015, 15, 3508–3520. doi: 10.1002/pmic.201400614.
[109] Krumsiek, J., Mittelstrass, K., Do, K.T., Stückler, F., Ried, J., Adamski, J. et al., Gender-specific pathway differences in the human serum metabolome. Metabolomics. 2015, 11, 1815–1833. doi: 10.1007/s11306-015-0829-0.
[110] Hill, C.B., Czauderna, T., Klapperstück, M., Roessner, U., Schreiber, F. Metabolomics, standards, and metabolic modeling for synthetic biology in plants. Front Bioeng Biotechnol. 2015, 3, 167. doi: 10.3389/fbioe.2015.00167.

[111] Wang, J.H., Byun, J., Pennathur, S. Analytical approaches to metabolomics and applications to systems biology. Semin Nephrol. 2010, 30, 500–511. doi: 10.1016/j.semnephrol.2010.07.007.
[112] Aldridge, B.B., Rhee, K.Y. Microbial metabolomics: Innovation, application, insight. Curr Opin Microbiol. 2014, 19, 90–96. doi: 10.1016/j.mib.2014.06.009.
[113] Pan, Z., Raftery, D. Comparing and combining NMR spectroscopy and mass spectrometry in metabolomics. Anal Bioanal Chem. 2007, 387, 525–527. doi: 10.1007/s00216-006-0687-8.
[114] McMahon, M.A., Prakash, T.P., Cleveland, D.W., Bennett, C.F., Rahdar, M. Chemically modified Cpf1-CRISPR RNAs mediate efficient genome editing in mammalian cells. Mol Ther. 2018, 26, 1228–1240. doi: 10.1016/j.ymthe.2018.02.031.
[115] Ter Kuile, B.H., Westerhoff, H.V. Transcriptome meets metabolome: Hierarchical and metabolic regulation of the glycolytic pathway. FEBS Lett. 2001, 500, 169–171.
[116] Oliver, S.G., Winson, M.K., Kell, D.B., Baganz, F. Systematic functional analysis of the yeast genome. Trends Biotechnol. 1998, 16, 373–378.
[117] Raamsdonk, L.M., Teusink, B., Broadhurst, D., Zhang, N., Hayes, A., Walsh, M.C., Berdon, J.A., Brindle, K.M., Kell, D.B. &#x0026 other authors A functional genomics strategy that uses metabolome data to reveal the phenotype of silent mutations. Nat Biotechnol. 2001, 19, 45–50.
[118] Dunn, W.B., Bailey, N.J., Johnson, H.E. Measuring the metabolome: Current analytical technologies. Analyst. 2005, 130, 606–625.
[119] Dettmer, K., Aronov, P.A., Hammock, B.D. Mass spectrometry-based metabolomics. Mass Spectrom Rev. 2007, 26, 51–58. doi: 10.1002/mas.20108.
[120] Lee, W.N., Go, V.L. Nutrient-gene interaction: Tracer-based metabolomics. J Nutr. 2005, 135, 3027S–3032S. doi: 10.1093/jn/135.12.3027S.
[121] Wiechert, W. 13C metabolic flux analysis. Metab Eng. 2001, 3, 195–206. doi: 10.1006/mben.2001.0187.
[122] Sauer, U. Metabolic networks in motion: 13C-based flux analysis. Mol Syst Biol. 2006, 2, 62. doi: 10.1038/msb4100109.
[123] Tang, Y.J., Hwang, J.S., Wemmer, D.E., Keasling, J.D. Shewanella oneidensis MR-1 fluxome under various oxygen conditions. Appl Environ Microbiol. 2007, 73, 718–729. doi: 10.1128/AEM.01532-06.
[124] Maphosa, F., Smidt, H., de Vos, W.M., Roling, W.H.F.M. Microbial community and metabolite dynamics of an anoxic dechlorinating bioreactor. Stud Environ Sci. 2010b, 44, 4884–4890. doi: 10.1021/es903721s.
[125] Wierckx, N.J.P., Ballerstedt, H., De Bont, J.A.M., DeWinde, J.H., Ruijssenaars, H.J., Wery, J. Transcriptome analysis of a phenolproducing Pseudomonas putida S12 construct: Genetic and physiological basis for improved production. J Bacteriol. 2008, 190, 2822–2830. doi: 10.1128/JB. 01379–07.
[126] Kukurugya, M.A., Mendonca, C.M., Solhtalab, M., RWilkes, R.A., Thannhauser, T.W., Aristilde, L. Multi-omics analysis unravels a segregated metabolic flux network that tunes co-utilization of sugar and aromatic carbons in Pseudomonas putida. J Biol Chem. 2019, 294, 8464–8479. doi: 10.1074/jbc.RA119.007885.
[127] Barberan, A., Bates, S.T., Casamayor, E.O., Fierer, N. Using network analysis to explore co-occurrence patterns in soil microbial communities. ISME J. 2012, 6, 343–351.
[128] Duran-Pinedo, A.E., Paster, B., Teles, R., Frias-Lopez, J. Correlation network analysis applied to complex biofilm communities. PLoS One. 2011, 6, e28438.
[129] Malik Seidu, B., Michael Megharaj, M., Naidu, R. The use of molecular techniques to characterize the microbial communities in contaminated soil and water. Environ Intl. 2008, 34, 265–276.

[130] Stenuit, B., Eyers, L., Schuler, L., Agathos, S.N., George, I. Emerging high-throughput approaches to analyze bioremediation of sites contaminated with hazardous and/or recalcitrant wastes. Biotechnol Adv. 2008, 26, 561–575.

[131] Laczi, K., Erdeiné Kis, Á., Szilágyi, Á., Bounedjoum, N., Bodor, A., Vincze, G.E., Kovács, T., Rákhely, G., Perei, K. New Frontiers of Anaerobic Hydrocarbon Biodegradation in the Multi-Omics Era. Front Microbiol. 2020, 11, 2886.

[132] Gutleben, J., Chaib De Mares, M., van Elsas, J.D., Smidt, H., Overmann, J., Sipkema, D. The multi-omics promise in context: From sequence to microbial isolate. Crit Rev Microbiol. 2018, 44(2), 212–229. DOI:10.1080/1040841X.2017.1332003.

[133] Wang, P.I., Hwang, S., Kincaid, R.P., Sullivan, C.S., Lee, I., Marcotte, E.M. RIDDLE: Reflective diffusion and local extension reveal functional associations for unannotated gene sets via proximity in a gene network. Genome Biol. 2012a, 13, R125.

[134] Forster, J., Gombert, A.K., Nielsen, J. A functional genomics approach using metabolomics and in silico pathway analysis. Biotechnol Bioeng. 2002, 79, 703–712.

Illathu Kandy Nidhin and Indranil Chattopadhyay

2 Emerging technologies in environmental microbiology

Microbes in the environment

Abstract: Microorganisms inhabit most of the ecosystems around us, including soil, rhizosphere, phyllosphere, coral reefs, insect guts, streams, ocean, and even atmospheric aerosols. Even remote ecosystems with extreme geographic and climatic conditions are inhabited by microorganisms. Bacteria, archaea, and fungi are the major microbial groups found. Proteobacteria, Acidobacteria, Actinobacteria, Thaumarchaeota, Euryarchaeota, Ascomycota, and Basidiomycota are some of the major microbial phyla found in most of the ecosystems. Metagenomic approaches enable efficient analysis of microbial compositions and abundances of various microbiomes. In this chapter, we discuss the abundance and composition of microbes present in various terrestrial and aquatic habitats.

Keywords: Microbes, metagenomic, environment, climate

2.1 Introduction

Flemming and Wuertz [1], in their review on the abundance of archaea and bacteria on the Earth, estimated that the total archaeal and bacterial cell count around the globe is roughly 1.2×10^{30} cells. Archaea, bacteria, and other microbes can aggregate and grow together into layers that may be attached to some substrates or interfaces to form the biofilms. These biofilms may have some new characteristics and properties developed due to this aggregation which enables them to perform the novel functions, which were unable to perform by any of the free-living members alone. Deep continental and oceanic subsurfaces, oceans, soils, and upper oceanic sediments are the major habitats, where the majority of the archaea and bacteria inhabit. Among these, deep oceanic subsurface is the largest microbial habitat, in terms of both area and abundance, and it is suggested that 20–80% of the microbial cells in this habitat belongs to biofilms. Biofilms are also found in the deep continen-

Illathu Kandy Nidhin, Dept. of Life Sciences, Central University of Tamil Nadu, Thiruvarur, Tamil Nadu, India
Indranil Chattopadhyay, Dept. of Life Sciences, Central University of Tamil Nadu, Thiruvarur, Tamil Nadu, India, e-mail: indranil@cutn.ac.in, indranil_ch@yahoo.com

https://doi.org/10.1515/9783110727227-002

tal subsurface, and here also 20–80% of microbial cells belong to biofilms. Till 50 cm depth of upper oceanic sediment layer, most of the microbial cells are found in biofilms. Biofilms play a crucial role in ecological processes and biogeochemical cycles in the soil layers, and almost 3×10^{29} archaeal and bacterial cells belong to biofilms. Even though in the ocean waters most of the microbial cells belong to free-living microbial groups, the biofilms are found in the ocean surface microlayer and associated with the coral reefs, planktons, organic debris, and seaweeds. Fungal cells, spores, and bacteria together form the microbial aerosols which constitute 74% of the atmospheric aerosol.

2.2 Terrestrial

Microbes occupy every microniche available in the terrestrial ecosystems all over the world, including deep soil layers and extreme habitats like permafrost soils. Microbes are seen abundant in soils, especially in the forest soils. Microbes are also found associated with other macroorganisms like plants and animals. Rhizosphere, phyllosphere, and gut microbiomes are the major examples of microbial groups in association with other higher organisms.

2.2.1 Soil

Baldrian [2] reviewed the forest soil microbial communities and ecosystem processes. Bacteria, archaea, molds, saprotrophic fungi, and ectomycorrhizal fungi play a crucial role in the ecosystem processes and dynamics. The archaeal groups are comparatively less abundant in forest soils compared to the fungal and bacterial groups, and the bacteria are the most abundant group. The impact of seasonal variations and litter input over the microbial communities is studied and reported, whereas the impact of circadian rhythm and the long-term processes over the microbial communities is less studied. Ectomycorrhizal fungi are the microbial group that is mostly affected by the seasonal variations because of their close interactions with the tree roots. The abundance of ectomycorrhizal fungi is found to be lower in the winter compared to that in the summer season. Bacteria and saprotrophic fungi are also affected by the seasonal variation in root exudates and rhizodeposition. The competition between saprotrophic fungi and ectomycorrhizal fungi is also affected by seasonality. Ectomycorrhizal fungi and rhizosphere-associated bacteria absorb carbon from the plant hosts. Molds and bacteria mainly use the decomposition of simple organic compounds to absorb carbon, whereas saprotrophic fungi depend on complex organic substances to obtain carbon.

2.2.1.1 Tropical forest soil

Pajares et al. [3] reported that Actinobacteria, Firmicutes, Proteobacteria, Acidobacteria, Chloroflexi, Bacteroidetes, Gemmatimonadetes, and Verrucomicrobia *were the other major phyla present in a seasonally dry tropical forest in* Yucatan Peninsula, Mexico. In a study, Pedrinho et al. [4] reported that bacteria were the major microbial group present in primary and secondary forests and pasture land in the Amazon forests. Proteobacteria and Actinobacteria were the most dominant bacterial phyla, followed by Acidobacteria and Firmicutes. It was observed that Bacteroidetes, Firmicutes, and Chloroflexi had increased abundance in the pasture lands, whereas Proteobacteria had a higher abundance in the secondary forests and Thaumarchaeota and Actinobacteria were major groups present in the primary forests. The major genera observed were *Mycobacterium*, *Rhodopseudomonas*, *Bradyrhizobium*, *Frankia*, *Nitrobacter*, and *Acidothermus*. The primary and secondary forests and pasture lands had various abundances of various genera. *Frankia*, *Mycobacterium*, and *Acidothermus* had higher abundance in the primary forests and *Bradyrhizobium*, *Burkholderia*, and *Rhodopseudomonas* were having higher abundance in secondary forests, whereas the pasture land had a higher abundance of many genera including *Geobacter*, *Bacillus*, and *Anaeromyxobacter*. Further, in a study, Benucci et al. [5] found that Proteobacteria, Actinobacteria, and Acidobacteria were the most abundant prokaryotic community, and other abundant phyla observed were Verrucomicrobia, Chloroflexi, Planctomycetes, Crenarchaeota, and Nitrospirae. Ascomycota, Basidiomycota, and Mucoromycota were the dominant fungal phyla and Glomeromycota, Chytridiomycota, and Rozellomycota were the other abundant fungi phyla present. *Glomus*, *Scutellospora*, *Rhizophagus*, *Septoglomus*, and *Acaulospora* were major arbuscular mycorrhizal fungi, and *Tomentella*, *Chloridium*, *Entoloma*, *Cenococcum*, *Inocybe*, and *Amanita* were the major ectomycorrhizal fungi present.

Thaumarchaeota was the most dominant archaeal phylum found, followed by phylum Euryarchaeota. Nitrososphaerales and Cenarchaeales were the abundant Thaumarchaeota orders present. Proteobacteria, Actinobacteria, Acidobacteria, and Thermoleophilia were the major bacterial phyla found. The major fungal phyla present were Ascomycota, Mucoromycota, and Basidiomycota. Mucoromycotina and Agaricomycetes were the most abundant classes belonging to the phyla Mucoromycota and Basidiomycota, respectively, whereas the major classes belonging to Ascomycota were Eurotiomycetes, Dothideomycetes, Sordariomycetes, and Leotiomycetes [6]. Kerfahi et al. [7] reported that Thaumarchaeota and Euryarchaeota were the major archaeal phyla found in the tropical rainforests in Brunei and Borneo [7]. The major bacterial phyla present in the evergreen rainforest soils of the Amazon were Proteobacteria, Acidobacteria, Firmicutes, Actinobacteria, and Chloroflexi [8]. Lacerda-Júnior et al. [9] found that Actinobacteria, Proteobacteria, Acidobacteria, Cyanobacteria, and Bacteroidetes were the major bacterial phyla present in the Brazilian dry forest soils.

2.2.1.2 Temperate forest soil

In the forests, bacteria, fungi, and archaea are found in various habitats, including soil, phyllosphere, rhizosphere, deadwood, barks, litter, and even on rock surfaces and in atmospheric aerosol. These microbes are also seen in association with invertebrates like insects. All these forest habitats have different compositions, richness, and abundance of microbes. Fungal spores and bacteria are the major microbes found in the forest atmosphere, and bacteria, lichens, and ectomycorrhizal fungi are major microbes found on the rock surfaces. The bacterial phyla including Proteobacteria, Actinobacteria, Firmicutes, and Acidobacteria, and the fungi including Leotiomycetes and Dothideomycetes are the major microbial groups present in the phyllosphere. In the streams and lakes, the major microbial groups observed were Bacteroidetes, Actinobacteria, Proteobacteria, Ascomycota, and protists. Soil habitats have a high abundance of many bacterial phyla, and saprotrophic and mycorrhizal fungi are also found. The rhizosphere microbial community has a high abundance of many symbiotic mycorrhizal fungi and associated bacteria and is significantly different from the soil microbial community. Many invertebrates including woodwasps, termites, and beetles have symbiotic microbes associated with them, especially in the guts. Actinobacteria, Bacteroidetes, Proteobacteria, and some fungi are the major gut microbiome observed in insects [10]. Miyashita et al. 11 studied the microbial composition in temperate forests in Japan and tropical forests in Southeast Asia. In all the forest samples, the most dominant bacterial phyla were Acidobacteria and Proteobacteria. The bacterial phyla Verrucomicrobia, Bacteroidetes, Planctomycetes, and Actinobacteria were also abundant in these forests. Alpha-Proteobacteria was the most abundant Proteobacteria class, followed by Beta-Proteobacteria and Delta-Proteobacteria. The valley and remnant forest samples of the tropical forests and the temperate forest samples had a high abundance of Alpha-Proteobacteria and a lower abundance of Beta-Proteobacteria. Acidobacteria_Gp1 and Gp2 were the most abundant classes of Acidobacteria, whereas Spartobacteria and subdivision 3 were the most abundant classes of Verrucomicrobia. Sphingobacteria, Planctomycetacia, and Actinobacteridae were the most abundant classes of the phyla Bacteroidetes, Planctomycetes, and Actinobacteria, respectively.

In a study by Osburn et al. [12] on the fungal and bacterial communities in disturbed temperate forests, they found that Proteobacteria and Acidobacteria were the most abundant bacterial phyla. Agaricomycetes was the most abundant fungal group. Increased abundance of bacterial phyla, Proteobacteria, Actinobacteria, Chloroflexi, and Nitrospirae, was observed in the disturbed forests. It was observed that the soil disturbances had significantly affected the relative abundances of the microbial communities, especially that of the bacterial phyla including Planctomycetes and Acido bacteria. The Mucoromycotina and Geminibasidiomycetes had lower abundance in disturbed soil, whereas Eurotiomycetes and Sordariomycetes had higher abundance in the disturbed soils. In another study, Nguyen et al. [13] reported that Ascomycota and Basidiomycota are the most abundant fungal groups present in a temperate forest. Agaricales, Russulales, Pezizales, and Helotiales are the major fungal orders present. A higher number of symbiotrophic orders were observed compared to the saprotrophic orders. Root NO_3 had a significant negative effect over the fungal orders Russulales, Boletales, and symbiotrophic Pezizales, whereas saprotrophic Pezizales was positively affected by root NO_3. Further, Miyake et al. [14] found that the most dominant arbuscular mycorrhizal fungal group was Glomeraceae, and the other abundant groups were Acaulosporaceae, Diversisporaceae, Claroideoglomeraceae, Archaeosporaceae, Gigasporaceae, Ambisporaceae, and Pacisporaceae. The plant taxa had a significant impact on the temperate forest arbuscular mycorrhizal composition, but the geographic distance had no relevant impact on the fungal composition. The samples from secondary forest and plantation sites differed from each other, but it was unable to make a conclusion based on these vegetation types because in one location the plantations had higher fungal abundance, whereas the secondary forests had higher abundance in another location.

2.2.1.3 Alpine forest soil

Wang et al. [59] studied the soil microbial composition of alpine wetland and alpine forests in the Tibetan Plateau. They found that alpine forests and alpine wetlands had a similar soil microbial composition. Proteobacteria was the dominant phylum found in both the alpine wetland and alpine forest soils. Acidobacteria was the second dominant phylum, followed by Actinobacteria, Chloroflexi, and Bacteroidetes. The other bacterial phyla found in lesser abundance were Gemmatimonadetes, Patescibacteria, Firmicutes, Verrucomicrobia, Rokubacteria, Planctomycetes, Cyanobacteria, and Latescibacteria.

2.2.1.4 Mangrove forest soil

In a study, Behra et al. [15] reported that Proteobacteria, Actinobacteria, Acidobacteria, Bacteroidetes, Firmicutes, Verrucomicrobia, and Ignavibacteria were the major group of bacteria found in the sediments of Bhitarkanika mangrove forest in Odisha. They found that the bacterial composition and abundance were different for the riverine and the bay soils [15]. Also, Huergo et al. [16] studied the influence of carbonaceous debris from the Sambaqui site over the mangrove soil microbiomes in Southern Brazil. Even though the Sambaqui site leads to altering soil pH in the mangrove area, the mangrove microbiome was observed to be resilient toward the change in soil pH. Proteobacteria, Bacteroidetes, Chloroflexi, Planctomycetes, Acidobacteria, Euryarchaeota, Deferribacteres, Bathyarchaeota, Latescibacteria, and Spirochetes were the most abundant phyla present in these mangrove soils. The predominant families found here were Desulfarculaceae, Desulfobacteraceae, and Desulfobulbaceae. It was observed that in the soils influenced by the Sambaqui, the abundance of Bacteroidetes and Chlorobi had decreased and that of Crenarchaeota had increased [16]. In another study by Li et al. [17] on the microbiome composition in various sites in the Dongzhai Harbor mangrove forest in China, they observed that the microbial composition was similar in the mangrove soils of the reserve and the soil near the sea but the microbial composition of the soil near the wastewater sewage was different from the other two samples. Chloroflexi was the most dominant in the soil near the wastewater sewage and Proteobacteria was the second dominant, whereas Proteobacteria was the most dominant in both the other samples, followed by Chloroflexi. Bacteroidetes had a similar abundance in all three sample locations. Actinobacteria, Firmicutes, Acidobacteria, Nitrospirae, and Ignavibacteriae were other major phyla found. Cabral et al. [18] studied the anthropogenic impacts on the mangrove microbiome, and found that *Aromatoleum, Bradyrhizobium, Geobacter* were the major bacterial genera present in the mangrove soils. In both the pristine and oil-contaminated mangrove soils, the genus *Sulforovum* was the most abundant. In the oil-contaminated site, *Rhodopseudomonas* was observed as an abundant genus. The oil spills and sludge or wastewater due to anthropogenic activity were observed to be actively detoxified by the microbial members of *Desulfococcus*, *Desulfitobacterium*, *Vibrio*, and *Desulfatibacillum*.

Ceccon et al. [19] studied the mangrove microbial composition in Brazil and reported that bacteria was the most dominant, followed by archaea. Proteobacteria was the most dominant bacterial phyla, and Actinobacteria, Chloroflexi, Acidobacteria, Bacteroidetes, and Cyanobacteria were the other abundant bacterial phyla present. Crenarchaeota and Euryarchaeota were the abundant archaeal phyla present. Delta-Proteobacteria, Alpha-Proteobacteria, and Gamma-Proteobacteria were the most abundant bacterial classes, and Thermoplasmata was the prevalent archaeal group. Desulfobacterales, Thiotrichales, Chromatiales, and Bacteroidales were the predominant microbial orders present. In another study, Jin et al. [20] described the composition of viruses in mangrove soils and found that Circoviridae, Microviridae, and Nanoviridae were the major viral families present. Phycodnaviridae, Parvoviridae, Geminiviridae, Mimiviridae, and Inoviridae were the other abundant viral families present. Comparative study of samples from the bay, port, and river samples from mangrove forests suggested that virus compositions were different for the three sample sites. Some mycobacteriophages and halophages were found only in the bay samples. The unique viruses found in the river site samples were enterobacteria phage, orthovirus, and cyclovirus. No such unique viruses were reported in the port site samples. The study suggested that compared to the freshwaters, marine waters had a higher influence on the virus composition of this mangrove area.

Shiau et al. [21] studied the soil methanotrophic community in the red mangroves and reported *Methylosarcina* was the abundant methanotroph in both the tidal and dwarf mangrove soils, whereas *Methylomonas* was abundant only in the tidal mangrove soils and *Methylobacter* was only found abundant in the dwarf mangrove soils. *Methylocystis* were found in higher abundance in the dwarf mangrove soils. Tidal mangrove soils had lesser methanotrophic activity compared to the dwarf mangrove soils. In another study, Guerrero-Cruz et al. [22] described that methane oxidation can be performed by the methanotrophic microbes like Proteobacteria and archaeal groups either aerobically or anaerobically. Proteobacteria are aerobic methanotrophs and archaeal groups are anaerobic methanotrophs. Members of bacterial families Methylothermaceae, Methylococcaceae (Gamma-Proteobacteria) Beijerinckiaceae, and Methylocystaceae (Alpha-Proteobacteria) are major aerobic methanotrophs. Verrucomicrobial methanotrophs belonging to the family Methylacidiphilaceae are even present in extreme environments including volcanic and geothermal environments. Some proteobacterial methanotrophs are even seen in thermal lakes, thermal springs, peat soil, and acidic wetlands. Anaerobic methanotrophic archaea are capable of methane oxidation in anaerobic conditions with the help of sulfate-reducing bacteria.

2.2.1.5 Desert soil

Baubin et al. [23] studied the bacterial composition of desert soils and reported Proteobacteria, Actinobacteria, and Bacteroidetes were the major abundant bacterial phyla present. In the dry seasons, Deinococcus-Thermus was found to be the abundant bacterial phylum especially in the sample collected from undershrub and sites near ant nests. Proteobacteria and Actinobacteria were the other abundant phyla found in these sites other than the Deinococcus-Thermus phylum. The bacterial orders Thermales, Rhodobacterales, and Acidimicrobiales were only detected in the DNA-based community. Also, Scola et al. [24] studied the soil microbial composition in various xeric zones in the Namib Desert. They reported that the microbial composition of the fog zone soil was significantly different from the rain zone soils. Soil chemistry and the spatial separation in xeric zones are major factors affecting bacterial compositions. The viral composition also had higher abundance and diversity in the rain zones, which suggests that both the viruses and the bacteria had a similar trend in the various xeric zones. Siphoviridae, Phycodnaviridae, and Mimiviridae were the major viral families observed.

In a study, Araya et al. [25] found that Ascomycota was the most abundant fungal phylum in the soil and rhizosphere microbiome of the Atacama flowering desert and reported that other abundant fungal phyla were Basidiomycota, Chytridiomycota, and Mortierellomycota. The fungi community composition was observed to be different among various plant rhizospheres and also between the bulk soil and rhizosphere.

The most abundant bacterial phyla were Proteobacteria and Actinobacteria, and the other abundant bacterial phyla were Firmicutes, Chloroflexi, Bacteroidetes, Planctomycetes, Acidobacteria, and Gemmatimonadetes. The archaeal groups were found in lower abundance. Thaumarchaeota and Euryarchaeota were the major archaeal phyla present. The Proteobacteria is observed to have a higher abundance in the dry year. Both the rhizosphere fungi and prokaryotic phyla exhibited different abundances in dry and rainy years. Actinobacteria was the most abundant rhizosphere phyla during the rainy year, and Bacteroidetes had decreased abundance in the dry year, whereas the dry and rainy years had made no significant difference in the Proteobacteria abundance in the rhizosphere. The rhizosphere and the bulk soil bacterial composition were observed to be similar in the dry year. Also, Andrew et al. [26] studied the microbial composition of the desert soil of the Sonoran Desert. The desert soil microbiome had the archaeal phylum Crenarchaeota, as the dominant group (18.7%). The bacterial phyla Proteobacteria, Bacteroidetes, Actinobacteria, Firmicutes, and Acidobacteria were dominant. The microbial phyla Chloroflexi, Euryarchaeota, Gemmatimonadetes, Thermomicrobia, Cyanobacteria, Nitrospira, and Verrucomicrobiae were also present in low abundance.

2.2.2 Phyllosphere

Actinobacteria, Bacteroidetes, Gamma-Proteobacteria, and Alpha-Proteobacteria were the major bacterial group present in the phyllosphere of the hardwood forest in Germany [27]. Yao et al. [28] studied the phyllosphere fungal composition of tropical mangrove forests and reported that endophytic fungi were significantly affected by the plant types but epiphytic fungi were not affected by the plant types. The fungal compositions of both the endophytic and epiphytic communities were different from each other. Tremellomycetes and Dothideomycetes were the major fungal groups present in both endophytic and epiphytic communities. Ascomycota was the most dominant, followed by Basidiomycota and Cryptomycota. Also, Laforest-Lapointe et al. [29] studied the phyllosphere bacterial composition of temperate forests and found that Proteobacteria, Bacteroidetes, Actinobacteria, and Acidobacteria were the most abundant bacterial phyla. Among the most dominant classes, four of them belonged to the Proteobacteria, namely Alpha-Proteobacteria, Gamma-Proteobacteria, Beta-Proteobacteria, and Delta-Proteobacteria. Cytophagia, Saprospirae, and Sphingobacteria were the abundant Bacteroidetes classes. The host plant species have a significant impact on the phyllosphere bacterial composition. Moreover, a core microbiome is commonly shared among the tree species. Methylocystaceae, Sphingomonas, Beijerinckia, Acidobacteriaceae, Acetobacteraceae, and Oxalobacteraceae were the major abundant bacterial groups present in the core microbiome.

2.2.3 Rhizosphere

Proteobacteria, Bacteroidetes, Planctomycetes, Euryarchaeota, and Crenarchaeota were the major bacterial and archaeal phyla present in the rhizosphere microbiomes of the Amazon rainforests. Ascomycota and Basidiomycota were the major fungal phyla present. Other bacterial phyla like Acidobacteria and Actinobacteria were also present [30].

Maquia et al. [31] found that Firmicutes, Bacteroidetes, Tenericutes, Proteobacteria, and Actinobacteria were the major bacterial phyla present in the rhizosphere soil microbiomes of subtropical woodlands of the African savanna. Also, Ma et al. [32] studied the microbial composition of temperate forest rhizosphere in Europe. Actinobacteria, Acidobacteria, and Proteobacteria were the abundant rhizosphere bacterial phyla. The major factors that influenced the bacterial abundance were plant species and soil acidity. Bacterial phyla Nitrospirae was positively affected by the soil acidity, whereas Acidobacteria and Chlamydiae were negatively affected. Both the recent forest sites and the ancient forest sites had similar rhizosphere bacterial composition, and there was no significant difference observed between the rhizosphere bacterial composition of the recent and ancient forests.

In a study by Alzubaidy et al. [33] on the rhizosphere microbial community of the *Avicennia marina* in the northern Red Sea mangroves, they found that the bacterial phyla Bacteroidetes, Proteobacteria, and Firmicutes were found in high abundance in the gray mangrove rhizosphere. Methanogens including Methanobacteria, Methanomicrobia, Methanococci, and Methanopyri and sulfate reducers like Delta-Proteobacteria were also dominant in the *Avicennia marina* rhizosphere. The dominant archaeal groups were Methanosarcinales, Methanococcales, Thermococcales, Halobacteriales, Methanobacteriales, and Nitrosopumilales.

2.2.4 Aerosol

The aerosol microbiome in the rainforests of the Amazon was mainly composed of the bacterial families Enterobacteriaceae, Beijerinckiaceae, Polyangiaceae, Bacillaceae, and Ktedonobacteraceae [34].

2.3 Aquatic

Dann et al. [35] studied the niche partitioning of aquatic microbial groups. *Shewanella* dominated the marine samples, and Enterobacteriaceae was dominant in the freshwater samples. *Fusobacterium* and *Arcobacter* were abundant in the marine inoculum, and *Pseudoalteromonas*, Bacteroidales, Spirochaetaceae, and Gamma-Proteobacteria were also present. Bacteroidales and Comamonadaceae were dominant in the freshwater inoculum, and Porphyromonadaceae and *Acinetobacter* were also present.

2.3.1 Freshwater

Zhu et al. [36] studied the spatial and seasonal effects on freshwater microbial communities in Lake Taihu, and reported that Proteobacteria, Cyanobacteria, Actinobacteria, Verrucomicrobia, and Planctomycetes were the bacterial phyla found to be dominant during all four seasons. Cyanobacteria were the most dominant phylum, and *Microcystis*, *Dolichospermum*, and *Synechococcus* were the predominant cyanobacterial genera present. The seasonal variations had a greater impact on the microbial compositions, and the abundant phyla were observed to be in their peak abundance during different seasons. The Proteobacteria was most abundant in the month of January and Cyanobacteria in December, whereas Actinobacteria and Verrucomicrobia were most abundant during April and October, respectively. The density of phytoplankton was the highest during the summer and the chlorophyll a was the highest during the autumn. Also, Eissler et al. [37] studied the aquatic microbiome structure of high-altitude wetlands in Chile, and reported that among the 52 phyla found, the majority were bacterial phyla and only 10 belonged to the archaea. The major archaeal phyla present were Euryarchaeota, Thaumarchaeota, Diapherotrites, and Woesearchaeota. Proteobacteria was the most dominant bacterial phyla. Bacteroidetes, Verrucomicrobia, and Cyanobacteria were other predominant phyla present. Alpha-Proteobacteria, Beta-Proteobacteria, and Gamma-Proteobacteria were the major abundant classes belonging to the phyla Proteobacteria. Ponds were observed to have higher microbial abundance compared to the springs and lagoons. The Bacteroidetes classes Flavobacteria and Cytophagia were abundant in the lagoon.

In a study, Santos-Junior et al. [38] found that the Amazon river microbiome was mainly composed of the genera *Limnohabitans*, *Polynucleobacter*, *Planktophila*, *Methylopumilus*, *Ramlibacter*, and *Fonsibacter*. Further, Findlay et al. [39] reported that 60% of the stream sediment microbial biomass belong to the bacterial groups. The stream microbiomes of tropical evergreen forests of Costa Rica were more similar to each other, whereas higher variations were observed among the stream microbiomes of coniferous forests of New Jersey.

2.3.2 Marine

Proteobacteria and Bacteroidetes were found to be the major microbial phyla present in the marine microbial population of the Antarctic and Arctic regions. Cyanobacteria, Actinobacteria, Crenarchaeota, Marinimicrobia, and Parcubacteria were other abundant phyla. Microbial phyla Chloroflexi, Chlamydiae, Verrucomicrobia were also present in this polar microbiota [40]. Carter-Gates et al. [41] studied the influence of the Atlantic Ocean on the Arctic microbiome, and found that Gamma-Proteobacteria was the most dominant bacterial group, followed by Alpha-Proteobacteria and Flavobacteria. Eight out of ten major bacterial groups including Flavobacteriia, Acidimicrobiia, Firmicutes, Verrucomicrobia, Alpha-, Gamma-, Beta-, and Delta-Proteobacteria exhibited a significant difference in abundance between the low influence and high influence regions, whereas the Atlantic water influence had little impact over Cytophagia and Epsilon-Proteobacteria. Among the eukaryotic groups Charophyta, Dinoflagellata, Chlorophyta, Excavata, Apicomplexa, Protalveolata, Cryptophyta, and Opisthokonta were also influenced by the Atlantic water influence. But no variation in abundance was observed in the eukaryotic groups Rhizaria, Picozoa, Stramenopiles, Centrohelida, Ciliophora, and Haptophyta. In a study by Wang et al. [59] on the surface water microbiome of the Southern Ocean, they found that Bacteroidetes and Proteobacteria were the major bacterial phyla present. Alpha-Proteobacteria, Flavobacteria, and Gamma-Proteobacteria were the predominant bacterial classes. *Halomonas*, *Polaribacter*, and *Candidatus Pelagibacter* were the most abundant bacterial genera present. In most of the sites, bacterial phyla like Actinobacteria, Verrucomicrobia, and Firmicutes were found to be rare. The abundances of the microbial groups varied with increasing distances between the sample locations. The microbial composition of sample site 334 km apart was the most similar and that of sample site 6,870 km was the least similar. The study suggests that the geographical distance leads to a differential abundance of the microbial group even in the well-connected ocean waters. In another study by De Corte et al. [42], they reported that Proteobacteria and Bacteroidetes were the major zooplankton-associated bacterial phyla in the North Atlantic Ocean. The major zooplankton-associated bacterial families were Rhodobacterales, Flavobacteriales, Pseudoalteromonadales, and Burkholderiales. In the ambient water, the major bacterial groups found were Flavobacteriales, Delta-Proteobacteria, Rickettsiales, Oceanospirillales, Rhodobacteriales, Acidimicrobiales, Alteromonadales, Marinimicrobia, and Chloroflexi. The common bacterial families to both the zooplankton-associated and the ambient water community were Rhodobacterales, Flavobacteriales, and Alteromonadales, whereas Acidimicrobiales and Rickettsiales were only found in the ambient water.

2.3.3 Brackish/estuarine

Bacteria dominated the microbial groups of estuary water samples of the Sundarbans, and only 4% of the community belonged to the archaea. Copiotrophic archaeal and bacterial communities cause eutrophication in the open marine water, whereas a high abundance of oligotrophic microbes was observed in the island of Sundarban mangroves. The bacterial and archaeal community compositions of the open marine water differed from that of the island of Sundarban mangrove samples. Flavobacteria,

Acidimicrobiia, and Alpha-Proteobacteria were the dominant bacterial classes found in the island of Sundarban mangroves, and Gamma-Proteobacteria was the dominant class in the open marine water samples. Bacterial family Flavobacteriaceae was dominant in water samples from the island of Sundarban mangroves, and the families Spongiibacteraceae and Oceanospirillaceae were dominant in the open marine waters, whereas Rhodobacteraceae had shown equal dominance in both the communities. Thaumarchaeota and Euryarchaeota were the major archaeal group found and the presence of Woesearchaeota was also reported. Euryarchaeota was most dominant in the open marine waters and Thaumarchaeota was most dominant in the island of Sundarban mangroves [43]. Also, Rocca et al. [44] studied the microbial composition and its response toward experimental mixing of marine and freshwaters. The major bacterial families found in the freshwater were Acetobacteraceae, Beijerinckiaceae, Paracaedibacteraceae, and Burkholderiaceae, whereas Alteromonadaceae, Saprospiraceae, Rhodobacteraceae, and Spirosomaceae were the major marine bacterial families. In the experimental mixing of marine and freshwater, the resultant microbial community had a strong similarity with the marine microbiome, and these brackish intermediates had a severe negative effect on the freshwater microbiome. It was observed that in marine microbiomes, the tolerant and less prevalent sister taxa had replaced the microbial group that had been lost due to the mixing. *Alteromonas* was the most dominant in the experimental mixes done using marine and brackish waters. The microbial enzymatic activities were highest in the freshwater and were reduced by the mixing with marine and brackish waters.

2.4 Extreme climatic microbial habitats

Microbes are found present even in ecosystems with extreme conditions. The presence of bacteria and other microbes are even reported in the hot springs with extremely high temperatures and also in the frozen permafrost soil with extremely low temperatures.

2.4.1 Hot springs

The major bacterial phyla found in the hot spring were Proteobacteria, Cyanobacteria, Deinococci, Aquificae, and Ignavibacteria. The cyanobacteria genus *Thermosynechococcus* was found abundant in the hot spring microbiome. The major archaeal phyla present in the hot spring were Woesearchaeota and Thermoprotei. *Acidilobus* and *Sulfolobus* were the major Thermoprotei genera present. The major eukaryotic groups found were the green algae Zygnematophyceae and Bdelloidea belonging to Rotifers [45]. Further, Guo et al. [46] studied the hot spring microbial composition in the Himalayan Mountain belt. The most dominant bacterial phylum was Proteobacteria followed by Aquificae. The other abundant bacterial phyla were Deinococcus-Thermus, Cyanobacteria, Firmicutes, Nitrospirae, Bacteroidetes, Thermotogae, and Thermodesulfobacteria. The hot spring microbial composition was observed to be mainly affected by the temperature. In the low-temperature hot springs, the most abundant phyla were Nitrospirae and Proteobacteria, whereas Deinococcus-Thermus, Aquificae, Thermodesulfobacteria, Cyanobacteria, and Thermotogae were the major phyla present in the moderate- to high-temperature hot springs. Crenarchaeota, Thaumarchaeota, Euryarchaeota, and Woesearchaeota were the major archaeal phyla present. A higher abundance of the phyla Crenarchaeota, Cyanobacteria, Deinococcus-Thermus, and Thermotogae were found in the high-temperature hot springs.

2.4.2 Permafrost

The most active bacterial groups in the permafrost soils were Beta-Proteobacteria, Gamma-Proteobacteria, Acidobacteria, Firmicutes, and Actinobacteria. The permafrost thaw mainly affected the phyla Bacteroidetes, Firmicutes, and Euryarchaeota. The species *Psychrobacter arcticus* and *Psychrobacter cryohalolentis* were able for DNA synthesis even at a low temperature of −15 °C. Even in the low temperature of −25 °C, species like *Planococcus halocryophilus* was found [47]. In another study by Liang et al. [48] on the bacterial composition in the Siberian permafrost layers, they found that Proteobacteria and Actinobacteria were the dominant bacterial phyla present in the younger sediments, whereas Firmicutes was the dominant phyla in the older sediments. The abundances of the bacterial phyla Acidobacteria, Actinobacteria, and Chloroflexi had a negative impact by the increasing geological age of permafrost.

2.5 Symbiotic and parasitic microbes

Microbes are also found associated with other macroorganisms. The gut microbiota of the insects and animals, and the microbes associated with the coral reefs and sponges are some examples of the symbiotic microbial association with other higher organisms.

2.5.1 Gut microbiome

The gut microbiota mainly helps their symbiotic hosts in the digestion process. Actinobacteria, Proteobacteria, and Bacteroidetes were the major gut microbes found in the beetles. The hindgut microbes of xylophagous beetles consist of the bacterial phyla Actinobacteria, Bacteroidetes, Proteobacteria, and Firmicutes. In the guts of wood-feeding woodwasps belonging to *Sirex* genera, the major microbes present were basidiomycete Amylostereum, Proteobacteria, and Actinobacteria [10]. Hannula et al. [49] reported that the microbial composition of the caterpillars was more similar to the microbial composition of the soil and was less similar to the host plant. The major bacterial phyla present in the caterpillar core microbiome were Acidobacteria, Proteobacteria, and Firmicutes. The caterpillars feed on intact plants had microbial compositions similar to the soil microbiome, whereas the caterpillars fed using the detached leaves had a different microbial composition. Actinobacteria, Eurotiomycetes, and some fungal groups were observed as common between the soil and caterpillars fed on intact plants. In the leaf microbiome, the most abundant phylum was Gamma-Proteobacteria. The major abundant phyla present in the roots were Gamma-Proteobacteria, Bacteroidetes, Firmicutes, Sordariomycetes, Glomeromycotina, and Agaricomycetes. Also, Arango et al. [50] studied the antimicrobial activity of the Actinobacteria isolated from the gut microbiome of the termite species *Reticulitermes flavipes* and *Reticulitermes tibialis*. Subterranean termite gut contains Actinobacteria with a wide range of antimicrobial activity against the subterranean fungi and yeast. These Actinobacteria mainly belong to *Streptomyces* and show various levels of inhibition toward the *Metarhizium* and *Trichoderma* mold species and three strains of *Serratia marcescens* and yeast species *Candida albicans*. The gut Actinobacteria not only inhibited the pathogenic fungi groups but also showed inhibition toward the competitive basidiomycetes species *Gleophyllum trabeum*, *Tyromyces palustris*, *Irpex lacteus*, and *Trametes versicolor*.

2.5.2 Coral microbiome

Bacteria, archaea, dinoflagellates, fungi, and viruses are the major members of the coral microbiome. The different parts of the corals including the skeleton, mucus, gastrovascular cavity, and tissue have distinct and separate microbial communities. The skeleton and the mucus layer are the major microhabitats of the coral microbiome, especially for the bacteria. The mucus layer is a nutrient-rich habitat due to the coral exudes in these layers. The coral colony as a whole may have specific core microbiomes. The highest bacterial diversity is observed in the coral skeleton. In the coral tissues, the bacterial aggregates are found associated with the tissue cells, known as the cell-associated microbial aggregates. Further, the bacterial communities and the coral-algal symbionts inhabit the gastrodermal layer [51]. The coral skeleton and tissues inhabit the majority of the coral microbiomes including archaeal and bacterial species. The skeleton has the highest diversity of the prokaryotes. The most abundant bacterial classes were Gamma-Proteobacteria and Alpha-Proteobacteria. In the live coral skeletons, the eukaryotic green algae are found in abundance. The *Ostreobium* genus of green algae is found common in most of the stony corals. Fungi are also found in the coral skeleton and form an integral part of the coral microbiomes. Many functionally different microbial groups like anaerobic and aerobic bacteria, decomposers, diazotrophs, and phototrophs inhabit the various microniches of the microbiomes [52].

2.6 Metagenomic approaches

Metagenomics enables us to find out the novel genes, taxonomic groups, and microbial functions. The metabolic process of microbial communities can also be unraveled, which in turn describes the complex dynamic process in the microbial communities. The metagenomic analysis based on the sequences is done either by using any phylogenetic marker that can taxonomically identify the DNA fragments or else random fragments are sequenced to find the gene of interest, and taxonomic markers will be found by sequencing the adjacent DNA regions. Metagenomics also improved the understanding of microbial evolutionary biology, and metagenomic analyses proved that most microbes were not genomic clones as assumed previously [53]. Metagenomics is an essential field of study to enable a much deeper study of soil microbial diversity and to examine functional and physiological roles and abilities of all the individual microbial taxa [54]. Quantification of soil microbial taxa can be done with advanced techniques called metagenomics, without the need for isolating the organism and even for the microbes that are difficult to culture or are nonculturable taxa [55]. The population dynamics of microbiomes at the species level can be determined using metagenomic techniques. The identification of subpopulations and genetic features is used to study the microbial strains or ecotypes. Detections of single nucleotide variants, type-defining genes, and horizontal gene transfer can be done even without the assemblage of genomes using the metagenomic approaches. A large number of species can be studied or detected using metagenomic sequencing [56]. Dubey et al. [57] explained that metagenomics is a branch of science dealing with the genomic analysis of DNA from environmental samples. Metagenomics studies the presence of various gene families and their functional attributes in a microbiome and helps in proper taxonomic cataloguing of microbes along with providing significant ecological, evolutionary, genetic, and adaptation aspects of the microbiome. Also, Berg et al. [58] explained that the functional potential of the microbial communities and individual microbes in the natural environment can be analyzed from the metagenome, which refers to the collection of genes and genomes from the microbiota groups and members.

2.7 Conclusion

Microbes are found abundant in most of the environments around us and are even present in ecosystems with extreme conditions, including the hot springs and permafrost soils. Bacteria is the major microbial group dominant in most of the ecosystems, followed by archaea and fungi. Protists and viruses are least represented in the microbiomes. Moreover, there are only a few studies found regarding the virus and protist groups, and the majority of studies concentrated on bacteria and fungi. Proteobacteria, Acidobacteria, and Actinobacteria were the most dominant bacterial phyla present in most of the ecosystems, whereas Thaumarchaeota and Euryarchaeota were the dominant archaeal phyla and the dominant fungal phyla were Ascomycota and Basidiomycota. The abundance, diversity, and composition of the microbes were observed to differ with various ecosystems due to the geographical and climatic influences.

Table 2.1: Microbial composition of various microbiomes.

Habitat type and location	Major microbial groups present	Reference
Tropical forest soil		
Yucatan Peninsula, Mexico	*Bacteria* Actinobacteria, Firmicutes, Proteobacteria, Acidobacteria, Chloroflexi, Bacteroidetes, Gemmatimonadetes, and Verrucomicrobia	Pajares et al. [3]
Amazon region	*Bacteria* *Most dominant* Proteobacteria and Actinobacteria *Abundant* Acidobacteria, Firmicutes. Bacteroidetes, Firmicutes, and Chloroflexi *Archaea* *Most dominant* Thaumarchaeota	Pedrinho et al. [4]
Fiji, South Pacific	*Bacteria* *Most dominant* Proteobacteria, Actinobacteria, and Acidobacteria *Abundant* Verrucomicrobia, Chloroflexi, Planctomycetes, and Nitrospirae *Archaea*	Benucci et al. [5]

Table 2.1 (continued)

Habitat type and location	Major microbial groups present	Reference
	Most dominant Crenarchaeota *Fungi* *Most dominant* Ascomycota, Basidiomycota, and Mucoromycota *Abundant* Glomeromycota, Chytridiomycota, and Rozellomycota	
Eastern China	*Bacteria* Proteobacteria, Actinobacteria, and Acidobacteria *Archaea* Thaumarchaeota and Euryarchaeota *Fungi* Ascomycota, Mucoromycota, and Basidiomycota	Ma et al. [6]
Brunei and Borneo	*Archaea* Thaumarchaeota and Euryarchaeota	Kerfahi et al. [7]
Amazon forest	*Bacteria* Proteobacteria, Acidobacteria, Firmicutes, Actinobacteria, and Chloroflexi	Buscardo et al. [8]
Brazilian dry forest soils	*Bacteria* Actinobacteria, Proteobacteria Acidobacteria, Cyanobacteria, and Bacteroidetes	Lacerda-Júnior et al. [9]
Temperate forest soil		
Temperate forests in Japan	*Bacteria* *Most dominant* Acidobacteria and Proteobacteria *Abundant* Verrucomicrobia, Bacteroidetes, Planctomycetes, and Actinobacteria	Miyashita et al. [11]
Disturbed temperate forests	*Bacteria* *Most dominant* Proteobacteria and Acidobacteria *Abundant* Actinobacteria, Chloroflexi, and Nitrospirae Fungi *Most dominant* Agaricomycetes	Osburn et al. [12]

Table 2.1 (continued)

Habitat type and location	Major microbial groups present	Reference
Temperate forests	*Fungi* *Most dominant* Ascomycota and Basidiomycota *Dominant* – Arbuscular mycorrhizal fungi – Glomeraceae *Abundant* – Arbuscular mycorrhizal fungi – Acaulosporaceae, Diversisporaceae, Claroideoglomeraceae, Archaeosporaceae, Gigasporaceae, Ambisporaceae, and Pacisporaceae	Nguyen et al. [13] Miyake et al. [14]
Alpine forest soil		
Tibetan plateau	*Bacteria* *Most dominant* Proteobacteria *Abundant* Acidobacteria, Actinobacteria, Chloroflexi, and Bacteroidetes.	Wang et al. [59]
Mangrove forest soil		
Bhitarkanika mangrove forest, Odisha	*Bacteria* Proteobacteria, Actinobacteria, Acidobacteria, Bacteroidetes, Firmicutes, Verrucomicrobia, and Ignavibacteria	Behra et al. [15]
Southern Brazil	*Bacteria* Proteobacteria, Bacteroidetes, Chloroflexi, Planctomycetes, Acidobacteria, Deferribacteres, Latescibacteria, and Spirochetes *Archaea* Euryarchaeota and Bathyarchaeota	Huergo et al. [16]
The Dongzhai Harbor mangrove forest in China	*Bacteria* Proteobacteria, Chloroflexi, Bacteroidetes Actinobacteria, Firmicutes, Acidobacteria, Nitrospirae, and Ignavibacteriae	Li et al. [17]
Brazil	*Bacteria* *Most dominant* Proteobacteria *Abundant* Actinobacteria, Chloroflexi, Acidobacteria, Bacteroidetes, and Cyanobacteria *Archaea* Crenarchaeota and Euryarchaeota	Ceccon et al. [19]

Table 2.1 (continued)

Habitat type and location	Major microbial groups present	Reference
Mangrove soil	*Viruses* *Dominant* Circoviridae, Microviridae, and Nanoviridae *Abundant* Phycodnaviridae, Parvoviridae, Geminiviridae, Mimiviridae, and Inoviridae	Jin et al. [20]
Desert soil		
Desert soil	*Bacteria* Proteobacteria, Actinobacteria, Bacteroidetes, and Deinococcus-Thermus	Baubin et al. [23]
Namib desert	*Virus* Siphoviridae, Phycodnaviridae, and Mimiviridae	Scola et al. [24]
Atacama flowering desert	*Bacteria* *Most dominant* Proteobacteria and Actinobacteria *Abundant* Firmicutes, Chloroflexi, Bacteroidetes, Planctomycetes, Acidobacteria, and Gemmatimonadetes *Archaea* Thaumarchaeota and Euryarchaeota *Fungi* *Most dominant* Ascomycota *Abundant* Basidiomycota, Chytridiomycota, and Mortierellomycota	Araya et al. [25]
Sonoran desert	*Bacteria* Proteobacteria, Bacteroidetes, Actinobacteria, Firmicutes, and Acidobacteria *Archaea* Crenarchaeota	Andrew et al. [26]
Phyllosphere		
Hardwood forest in Germany	*Bacteria* Actinobacteria, Bacteroidetes, Gamma-Proteobacteria, and Alpha-Proteobacteria	Herrmann et al. [27]

Table 2.1 (continued)

Habitat type and location	Major microbial groups present	Reference
Tropical mangrove forests	*Fungi* *Most dominant* Ascomycota *Abundant* Basidiomycota and Cryptomycota	Yao et al. [28]
Temperate forests	Bacteria Proteobacteria, Bacteroidetes, Actinobacteria, and Acidobacteria	Laforest-Lapointe et al. [29]
Rhizosphere		
Amazon rainforests	*Bacteria* Proteobacteria, Bacteroidetes, and Planctomycetes *Fungi* Euryarchaeota and Crenarchaeota	Fonseca et al. [30]
Subtropical woodlands, African Savannah	*Bacteria* Firmicutes, Bacteroidetes, Tenericutes, Proteobacteria, and Actinobacteria	Maquia et al. [31]
Red Sea mangroves	*Bacteria* Bacteroidetes, Proteobacteria, and Firmicutes	Alzubaidy et al. [33]
Temperate forest rhizosphere in Europe	*Bacteria* Actinobacteria, Acidobacteria, and Proteobacteria	Ma et al. [32]
Aerosol		
Amazon rainforest	Bacteria Enterobacteriaceae, Beijerinckiaceae, Polyangiaceae, Bacillaceae, and Ktedonobacteraceae	Souza et al. [34]
Freshwater		
Lake Taihu	*Bacteria* Proteobacteria, Cyanobacteria, Actinobacteria, Verrucomicrobia, and Planctomycetes	Zhu et al. [36]
Chile	*Bacteria* *Most dominant* Proteobacteria *Abundant* Bacteroidetes, Verrucomicrobia, and Cyanobacteria	Eissler et al. [37]

Table 2.1 (continued)

Habitat type and location	Major microbial groups present	Reference
	Archaea Euryarchaeota, Thaumarchaeota, Diapherotrites, and Woesearchaeota	
Amazon River	Limnohabitans, Polynucleobacter, Planktophila, Methylopumilus, Ramlibacter, and Fonsibacter	Santos-Junior et al. [38]
Marine		
Antarctic and Arctic	*Bacteria* *Most dominant* Proteobacteria and Bacteroidetes *Abundant* Cyanobacteria, Actinobacteria, Crenarchaeota, Marinimicrobia, and Parcubacteria	Cao et al. [40]
ARCTIC	*Bacteria* *Most dominant* Gamma-Proteobacteria *Abundant* Alpha-Proteobacteria and Flavobacteria, Acidimicrobiia, Firmicutes, and Verrucomicrobia	Carter-Gates et al. [41]
Southern Ocean	*Bacteria* Bacteroidetes, Flavobacteria, and Proteobacteria	Wang et al. [60]
North Atlantic Ocean – zooplankton associated	*Bacteria* Proteobacteria and Bacteroidetes	De Corte et al. [42]
Brackish/estuarine		
Indian Sundarban mangrove areas	*Bacteria* Flavobacteria, Acidimicrobiia, and Alpha-Proteobacteria *Archaea* Thaumarchaeota and Euryarchaeota	Dhal et al. [43]
Hot springs		
Ancient rock	*Bacteria* Proteobacteria, Cyanobacteria, Deinococci, Aquificae, and Ignavibacteria *Archaea* Woesearchaeota and Thermoprotei	Schuler et al. [45]

Table 2.1 (continued)

Habitat type and location	Major microbial groups present	Reference
Eastern Tibetan Plateau geothermal belt, Western China	*Bacteria* Proteobacteria, Aquificae, Deinococcus-Thermus, Cyanobacteria, Firmicutes, Nitrospirae, Bacteroidetes, Thermotogae, and Thermodesulfobacteria *Archaea* Crenarchaeota, Thaumarchaeota, Euryarchaeota, and Woesearchaeota	Guo et al. [46]
Permafrost		
Frozen and thawing soils	*Bacteria* Beta-Proteobacteria, Gamma-Proteobacteria, Acidobacteria, Firmicutes, and Actinobacteria.	Nikrad et al. [47]
SIBERIAN PERMAFROST LAYERS	*Dominant bacteria* *–Younger sediment* Proteobacteria and Actinobacteria *–Older sediment* Firmicutes	Liang et al. [48]
Gut microbiome		
Beetles	*Bacteria* Actinobacteria, Proteobacteria, and Bacteroidetes	Baldrian [2]
Woodwasps belonging to SIREX	*Bacteria* Proteobacteria and Actinobacteria	Baldrian [2]
Xylophagous beetles	*Bacteria* Actinobacteria, Bacteroidetes, Proteobacteria, and Firmicutes	Baldrian [2]
Subterranean termite	*Bacteria* *Most dominant* Actinobacteria	Arango et al. [50]
Caterpillar	*Bacteria* Acidobacteria, Proteobacteria, and Firmicutes	Hannula et al. [49]
Coral microbiome	Bacteria Gamma-Proteobacteria and Alpha-Proteobacteria	Ricci et al. [52]

References

[1] Flemming, H.C., Wuertz, S. Bacteria and archaea on Earth and their abundance in biofilms. Nat Rev Microbiol. 2019, 17(4), 247–260.
[2] Baldrian, P. Microbial activity and the dynamics of ecosystem processes in forest soils. Curr Opin Microbiol. 2017, 37, 128–134.
[3] Pajares, S., Campo, J., Bohannan, B.J., Etchevers, J.D. Environmental controls on soil microbial communities in a seasonally dry tropical forest. Forest Appl Environ Microbiol. 2018, 84, 17.
[4] Pedrinho, A., Mendes, L.W., Merloti, L.F., Andreote, F.D., Tsai, S.M. The natural recovery of soil microbial community and nitrogen functions after pasture abandonment in the Amazon region. FEMS Microbiol Ecol. 2020, 96(9), fiaa149.
[5] Benucci, G.M.N., Bonito, V., Bonito, G. Fungal, bacterial, and archaeal diversity in soils beneath native and introduced plants in Fiji, South Pacific. Microb Ecol. 2019, 78(1), 136–146.
[6] Ma, B., Dai, Z., Wang, H., Dsouza, M., Liu, X., He, Y., Wu, J., Rodrigues, J.L., Gilbert, J.A., Brookes, P.C., Xu, J. Distinct biogeographic patterns for archaea, bacteria, and fungi along the vegetation gradient at the continental scale in Eastern China. mSystems. 2017, 2(1), e00174–16.
[7] Kerfahi, D., Tripathi, B.M., Slik, J.W.F., Sukri, R.S., Jaafar, S., Adams, J.M. Distinctive soil archaeal communities in different variants of tropical equatorial forest. Microb Ecol. 2018, 76 (1), 215–225.
[8] Buscardo, E., Geml, J., Schmidt, S.K., Freitas, H., da Cunha, H.B., Nagy, L. Spatio-temporal dynamics of soil bacterial communities as a function of Amazon forest phenology. Sci Rep. 2018, 8(1), 4382.
[9] Lacerda-Júnior, G.V., Noronha, M.F., Cabral, L., Delforno, T.P., de Sousa, S.T.P., Fernandes-Júnior, P.I., Melo, I.S., Oliveira, V.M. Land use and seasonal effects on the soil microbiome of a Brazilian dry forest. Front Microbiol. 2019, 10, 648.
[10] Baldrian, P. Forest microbiome: Diversity, complexity and dynamics. FEMS Microbiol Rev. 2017, 41(2), 109–130.
[11] Miyashita NT, Iwanaga H, Charles S, Diway B, Sabang J, Chong L. Soil bacterial community structure in five tropical forests in Malaysia and one temperate forest in Japan revealed by pyrosequencing analyses of 16S rRNA gene sequence variation. Genes Genet Syst. 2013; 88(2):93-103. doi: 10.1266/ggs.88.93.
[12] Osburn, E.D., McBride, S.G., Aylward, F.O., Badgley, B.D., Strahm, B.D., Knoepp, J.D., Barrett, J.E. Soil bacterial and fungal communities exhibit distinct long-term responses to disturbance in temperate forests. Front Microbiol. 2019, 10, 2872.
[13] Nguyen, D.Q., Schneider, D., Brinkmann, N., Song, B., Janz, D., Schöning, I., Daniel, R., Pena, R., Polle, A. Soil and root nutrient chemistry structure root-associated fungal assemblages in temperate forests. Environ Microbiol. 2020, 22(8), 3081–3095.
[14] Miyake, H., Ishitsuka, S., Taniguchi, T., Yamato, M. Communities of arbuscular mycorrhizal fungi in forest ecosystems in Japan's temperate region may be primarily constituted by limited fungal taxa. Mycorrhiza. 2020, 30(2–3), 257–268.
[15] Behera, P., Mohapatra, M., Kim, J.Y., Adhya, T.K., Pattnaik, A.K., Rastogi, G. Spatial and temporal heterogeneity in the structure and function of sediment bacterial communities of a tropical mangrove forest. Environ Sci Pollut Res Int. 2019, 26(4), 3893–3908.
[16] Huergo, L.F., Rissi, D.V., Elias, A.S., Gonçalves, M.V., Gernet, M.V., Barreto, F., Dahmer, G.W., Reis, R.A., Pedrosa, F.O., Souza, E.M., Monteiro, R.A., Baura, V.A., Balsanelli, E., Cruz, L.M. Influence of ancient anthropogenic activities on the mangrove soil microbiome. Sci Total Environ. 2018, 645, 1–9.

[17] Li, P., Wu, S., Yin, H., Wu, H., Peng, Q., Zhang, W., Wang, R. Bacterial community diversity and dynamics of Dongzhai harbor mangrove soils in China. Can J Microbiol. 2019, 65(10), 703–712.

[18] Cabral, L., Pereira de Sousa, S.T., Júnior, G.V.L., Hawley, E., Andreote, F.D., Hess, M., de Oliveira, V.M. Microbial functional responses to long-term anthropogenic impact in mangrove soils. Ecotoxicol Environ Saf. 2018, 160, 231–239.

[19] Ceccon, D.M., Faoro, H., Lana, P.D.C., Souza, E.M., Pedrosa, F.O. Metataxonomic and metagenomic analysis of mangrove microbiomes reveals community patterns driven by salinity and pH gradients in Paranaguá Bay, Brazil. Sci Total Environ. 2019, 694, 133609.

[20] Jin, M., Guo, X., Zhang, R., Qu, W., Gao, B., Zeng, R. Diversities and potential biogeochemical impacts of mangrove soil viruses. Microbiome. 2019, 7(1), 58.

[21] Shiau, Y.J., Cai, Y., Lin, Y.T., Jia, Z., Chiu, C.Y. Community structure of active aerobic methanotrophs in red mangrove (Kandelia obovata) soils under different frequency of tides. Microb Ecol. 2018, 75(3), 761–770.

[22] Guerrero-Cruz, S., Vaksmaa, A., Horn, M.A., Niemann, H., Pijuan, M., Ho, A. Methanotrophs: Discoveries, environmental relevance, and a perspective on current and future applications. Front Microbiol. 2021, 12, 1057.

[23] Baubin, C., Farrell, A.M., Šťovíček, A., Ghazaryan, L., Giladi, I., Gillor, O. Seasonal and spatial variability in total and active bacterial communities from desert soil. Pedobiologia. 2019, 74, 7–14.

[24] Scola, V., Ramond, J.B., Frossard, A., Zablocki, O., Adriaenssens, E.M., Johnson, R.M., Seely, M., Cowan, D.A. Namib desert soil microbial community diversity, assembly, and function along a natural xeric gradient. Microb Ecol. 2018, 75(1), 193–203.

[25] Araya, J.P., González, M., Cardinale, M., Schnell, S., Stoll, A. Microbiome dynamics associated with the Atacama flowering desert. Front Microbiol. 2020, 10, 3160.

[26] Andrew, D.R., Fitak, R.R., Munguia-Vega, A., Racolta, A., Martinson, V.G., Dontsova, K. Abiotic factors shape microbial diversity in Sonoran Desert soils. Appl Environ Microbiol. 2012, 78 (21), 7527–7537.

[27] Herrmann, M., Geesink, P., Richter, R., Küsel, K. Canopy position has a stronger effect than tree species identity on phyllosphere bacterial diversity in a floodplain hardwood forest. Microb Ecol. 2021, 81(1), 157–168.

[28] Yao, H., Sun, X., He, C., Maitra, P., Li, X.C., Guo, L.D. Phyllosphere epiphytic and endophytic fungal community and network structures differ in a tropical mangrove ecosystem. Microbiome. 2019, 7(1), 57.

[29] Laforest-Lapointe, I., Messier, C., Kembel, S.W. Host species identity, site and time drive temperate tree phyllosphere bacterial community structure. Microbiome. 2016, 4(1), 27.

[30] Fonseca, J.P., Hoffmann, L., Cabral, B.C.A., Dias, V.H.G., Miranda, M.R., de Azevedo Martins, A.C., Boschiero, C., Bastos, W.R., Silva, R. Contrasting the microbiomes from forest rhizosphere and deeper bulk soil from an Amazon rainforest reserve. Gene. 2018, 642, 389–397.

[31] Maquia, I.S., Fareleira, P., Videira e Castro, I., Brito, D.R., Soares, R., Chaúque, A., Ferreira-Pinto, M.M., Lumini, E., Berruti, A., Ribeiro, N.S., Marques, I. Mining the microbiome of key species from African savanna woodlands: Potential for soil health improvement and plant growth promotion. Microorganisms. 2020, 8(9), 1291.

[32] Ma, S., De Frenne, P., Boon, N., Brunet, J., Cousins, S.A.O., Decocq, G., Kolb, A., Lemke, I., Liira, J., Naaf, T., Orczewska, A., Plue, J., Wulf, M., Verheyen, K. Plant species identity and soil characteristics determine rhizosphere soil bacteria community composition in European temperate forests. FEMS Microbiol Ecol. 2019, 95(6), fiz063.

[33] Alzubaidy, H., Essack, M., Malas, T.B., Bokhari, A., Motwalli, O., Kamanu, F.K., Jamhor, S.A., Mokhtar, N.A., Antunes, A., Simões, M.F., Alam, I., Bougouffa, S., Lafi, F.F., Bajic, V.B., Archer, J.A. Rhizosphere microbiome metagenomics of gray mangroves (Avicennia marina) in the Red Sea. Gene. 2016, 576(2 Pt 1), 626–636.
[34] Souza, F.F., Mathai, P.P., Pauliquevis, T., Balsanelli, E., Pedrosa, F.O., Souza, E.M., Baura, V.A., Monteiro, R.A., Cruz, L.M., Souza, R.A., Andreae, M.O. Influence of seasonality on the aerosol microbiome of the Amazon rainforest. Sci Total Environ. 2021, 760, 144092.
[35] Dann, L.M., Clanahan, M., Paterson, J.S., Mitchell, J.G. Distinct niche partitioning of marine and freshwater microbes during colonisation. FEMS Microbiol Ecol. 2019, 95(8), fiz098.
[36] Zhu, C., Zhang, J., Nawaz, M.Z., Mahboob, S., Al-Ghanim, K.A., Khan, I.A., Lu, Z., Chen, T. Seasonal succession and spatial distribution of bacterial community structure in a eutrophic freshwater Lake, Lake Taihu. Sci Total Environ. 2019, 669, 29–40.
[37] Eissler, Y., Gálvez, M.J., Dorador, C., Hengst, M., Molina, V. Active microbiome structure and its association with environmental factors and viruses at different aquatic sites of a high-altitude wetland. MicrobiologyOpen. 2019, 8(3), e00667.
[38] Santos-Júnior, C.D., Sarmento, H., de Miranda, F.P., Henrique-Silva, F., Logares, R. Uncovering the genomic potential of the Amazon River microbiome to degrade rainforest organic matter. Microbiome. 2020, 8(1), 151.
[39] Findlay, R.H., Yeates, C., Hullar, M.A., Stahl, D.A., Kaplan, L.A. Biome-level biogeography of streambed microbiota. Appl Environ Microbiol. 2008, 74(10), 3014–3021.
[40] Cao, S., Zhang, W., Ding, W., Wang, M., Fan, S., Yang, B., Mcminn, A., Wang, M., Xie, B.B., Qin, Q.L., Chen, X.L., He, J., Zhang, Y.Z. Structure and function of the Arctic and Antarctic marine microbiota as revealed by metagenomics. Microbiome. 2020, 8(1), 47.
[41] Carter-Gates, M., Balestreri, C., Thorpe, S.E., Cottier, F., Baylay, A., Bibby, T.S., Moore, C.M., Schroeder, D.C. Implications of increasing Atlantic influence for Arctic microbial community structure. Sci Rep. 2020, 10(1), 19262.
[42] De Corte, D., Srivastava, A., Koski, M., Garcia, J.A.L., Takaki, Y., Yokokawa, T., Nunoura, T., Elisabeth, N.H., Sintes, E., Herndl, G.J. Metagenomic insights into zooplankton-associated bacterial communities. Environ Microbiol. 2018, 20(2), 492–505.
[43] Dhal, P.K., Kopprio, G.A., Gärdes, A. Insights on aquatic microbiome of the Indian Sundarbans mangrove areas. PLoS One. 2020, 15(2), e0221543.
[44] Rocca, J.D., Simonin, M., Bernhardt, E.S., Washburne, A.D., Wright, J.P. Rare microbial taxa emerge when communities collide: Freshwater and marine microbiome responses to experimental mixing. Ecology. 2020, 101(3), e02956.
[45] Schuler, C.G., Havig, J.R., Hamilton, T.L.. Hot spring microbial community composition, morphology, and carbon fixation: Implications for interpreting the ancient rock record. Front Earth Sci 2017, 5, 97.
[46] Guo, L., Wang, G., Sheng, Y., Sun, X., Shi, Z., Xu, Q., Mu, W. Temperature governs the distribution of hot spring microbial community in three hydrothermal fields, Eastern Tibetan Plateau Geothermal Belt, Western China. Sci Total Environ. 2020, 720, 137574.
[47] Nikrad, M.P., Kerkhof, L.J., Häggblom, M.M. The subzero microbiome: Microbial activity in frozen and thawing soils. FEMS Microbiol Ecol. 2016, 92(6), fiw081.
[48] Liang, R., Lau, M., Vishnivetskaya, T., Lloyd, K.G., Wang, W., Wiggins, J., Miller, J., Pfiffner, S., Rivkina, E.M., Onstott, T.C. Predominance of anaerobic, spore-forming bacteria in metabolically active microbial communities from ancient Siberian permafrost. Appl Environ. 2019, 85(15), e00560–19.
[49] Hannula, S.E., Zhu, F., Heinen, R., Bezemer, T.M. Foliar-feeding insects acquire microbiomes from the soil rather than the host plant. Nat Commun. 2019, 10(1), 1254.

[50] Arango, R.A., Carlson, C.M., Currie, C.R., McDonald, B.R., Book, A.J., Green, F., Lebow, N.K., Raffa, K.F. Antimicrobial activity of actinobacteria isolated from the guts of subterranean termites. Environ Entomol. 2016, 45(6), 1415–1423.
[51] van Oppen, M.J.H., Blackall, L.L. Coral microbiome dynamics, functions and design in a changing world. Nat Rev Microbiol. 2019, 17(9), 557–567.
[52] Ricci, F., Rossetto Marcelino, V., Blackall, L.L., Kühl, M., Medina, M., Verbruggen, H. Beneath the surface: Community assembly and functions of the coral skeleton microbiome. Microbiome. 2019, 7(1), 159.
[53] Alves, L.F., Westmann, C.A., Lovate, G.L., de Siqueira, G.M.V., Borelli, T.C., Guazzaroni, M.E. Metagenomic approaches for understanding new concepts in microbial science. Int J Genomics. 2018, 2018, 2312987.
[54] Singh, B.K., Bardgett, R.D., Smith, P., Reay, D.S. Microorganisms and climate change: Terrestrial feedbacks and mitigation options. Nat Rev Microbiol. 2010, 8(11), 779–790.
[55] Chang, H.X., Haudenshield, J.S., Bowen, C.R., Hartman, G.L. Metagenome-wide association study and machine learning prediction of bulk soil microbiome and crop productivity. Front Microbiol. 2017, 8, 519.
[56] Van Rossum, T., Ferretti, P., Maistrenko, O.M., Bork, P. Diversity within species: Interpreting strains in microbiomes. Nat Rev Microbiol. 2020, 18(9), 491–506.
[57] Dubey, A., Malla, M.A., Khan, F., Chowdhary, K., Yadav, S., Kumar, A., Sharma, S., Khare, P.K., Khan, M.L. Soil microbiome: A key player for conservation of soil health under changing climate. Biodivers Conserv. 2019, 28(8), 2405–2429.
[58] Berg, G., Rybakova, D., Fischer, D., Cernava, T., Vergès, M.C.C., Charles, T., Chen, X., Cocolin, L., Eversole, K., Corral, G.H., Kazou, M. Microbiome definition re-visited: Old concepts and new challenges. Microbiome. 2020, 8(1), 1–22.
[59] Wang, X., Zhang, Z., Yu, Z., Shen, G., Cheng, H., Tao, S. Composition and diversity of soil microbial communities in the alpine wetland and alpine forest ecosystems on the Tibetan Plateau. Sci Total Environ. 2020, 747, 141358.
[60] Wang, Z.B., Sun, Y.Y., Li, Y., Chen, X.L., Wang, P., Ding, H.T., Chen, B., Zhang, X.Y., Song, X.Y., Wang, M., McMinn, A., Zhang, Y.Z., Qin, Q.L. Significant bacterial distance-decay relationship in continuous, well-connected Southern Ocean surface water. Microb Ecol. 2020, 80(1), 73–80.

Nelson Libardi Junior, Nathan Pacheco Amin Vieira da Costa, Rejane Helena Ribeiro da Costa, Carlos Ricardo Soccol and Luciana Porto de Souza Vandenberghe

3 Pollutants removal using aerobic granular sludge technology

Abstract: Aerobic granular sludge (AGS) is a promising wastewater treatment technology with advantages compared to the centenary conventional activated sludge process. This technology has been developed and tested for removal of nutrients as well as other pollutants present in both domestic and industrial wastewaters. A unified theory that completely describes the formation of aerobic granules was not yet consolidated due to several knowledge gaps. However, the hydrodynamic shear forces and the feast–famine feeding regime applied in sequencing batch reactors (SBR) have been successfully used for the development of granular biomass. Dense microbial aggregates are formed by the production of extracellular polymeric substances (EPS) naturally induced in SBRs. EPS plays a key role in the microbial aggregation, granule formation, and structural stability, contributing to the formation of aerobic, anoxic, and anaerobic microbial layers, allowing the simultaneous removal of carbon, nitrogen, and phosphorus in a single reactor, resulting in smaller constructed areas and lower energy demands. Aerobic granules have an excellent settleability, have a dense and strong microbial structure, are capable of high biomass retention, can withstand high organic loadings, and are tolerant to toxic compounds. Despite the advantages offered by AGS, there are still issues in the large-scale application of this technology, like the low-strength wastewaters, the low granule stability, and long time required for granule formation, resulting in reduced treatment efficiency. Recent studies are overcoming these limitations by the use of additives to improve granule stability and to reduce the time for granule formation and testing operational configurations to adapt AGS to diluted wastewater and subtropical conditions. In addition to the treatment of domestic wastewater, AGS has been tested for the treatment of petroleum wastewater, pulp mill effluents, rubber wastewater, food industry effluents, and others. While the potential of this technology has been extensively reviewed, there are still some challenges that are currently being overcome by researchers across the world.

Keywords: aerobic granular sludge, wastewater, pollutants

Nelson Libardi Junior, Departamento de Engenharia Sanitária e Ambiental, Universidade Federal de Santa Catarina – UFSC, Florianópolis-SC 88040-970, Brazil, e-mail: nelson.libardi@ufsc.br
Nathan Pacheco Amin Vieira da Costa, Rejane Helena Ribeiro da Costa, Departamento de Engenharia Sanitária e Ambiental, Universidade Federal de Santa Catarina – UFSC, Florianópolis-SC 88040-970, Brazil
Carlos Ricardo Soccol, Luciana Porto de Souza Vandenberghe, Departamento de Engenharia de Bioprocessos e Biotecnologia, Universidade Federal do Paraná – UFPR, Curitiba-PR 81531-980, Brazil

https://doi.org/10.1515/9783110727227-003

3.1 Introduction

The aggregation of microbial cells is a natural phenomenon observed in soil clumps, in kefir grains, as well as in the fungal spherical pellets commonly found in industrial bioprocess for antibiotics and enzymes' production in shaking flasks or stirred fermenters. The microbial aggregation in biological wastewater treatment processes is found in both anaerobic and aerobic conditions. The adhesion of biofilms in moving or fixed supporting materials has been used as a strategy to overcome the microbial washout from the reactors, resulting in a separation of the hydraulic and sludge retention time, increasing the microbial age. Examples are the membrane bioreactors, biological aerated filters, and other systems. The microbial self-aggregation (without the use of supporting materials) has been reported in anaerobic (upflow anaerobic sludge blanket reactors – UASB) and aerobic (sequencing batch reactor – SBR) wastewater treatment processes. Aerobic granular sludge (AGS) is the aggregated form of the centenary activated sludge process, whereas the microbial flocs are highly compact and display better settling properties.

The increasing demand for compact wastewater treatment plants (WWTP) pushed the development of granular biomass systems to maintain the biomass inside the reactor without settling tanks and to increase the surface area per volume of the reactor. The removal of carbon, nitrogen, and phosphorus was also improved and able to be performed in a single tank, since aerobic, anoxic, and anaerobic environments are present due to the microbial layers of the granules. The biopolymer layers produced by the microbial biomass display a protective effect against toxic pollutants present in domestic and industrial wastewater. This protection makes the aerobic granules less susceptive to phenolic compounds, dyes, and emerging contaminants, which makes aerobic granulation a suitable biological wastewater treatment technology to handle with diverse effluents.

A recent increase in research articles on isolation and characterization of biopolymers produced during the aerobic granule formation reveals the importance of discovering their characteristics for specific applications. Secondary uses for this "by-product" have increased the attention of researchers due to the applicability of this material since it has similar characteristics to the marine seaweed alginate for their contribution to wastewater sector toward circular economy and resource recovery.

3.2 Historical background

From the first observation in 1997 [1], AGS in SBR research has been developed, reaching the commercial application named as Nereda®, developed by Royal Haskoning DHV, in collaboration with Dutch sanitation companies and TU Delft. In

1998, an international patent was submitted and granted [2], followed by its extension in 2004, including the description of anaerobic feeding [3]. The first Nereda® prototype for industrial wastewater treatment was implemented in 2005 in the Netherlands. In parallel, two more prototype plants were built (South Africa and Portugal) for nonindustrial wastewater treatment, as well as the design and construction of the full-scale sewage treatment plant in the city of Epe, the Netherlands, in 2010–2011 [4]. Although AGS is a patented technology with full-scale treatment plants in operation, there is still many knowledge gaps [5] in which researchers across the world have been working on. In spite of the fact that there are increasing number of publications and patents, only few authors focused on the granule formation for the treatment of real domestic wastewater. Most of the published data are related to the use of synthetic wastewater in bench-scale experiments, hindering the pilot- and real-scale problems like low nutrient concentration, toxic effects, mechanical problems, granule disintegration, and others.

The research on AGS in Brazil started with the Liquid and Gaseous Effluent Laboratory of the Federal University of Santa Catarina. Granulation studies started on a pilot scale using synthetic wastewater [6], followed by the variation of its organic load [7] and with sanitary wastewater [8, 9] for different operating conditions and mathematical modeling and nitrogen dynamics with N_2O measurements [10]. Some authors have already tested the influence of the local climate in the AGS process like the coastal climate [11] and tropical climate [12]. Recently, the application of additives in order to overcome granule disintegration has been proposed [13]. The next frontier in AGS technology is the recovery and reuse of the process by-products like the spent granules and biopolymers. Extracted biopolymers have been reported as biosorbents for pollutants' removal [14, 15] as well as self-extinguishing biopolymer [16] or surface coating materials [17]. Sarma and Tay [5] proposed some trend topics in AGS research like the quorum-sensing mechanism and the related quorum-quenching enzymes, granule bioaccumulation, and biosorption of emerging contaminants.

3.3 AGS – general properties

3.3.1 AGS formation theories

A unified theory that completely describes the formation of the aerobic granules was not yet consolidated due to several knowledge gaps. Different factors for aerobic granulation are feeding pattern, shear forces, feast–famine regime, detachment processes, influence of substrate, selective pressure by settling rate, surface loading, hydrophobicity, biopolymers production, and others [18]. According to Sarma et al. [5], EPS synthesis, quorum sensing, cell surface hydrophobicity, and ionic bridges are the primary factors responsible for both aerobic and anaerobic granule formation.

Although the microbial multiple-step self-immobilization process describing the aerobic granulation was already proposed [18, 19], a unified theory is still a point of discussion [5]. The authors proposed that a granulation process is comprised of four steps, including (i) microbial movement to the initial cell to cell attachment to start the aggregation process resulting from the gravity, diffusion, and hydrodynamic forces; (ii) physical attractive forces development of microaggregates by these self-attached cells including van der Waals, opposite charge attraction, thermodynamic, and hydrophobicity effects; (iii) microbial metabolism resulting in the biosynthesis of extracellular biopolymers acting as a glue for the microbial aggregation; and (iv) granule maturation in response to the external hydrodynamics applied by the reactor configuration and operating conditions.

3.3.2 What is considered an aerobic granule?

The general concept for AGS was proposed during the first AGS meeting in 2004: "Aggregates of microbial origin which do not coagulate under low hydrodynamic shear conditions and which settle significantly faster when compared to activated sludge flocs." To make this concept more useful, some researchers established quantitative parameters to describe an aerobic granule and its development degree. Kreuk et al. [20] and Liu et al. [21] proposed that aerobic biomass is considered granular when at least 50% of the microbial particles have diameters greater than 200 µm. The authors also consider a complete granulation process when 80% of the particles' diameter reaches 200 µm. Beun et al. [22] and Toh et al. [23] proposed the use of the sludge volumetric index (SVI). According to these authors, the granular biomass is related to SVI_{30} lower than 50 mL/g, while flocculent biomass has SVI_{30} higher than 120 mL/g. Closer SVI_{30}/SVI_{10} ratio means better sedimentation properties, whereas Liu and Tay [24] indicate 0.9 as a ratio for granulation process.

3.4 Aerobic granular sludge – process conditions

The AGS process comprehends the interrelation of physicochemical and biological phenomena. Microbial self-immobilization is induced by environmental stress conditions, involving physical shear, physicochemical or biochemical interactions, and microbial attraction mechanisms [25]. In the AGS process, the round-shaped biomass is the central player for metabolic reactions and the reactor is where the pollutants' removal reactions take place. The reactor is where controllable parameters are set, and the microbial community responds changing its character from floccule to the granular form. Since AGS is a wastewater treatment technology, the treated wastewater is the desired product or service, aiming to achieve legislation standards. However, due to the challenging formation of high-quality granules, most of the published data focuses in the formation of aerobic granules. This aspect, in addition to the granule's extractable products, changed its status from biomass to a biotechnological product. So, although some AGS process conditions were already established, the complexity of the microbial interaction with the physicochemical parameters is still a matter of discussion.

The aerobic granulation process is affected by controllable and uncontrollable parameters, such as substrate composition, organic loading rate, hydrodynamic shear force, feast–famine regime, feeding strategy, dissolved oxygen concentration, reactor configuration, additives, solids retention time, SBR cycle time, settling time, and volume exchange ratio [26]. The high shear, short settling time, and high DO concentrations are considered critical for AGS [27].

3.4.1 Sequencing batch reactors (SBR)

SBR have been successfully used for the development of granular biomass [28–30]. The use of SBR becomes advantageous since it is possible to perform all treatment steps (filling, reaction, settling, and discharge) in the same tank, separated only by time intervals [31]. The column-type SBRs with upflow aeration allows a homogeneous circular flow inducing the microbial aggregation in round shape [18]. This same pattern is observed for fungal pellets formation in bubble column reactors [32]. Most of the bench-scale AGS reactors have high height-to-diameter (H/D) ratio, which induces the oxygen transfer, and the shear forces provided by the circular flow as described above. Liu and Tay [18] observed that a higher H/D ratio provides longer circular flowing trajectory, which creates an effective hydraulic attrition for microbial aggregation. Many studies about AGS were mainly established on well-controlled lab-scale SBR with an H/D ratio over 10. The transition to pilot-scale and full-scale plants requires the use of lower H/D ratios due to the constructive feasibility. To overcome the reduced shear stress necessary for AGS formation caused by lower H/D ratios, increased aeration is necessary, resulting in increased implementation and operation costs [33].

3.4.2 Reactor feeding

The reactor feeding is performed in a single aerobic reaction phase or an aerobic feeding/reaction phase followed by an aerobic reaction phase [34]. The anaerobic feeding was introduced to increase the biological phosphorus removal [35]. Intermittent feeding strategy was also proposed to enhance aerobic granulation which results in feast–famine conditions favoring the formation of compact and dense granule and the nitrification/denitrification [36].

3.4.3 Settling as selective pressure

The microbial particle settling is considered as the main selective pressure for AGS. Only the particles with settling velocities higher than the upward velocity of the fluid will not be washed out from the reactor, as in UASB or other fluid bed systems [37]. Under long settling time, flocculent sludge with poor settling ability cannot be washed out from the reactor. Short settling times allow maintaining only high settling microbial aggregates formed by lower growth rate microorganisms, responsible for the formation of dense microbial granules. The main example is the dense nitrifier aggregates compared to the heterotrophic flocs [38]. Together with the settling time, the volumetric exchange ratio and the discharge time are the controllable parameters in an SBR that selects dense microbial aggregates [39].

3.4.4 Extracellular polymeric substances

The microbial self-aggregation found in AGS is mainly related to the "biological glue" function displayed by the excreted microbial biopolymers. The extracellular polymeric substances (EPS) contribute mainly to the AGS formation and stability. The EPS consist of a mixture of polymeric units of different macromolecules like carbohydrates, proteins, lipids, and nucleic acids that stick together to join the cell and enclose the bacterial community [26, 33].

The EPS are related to many functions that include cell adherence and aggregation, biofilm formation, granule stability, cell protection against toxic compounds and organic load shocks, biosorption of organic compounds, nutrient accumulation, and water retention [40]. Many authors are investigating the relation of the EPS production and the formation of granules during the process of wastewater treatment in SBR or an event in continuous flow systems [41]. The recently described EPS isolated from wastewater treatment systems [42] include granular [43] acid-soluble polysugars [44], glycosylated proteins [45], the alginate-like exopolysaccharides (ALE) [46], and sulfated polysaccharides [47]. In a similar manner, the EPS matrix formed around the cells of an aerobic granule would prevent them to lose the secreted enzymes into the bulk liquid phase [5].

3.5 Treatment overview

3.5.1 Carbonaceous organic matter removal

Removal of nutrients by AGS systems is comparable to other biological wastewater treatment processes. The use of aerobic granules for carbon removal was historically developed in SBRs [22, 49]. Among organic carbon by-products, biomass and methane are the major end products of wastewater treatment by aerobic granules. Compared to the activated sludge process, AGS produces less sludge. The ability of aerobic granules to remove a large amount of carbon as soluble and gaseous inorganic carbon is the most probable reason behind reduced biomass production. Inorganic carbon such as carbonates, carbonic acid, bicarbonate anion, and carbon dioxide are the major metabolic end products [49].

3.5.2 Nitrogen removal

It has been claimed that aerobic granules are capable of simultaneous nitrification and denitrification in a single step process involving only one reactor [5]. In addition, assimilation and anaerobic ammonium oxidation are also possible in aerobic granules. Nitrogen removal is commonly known to require both aerobic and anaerobic/anoxic conditions, if nitrification and denitrification wants to be achieved [28, 48]. This removal efficiency depends not only on the process previously mentioned but also on the capacity of the granular biomass to incorporate the nutrient [27]. According to Wang et al. [50], the treatment efficiency using AGS could achieve 99.7% of ammonia-N removal and 78.4% of total-N, which is still a challenge for this technology. Conventional nitrogen removal requires both aerobic and anaerobic/anoxic conditions for ammonium nitrification and denitrification of nitrate or nitrite, respectively [28]. In sewage treatment, the nitrogen removal efficiency depends on the incorporation of nitrogen in the growing biomass, and the nitrification and denitrification capacity of the microbial community [27].

In order to improve N removal, an anoxic phase was introduced immediately after the aerobic phase, which led to the development of alternating nitrification and denitrification strategy [51]. Simultaneous nitrification and denitrification, and assimilation together with anaerobic ammonium oxidation are different nitrogen removal mechanisms of aerobic granule-based technology for wastewater treatment [48]. Lochmatter et al. [36] have reported that during simultaneous nitrification and denitrification, nitrogen removal efficiency of around 95% could be achieved by aerobic granule-based technology. The success of heterotrophic denitrification processes depends on the concentration of volatile fatty acids in the wastewater. According to De Kreuk et al. [35], the diameter of aerobic granule is a factor to influence nitrogen removal efficiency of aerobic granule-based technology. Apart from nitrification and denitrification, nitrogen is assimilated by the synthesis of new biomass. According to Sharma et al. [33], the oxidation of ammonia into nitrite or nitrate is accomplished by ammonia oxidizing bacteria (AOB) found in sludge. In the mature granule, diversification of these nitrifying bacteria is filamentous shaped on the outer surface and cocci shaped in the inner side. During nitrification, ammonia is oxidized to nitrite and nitrate, whereas denitrification involves reduction of nitrite or nitrate to molecular nitrogen. Denitrification of nitrite is 1.5–2 times faster than that of nitrate [33].

3.5.3 Phosphorus removal

The phosphorus in domestic wastewater is present in dissolved organic form, particulate organic phosphorus (1–5 mg/L), and inorganic phosphorus (3–10 mg/L) such as orthophosphate and condensed inorganic phosphates (polyphosphate, metaphosphate, and pyrophosphate) [48]. The mechanism behind phosphorus removal by enhanced biological phosphorus removal (EBPR) process used by modern WWTP is well established. Apart from the phosphorus removal mechanism similar to that of EBPR process, precipitation of phosphorus as inorganic phosphate such as $Ca_3(PO_4)_2$ and $MgHPO_4$ is also a phosphorus removal mechanism of aerobic granule-based technology. De Kreuk et al. [52] demonstrated that phosphorus removal efficiency as high as 94% could be achieved by aerobic granule-based technology, and biologically induced precipitation was one of the mechanisms behind phosphorus removal. The phosphorus accumulation by EPS is the third phosphorus removal mechanism, where $(PO_4)^{3-}$, Ca^{2+}, and Mg^{2+} are accommodated and concentrated in the EPS matrix and finally precipitate as calcium phosphate and hydroxyapatite [48].

3.6 Toxic pollutants degradation

AGS technology has been widely studied for the treatment of domestic wastewater but most of the presented data are related to bench-scale and synthetic wastewater studies. The studies testing the aerobic granular biomass formation and treatment efficiency using real domestic wastewater, which has a very complex composition compared to the synthetic one, are scarce. Recently, many researchers are exploiting the use of AGS to treat recalcitrant pollutants in both domestic and industrial wastewater to expand the applicability of AGS. Some microbial wastewater treatment technologies like the activated sludge are unable to face some toxic compounds found in industrial or even domestic wastewater.

The biodegradation of toxic pollutant is still a relatively less explored area of aerobic granulation technology; however, the better knowledge of these biodegradation mechanisms could increase the applicability of AGS for industrial wastewater treatment [53]. AGS is less sensitive to fluctuation than activated sludge, due to a higher tolerance to toxicity as well as a higher biomass concentration [54]. The granular structure is thought to enhance the sludge adaptation to toxic environment and shock loadings when compared to activated sludge. EPS are the first barrier avoiding microbial cells that directly contact and interact with the toxic substance. The mass transfer limitations can reduce the amount of a potentially toxic compound that bacteria deeper in the granule are exposed to. The dilution of toxins allows the protected bacteria to continue to function or even degrade the toxin [55]. The microbial community may respond to the presence of toxic compounds producing more EPS to protect the cells against the potential damage of compounds like micropollutants. The production of polysaccharides and proteins has been related to the microbial exposure to toxic materials or additives to improve granule stability like ZnO nanoparticles [56].

The strategies to develop or increase microbial activity in order to degrade specific toxic pollutants have been employed for the development of aerobic granules adapted to toxic environments and able to metabolize these toxic compounds. The microbial community and adaptation by the use of pollutants helps to enhance the microbial degradation. Jiang et al. [57] added aniline, a common by-product of chemical and petroleum industry, in the effluent composition of aerobic granular reactors. The addition resulted in change of microbial population, increasing the presence of aniline-degrading microorganism like *Pseudomonas* and other EPS-producing strains.

The biopolymer layer prevents the loss of extracellular enzymes involved in toxic pollutant degradation to the bulk liquid (Figure 3.1). The concentration of these enzymes would increase the efficiency of biodegradation of pollutants [53]. By the other side, adsorption plays a central role in degradation of pollutants by AGS. Adsorption plays a significant role in removing emerging pollutants due to its capacity to immobilize the toxins and facilitate the transport phenomenon associated with the degradation of these compounds. Classical adsorption processes are well known to be cost-

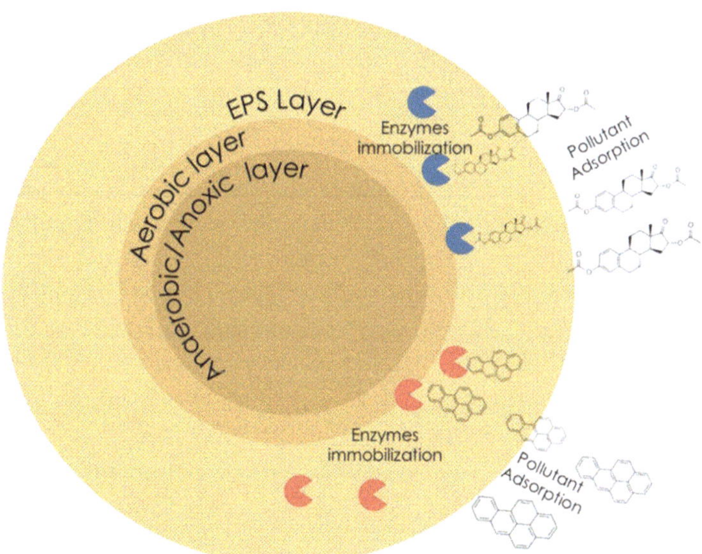

Figure 3.1: Immobilization and concentration of microbial enzymes to catalyze the degradation of adsorbed toxic pollutants in the extracellular polymeric layer of the aerobic granule.

effective, simply insensitive to toxic pollutants, and post-adsorption matrix is even biodegradable [58, 59]. The adsorption could enhance the contact between the bacterial cells and extracellular enzymes with the chemical to be degraded. The biopolymer layers immobilize the microbial extracellular enzymes that are concentrated in comparison to that excreted to the mixed liquor by activated sludge flocs.

A wide range of toxic pollutants can be found in domestic and industrial wastewaters. Besides the common nutrients present in the domestic wastewater, as carbon, nitrogen, and phosphorus, human or veterinary antibiotics, endocrine-disrupting compounds, herbicides, and pesticides can be present, even in very low concentration [53].

3.6.1 Phenols

The phenol degradation using AGS technology has been extensively described due to its wide dispersion in effluents. Phenolic compounds are found in diverse industrial processes such as petroleum refinement, coal tar processing and petrochemicals manufacturing, paints and resins production, as well as edible oil industry. The inhibitory or toxic effect of phenolic compounds affects nitrification mechanism of biological wastewater treatment processes [60]. So, the concentration gradient due to the diffusional resistance displays a protection effect on microorganisms by reducing the concentration of pollutants below some threshold value to avoid inhibition [61]. The microbial strains found in AGS are able to tolerate higher phenol concentrations compared to activated sludge flocs [61, 62].

3.6.2 Endocrine-disrupting chemicals

Endocrine-disrupting chemicals (EDC) are exogenous compounds or mixtures that have the potential to interfere with hormonal and endocrine system, even though at few nanograms per liter these micropollutants can cause serious damage to the environmental as well as human health. The known anthropogenic EDCs are pesticides, plastic adhesives, polychlorinated biphenyls, polycyclic aromatic hydrocarbons, organic compounds, alkylphenols, and industrial by-products such as dioxins and furans [63]. Typical EDCs of anthropogenic origin with an estrogen-like action include technical nonylphenol, bisphenol A (BPA), and 17α-ethinylestradiol [64]. BPA, as one of the most important man-made organic compounds with annual production exceeding 3.8 million tons, was widely used in the synthesis of polycarbonates, epoxy resins, and thermal paper [58].

BPA removal from municipal WWTP is mainly due to microbial degradation and adsorption processes. The BPA biodegradation is highly affected by the aerobic/anoxic/anaerobic conditions, temperature, initial BPA concentration, mixed liquor suspended solids, initial COD concentration, and pH [65]. According to Keskinkan and Balci [66], BPA can be toxic to activated sludge at high concentrations and can increase production of toxicity-resisting membrane proteins or enzymes such as superoxide dismutase or alkyl hydroperoxide reductase in microbial cells. The chemical composition of EPS samples in an AGS in the presence of BPA was evaluated by Dong Wei et al. [67]. The addition of BPA in a synthetic culture medium resulted in changes in the EPS structure, including inhibition on the activities of heterotrophic bacteria and mainly in nitrifying bacteria. Due to the EPS protection, BPA can be 10 times less toxic to immobilized biomass than to activated sludge. A wide range in BPA concentration in raw municipal wastewater has been presented. Concentrations from 3 to 62,000 ng/L with an average concentration of 2,031 ng/L were related by Wang et al. [65]. According to Li et al. [58], BPA has an inhibitory effect on heterotrophic and nitrifying microorganisms of aerobic granules, with the potential to change the EPS composition of the aerobic granules. AOB are able to synthesize the enzyme ammonia monooxygenase, which creates hydroxyl radicals that can break down resistant non-specific substrates such as steroidal estrogens [68].

3.6.3 Pharmaceuticals and personal care products (PPCPs)

The term "pharmaceuticals and personal care products (PPCPs)" is referred to products with healthcare or medical purposes, including prescription drugs, such as antibiotics, anti-inflammatory drugs, sedatives, and nonprescription drugs such as therapeutics, fragrances, sun screens, cosmetics, and vitamins. In the last decades, PPCPs are emerging due to its persistence in aquatic environments [69]. Endocrine-disrupting chemicals could be classified as PPCPs as a pharmaceutical ingredient. EDCs like steroidal estrogens could be discharged in the wastewater through feces and urine excretion [70].

PPCPs seem to influence the microbial community and EPS in AGS. Zhao et al. [71] observed an increase of the polysaccharide and protein content in EPS after the addition of prednisolone, ibuprofen, naproxen sulfamethoxazole, and norfloxacin in an AGS reactor. The PPCPs would stimulate the microbial metabolism, and the EPS production may be a microbial protection against these toxic pollutants. The research on PPCPs removal by AGS is still scarce, and the establishment of analytical procedures to identify these toxic compounds in a complex matrix like sanitary wastewater is an important research step.

Antibiotic resistance in human, animal, and food isolates has become a public health problem due to the transfer to the environment. Microbial antibiotic-resistant genes (ARG) have been found in wastewater from swine farms even after treatment like composting, lagoon storage, anaerobic digestion, and constructed wetlands. These treatment units are not designed to remove ARGs. This problem becomes even more serious since agroindustrial wastewater or manure is used as fertilizer in agriculture

[72]. More than 30% of antibiotic components have been found in animal wastes [73]. Zhou et al. [74] reported the presence of several antibiotics in swine and cattle farms in China such as chlortetracycline, bacitracin, lincomycin, and tetracycline ranging from 1.19 to 11.6 mg/day/swine. He et al. [72] used total DNA extraction and quantitative PCR to evaluate the presence of sulfonamide, tetracycline, chloramphenicol, and erythromycin-resistant genes in swine wastewater. Wang et al. [65] evaluated ampicillin, erythromycin, tetracycline, kanamycin, and ciprofloxacin degradation in piggery wastewater in a 30 L AGS SBR. The authors found that 94.8% of the total antibiotics were removed from wastewater after long-term (115 days) treatment with AGS system. From this amount, 62.5% of the total antibiotics was degraded and 32.3% was adsorbed by AGS. Liao et al. [75] achieved removal efficiencies of 83%, 85%, and 91% for sulfamerazine, sulfamethazine, and sulfamethoxazole from piggery slurry treated in a 5 L AGS reactor. While there is a biomass-adaptative phase for the removal of conventional pollutants, for antibiotics removal, no adaptative phases are observed since the removal efficiencies of antibiotics between the initial and final phases are similar. The EPS matrix may concentrate on the antibiotic in the granular structure to facilitate their degradation. Also, longer hydraulic retention times in AGS reactors may dilute the concentration of antibiotics facilitating the enzymatic hydrolysis efficiency. Stable COD and ammonia nitrogen removal seems to be independent from the presence of antibiotics [65, 75].

3.6.4 Dyes

Dyes are extensively used in the textile, leather, paper, pulp, cosmetic, drug, plastic, electroplating, and printing industry [76, 77]. Between 10% and 20% of the annually produced dyes are lost during processing steps and are released into the environment [78]. Aerobic granulation technology has been used to treat different industrial wastewaters containing synthetic dyes. Reactive blue 59, acid red 18, Congo red, and Eriochrome black T are some of the dyes that have been successfully treated by aerobic granulation technology [53].

Azo dye structure has azo groups (R–N = N–R′) flanked by aryl and/or alkyl moieties. They represent around 65% of the industrial dye stuff [79]. Azo dye molecules are first anaerobically degraded through the cleavage of their chromophores (–N = N–) forming colorless aromatic amines. An aerobic step is then necessary to degrade these carcinogenic and mutagenic amines to CO_2, H_2O, and NH_3 [80]. Simultaneous aerobic and anaerobic conditions are achieved by AGS systems. The simultaneous occurrence of reductive and oxidative biological conversions under microaerophilic conditions was also tested in AGS system, showing its potential to improve dye removal [81]. The use of synthetic dye-containing wastewater has been used to test the AGS ability to treat this pollutant. Mordant Orange [82], Sumifix

Black EXA, Sumifix Navy Blue EXF, Synozol Red K-4B [83], reactive blue 59 [84], and reactive yellow 15 [81] were tested diluted in synthetic wastewater.

Sarvajith et al. [81] reported azo dye removal efficiencies of 89–100%, followed by total organic carbon (79–95%) and ammoniacal nitrogen (92–100%) in an AGS reactor operated for 80 days. Up to date, most of the dye-removal studies using AGS systems use synthetic wastewater, which underestimates the real textile wastewater effects. Lotito et al. [85] have applied a hybrid system comprising AGS and biofilms for treating mixed municipal–textile wastewater. The authors evaluated, between the typical physical and chemical parameters, the decolorization percentage, measuring the absorbance of the samples in three wavelengths (426, 558, and 660 nm), achieving decolorization efficiencies of 33%, 41%, and 52%, respectively.

3.7 Additives to avoid granule disintegration

The long time required for granular biomass formation, unpredictable granule morphology, and granule disintegration are some of the main drawbacks observed for pilot- and full-scale application of this treatment technology [86, 87]. Some of these problems are observed only in long-term studies using real wastewater [88]. According to Rusanowska et al. [89], disturbances in the microbial EPS synthesis may cause biomass deflocculation, reducing the treated wastewater quality. Adav et al. [87] proposed that granule disintegration is related to the hydrolysis of β-polysaccharides, which are essential for granule stability since they form a matrix that favors the deposition of proteins, lipids, polysaccharides, and cells. However, Zhu et al. [90] came to a different conclusion, in which stable and high PN/PS ratio is crucial for maintaining the long-term stability of AGS. Disintegration of granules due to unknown process conditions is a challenge, which needs further investigation. Hamza et al. [91] studied the impact of food to microorganisms (F/M) ratio on the long-term stability of aerobic granules. Their goal was to determine stability boundaries of F/M ratio in an AGS reactor treating high-strength organic wastewater.

The addition of substances to accelerate and stabilize AGS, and therefore, the overall system function has been proposed as an alternative to overcome these problems. These substances may vary from loose ions dispersed inside the reactor, such as Ca^{2+} [92, 93], Mg^{2+} [94], Na^{+} [95, 96], and multi-ion matrices (e.g., seawater) [97], or the application of binders such as poly-aluminum chloride [98], static magnetic field [099], magnetite nanoparticle [100], crushed aerobic granules [101], anaerobic granules [102], and microbial flocculant [103].

The presence of divalent metal ions, such as Ca^{2+} [104] and Mg^{2+} [105], during the treatment process was proved to enhance the sludge granulation in an SBR [106].

Li et al. [107] proposed the utilization of another divalent ion. Their goal was to enhance the AGS process through the augmentation of Mg^{2+} into an SBR. Same as

Jiang et al. [108], the proposition was to use two identical reactors, working in parallel, one with the binder and the other without the binder; however, they adopted an SBR model. It was found out that the addition of Mg^{2+} enhanced the granulation process, cultivating larger sized granules, and also increased the biomass concentrations. Despite the fact that a meaningful impact for the COD removal was not noticed, neither did it bring up any difference in microbial morphology.

References

[1] Morgenroth, E., Sherden, T., Van Loosdrecht, M.C.M., Heijnen, J.J., Wilderer, P.A. Aerobic granular sludge in a sequencing batch reactor. Water Res. 1997, 31, 3191–3194.
[2] Heijnen, J.J., Van Loosdrecht, M.C.C. 1998. Method for Acquiring Grain-shaped Growth of a Microorganism in a Reactor, Patent Cooperation Treaty (PCT), pp. 13. Technische Universiteit Delft, US, European patent.
[3] Van Loosdrecht, M.C.M., De Kreuk, M.K. Method for the Treatment of Wastewater with Sludge Granules, Bureau voor Industriele Eigendom. The Netherlands, International, Technische Universiteit Delft, 2004.
[4] Franca, R.D.G., Pinheiro, M.H., Van Loosdrecht, M.C.M., Lourenço, N.D. Stability of aerobic granules during long-term bioreactor operation. Biotechnol Adv. 2018, 36, 228–246.
[5] Sarma, S.J., Tay, J.H., Chu, A. Finding knowledge gaps in aerobic granulation technology. Trends Biotechnol. 2017, 35, 66–78.
[6] Jungles, M.K., Figueroa, M., Morales, N., Val Del Rio, A., Da Costa, R.H.R., Campos, J.L., Mosquera-Corral, A., Méndez, R. Start up of a pilot scale aerobic granular reactor for organic matter and nitrogen removal. J Chem Technol Biotechnol. 2011, 86, 763–768.
[7] Jungles, M.K., Campos, J.L., COSTA, R.H.R. Sequencing batch reactor operation for treating wastewater with aerobic granular sludge. Brazilian J Chem Eng. 2014, 31, 27–33.
[8] Wagner, J., Weissbrodt, D.G., Manguin, V., Da Costa, R.H.R., Morgenroth, E., Derlon, N. Effect of particulate organic substrate on aerobic granulation and operating conditions of sequencing batch reactors. Water Res. 2015, 85, 158–166.
[9] Wagner, J., Da Costa, R.H.R. Aerobic Granulation in a Sequencing Batch Reactor Using Real Domestic Wastewater. J Environ Eng. 2013, 139, 1391–1396.
[10] Velho, V.F., Magnus, B.S., Xavier, J.A., Guimarães, L.B., Da Costa, R.H.R. Effect of COD/N ratio on N_2O production during nitrogen removal by aerobic granular sludge. Water Sci Technol. 2017, 76, 3452–3460.
[11] Guimarães, L.B., Wagner, J., Akaboci, T.R.V., Daudt, G.C., Nielsen, P.H., Van Loosdrecht, M.C.M., Weissbrodt, D.G., Da Costa, R.H.R. Elucidating performance failures in use of granular sludge for nutrient removal from domestic wastewater in a warm coastal climate region. Environ Technol. 2018, 1–33.
[12] Bassin, J.P., Tavares, D.C., Borges, R.C., Dezotti, M. Development of aerobic granular sludge under tropical climate conditions: The key role of inoculum adaptation under reduced sludge washout for stable granulation. J Environ Manage. 15, 168–182.
[13] Domingos, D.G., Henriques, R.O., Xavier, J.A., Libardi, N., Da Costa, R.H.R. Increasing activated sludge aggregation by magnetite nanoparticles addition. Water Sci Technol. 2019, 79, 993–999.
[14] Ladnorg, S., Libardi, N., Dall'Agnol, P., Domingos, D.G., Magnus, B.S., Wichern, M., Gehring, T., Da Costa, R.H.R. Alginate-like exopolysaccharide extracted from aerobic granular sludge

as biosorbent for methylene blue: Thermodynamic, kinetic and isotherm studies. J Environ Chem Eng. 2019, 7, 1–9.
[15] Dall'Agnol, P., Libardi, N., Muller, J.M., Xavier, J.A., Domingos, D.G., Da Costa, R.H.R. A comparative study of phosphorus removal using biopolymer from aerobic granular sludge: A factorial experimental evaluation. J Environ Chem Eng. 2019, 8, 103541.
[16] Lin, Y., Nam, K. (2019): Self-extinguishing property of biopolymers recovered from waste aerobic granular sludge. [3rd IWA Resource Recovery Conference, Venice, Italy, 8–12 September 2019] In: IWA (Eds.), 3rd IWA Resource Recovery Conference – Book of Abstracts, 8–12 September 2019, Venice, Italy, p.66
[17] Lin, Y.M., Nierop, K.G.J., Girbal-Neuhauser, E., Adriaanse, M., Van Loosdrecht, M.C.M. Sustainable polysaccharide-based biomaterial recovered from waste aerobic granular sludge as a surface coating material. Sustain Mat Technol. 2015, 4, 24–29.
[18] Liu, Y., Tay, J.H. The essential role of hydrodynamic shear force in the formation of biofilm and granular sludge. Water Res. 2002, 36, 1653–1665.
[19] Zhang, Q., Hu, J., Lee, D. Aerobic granular processes: Current research trends. Bioresour Technol. 2016, 210, 74–80.
[20] De Kreuk, M.K., Pronk, M., Van Loosdrecht, M.C.M. Formation of aerobic granules and conversion processes in an aerobic granular sludge reactor at moderate and low temperatures. Water Res. 2005, 39, 4476–4484.
[21] Liu, Y.Q., Moy, B., Kong, Y.H., Tay, J.H. Formation, physical characteristics and microbial community structure of aerobic granules in a pilot-scale sequencing batch reactor for real wastewater treatment. Enzyme Microb Technol. 2010, 46, 520–525.
[22] Beun, J.J., Van Loosdrecht, M.C.M., Heijnen, J.J. Aerobic granulation in a sequencing batch airlift reactor. Water Res. 2002, 36, 702–712.
[23] Toh, S., Tay, J., Moy, B., Ivanov, V., Tay, S. Size-effect on the physical characteristics of the aerobic granule in a SBR. Appl Microbiol Biotechnol. 2003, 60, 687–695.
[24] Liu, Y.Q., Tay, J.H. Influence of cycle time on kinetic behaviors of steady-state aerobic granules in sequencing batch reactors. Enzyme Microb Technol. 2007, 4, 516–522.
[25] Franca, R.D.G., Ortigueira, J., Pinheiro, H., Lourenco, N.D. Effect of SBR feeding strategy and feed composition on the stability of aerobic granular sludge in the treatment of a simulated textile wastewater. Water Sci Technol. 2017, 76, 1188–1195.
[26] Rollemberg, S., Barros, A.R.M., Firmino, P.I., Dos Santos, A. Aerobic granular sludge: Cultivation parameters and removal mechanisms. Bioresour Technol. 2018, 270, 678–688.
[27] Pronk, M., De Kreuk, M.K., De Bruin, B., Kamminga, P., Kleerebezem, R., Loosdrecht, M.C.M. Full scale performance of the aerobic granular sludge process for sewage treatment. Water Res. 2015, 84, 207–217.
[28] Nancharaiah, Y.V., Reddy, K.K. Aerobic granular sludge technology: Mechanisms of granulation and biotechnological application. Bioresour Technol. 2018, 247, 1128–1143.
[29] De Kreuk, M., Kishida, N., Loosdrecht, M.C.M. Aerobic granular sludge state of the art. Water Sci Technol. 2007, 55, 75–81.
[30] Zhang, Y., Dong, X., Nuramkhaan, M., Lei, Z., Shimizu, K., Zhang, Z., Adachi, Y., Lee, D., Tay, J. Rapid granulation of aerobic granular sludge: A mini review on operation strategies and comparative analysis. Bioresour Technol Rep. 2019, 7, 1–12.
[31] Wang, L.K., Li, Y. Sequencing Batch Reactors. In: Wang, L.K., Pereira, N.C., Hung, Y.T. Ed., Biological Treatment Processes. Humana Press, New Jersey, 2009, Vol. VIII, 459–511.
[32] Libardi, N., Soccol, C., Carvalho, J.C., Vandenberghe, L. Simultaneous cellulase production using domestic wastewater and bioprocesso efflluent treatment – A biorefinery approach. Bioresour Technol. 2018, 276, 42–50.

[33] Sharma, S., Sarma, S.J., Tay, J.T. Aerobic granulation in wastewater treatment: A general overview. In: Shah, M.P., Rodriguez-Couto, S. ed., Microbial Wastewater Treatment. Elsevier, Netherlands, 2019, 57–81.
[34] Carrera, P., Campo, R., Méndez, R., Di Bella, G., Campos, J.L., Mosquera-Corral, A., Val Del, R.A. Does the feeding strategy enhance the aerobic granular sludge stability treating saline effluents? Chemosphere. 2019, 226, 865–873.
[35] De Kreuk, M., Van Loosdrecht, N.C.M. Selection of slow growing organisms as a mean for improving aerobic granular sludge stability. Water Sci Technol. 2004, 49, 9–17.
[36] Lochmatter, S., Gonzalez-Gil, G., Holliger, C. Optimized aeration strategies for nitrogen and phosphorus removal with aerobic granular sludge. Water Res. 2013, 47, 6187–6197.
[37] Lettinga, G., Van Velsen, A.F.M., Hobma, S.W., De Zeeuw, W., Klapwijk, A. Use of upflow sludge blanket (USB) reactor concept for biological wastewater treatment, especially for anaerobic treatment. Biotechnol Bioeng. 1980, 22, 699–734.
[38] Van Loosdrecht, M.C.M., Tijhuis, L., Wijdieks, A.M.S., Heijnen, J.J. Population distribution in aerobic biofilms on small suspended particles. Water Sci Technol. 1995, 31, 163–171.
[39] Nancharaiah, Y.V., Sarvajith, M., Mohan, T.V.K. Aerobic granular sludge: The future of wastewater treatment. Curr Sci. 2019, 117, 395–404.
[40] Laspidou, C.S., Tirrmann, B. A unified theory for extracelular polymeric substances, soluble microbial products, and active and inert biomass. Water Res. 2002, 36, 2711–2720.
[41] Xin, X., Lu, H., Yao, L., Leng, L., Guan, L. Rapid Formation of Aerobic Granular Sludge and Its Mechanism in a Continuous-Flow Bioreactor. Appl Biochem Biotechnol. 2017, 181, 424–433.
[42] Seviour, T., Derlon, N., Dueholm, M.S., Flemming, H.-C., Girbal-Neuhauser, E., Horn, H., Kjelleberg, S., van Loosdrecht, M.C.M., Lotti, T., Malpei, M.F., Nerenberg, R., Neu, T.R., Paul, E., Yu, H., Lin, Y. Extracellular polymeric substances of biofilms: Suffering from an identity crisis. Water Res. 2018, 151, 1-.
[43] Seviour, T., Malde, A.K., Kjelleberg, S., Yuan, Z., Mark, A.E. Molecular dynamics unlocks atomic level self-assembly of the exopolysaccharide matrix of water-treatment granular biofilms. Biomacromolecules. 2012, 13, 1965–1972.
[44] Pronk, M., Neu, T.R., van Loosdrecht, M.C.M., Lin, Y.M. The acid soluble extracellular polymeric substance of aerobic granular sludge dominated by *Defluviicoccus* sp. Water Res. 2017, 122, 148–158.
[45] Lin, Y., Reino, C., Carrera, J., Pérez, J., van Loosdrecht, M.C.M. Glycosylated amyloid-like proteins in the structural extracellular polymers of aerobic granular sludge enriched with ammonium-oxidizing bacteria. Microbiol Open. 2018, 7, 1–13.
[46] Lin, Y., de Kreuk, M., van Loosdrecht, M.C.M., Adin, A. Characterization of alginate-like exopolysaccharides isolated from aerobic granular sludge in pilot-plant. Water Res. 2010, 44, 3355–3364.
[47] Xue, W., Zeng, Q., Lin, S., Zan, F., Hao, T., Lin, Y., Van Loosdrecht, M.C.M., Chen, G. Recovery of high-value and scarce resoures from biological wastewater treatment: Sulfated polysaccharides. Water Res. 2019, 163, 1–10.
[48] Sarma, S.J., Tay, J. Carbon, nitrogen and phosphorus removal mechanisms of aerobic granules. Crit Rev Biotechnol. 2018, 38, 1–12.
[49] Bathe, S., De Kreuk, M., McSwain, B., Schwarzenbeck, N. Aerobic Granular Sludge. London, IWA Publishing, 2005, 1–178.
[50] Wang, J., Wang, X., Zhao, Z., Li, J. Organics and nitrogen removal and sludge stability in aerobic granular sludge membrane bioreactor. Appl Microbiol Biotechnol. 2008, 79, 679–685.
[51] Adav, S.S., Lee, D., Lai, J. Aerobic granulation in sequencing batch reactors at different settling times. Bioresour Technol. 2009, 100, 5359–5361.

[52] De Kreuk, M., Heijnen, J., Van Loosdrecht, M.C.M. Simultaneous COD, nitrogen, and phosphate removal by aerobic granular sludge. Biotechnol Bioeng. 2005, 90, 761–769.
[53] Sarma, S.J., Tay, J. Biodegradation of toxic compounds by aerobic granulation technology. SM J Environ Toxicol. 2016, 2, 1–7.
[54] Kong, Q., Wang, Z., Shu, L., Miao, M. Characterization of the extracellular polymeric substances and microbial community of aerobic granulation sludge exposed to cephalexin. Int Biodeterior Biodegradation. 2015, 102, 375–382.
[55] Tay, S.T., Moy, B.Y., Jiang, H., Tay, J. Rapid cultivation of stable aerobic phenol-degrading granules using acetate-fed granules as microbial seed. J Biotechnol. 2005, 115, 387–395.
[56] He, Q., Yuan, Z., Zhang, J., Zhang, S., Zhang, W., Zou, Z., Wang, H. Insight into the impact of ZnO nanoparticles on aerobic granular sludge under shock loading. Chemosphere. 2017, 173, 411–416.
[57] Jiang, Y., Wei, L., Yang, K., Shi, X., Wang, H. Rapid formation of aniline-degrading aerobic granular sludge and investigation of its microbial community succession. J Clean Prod. 2017, 166, 1235–1243.
[58] Li, K., Wei, D., Zhang, G., Shi, L., Wang, Y., Wang, B., Du, B., Wei, Q. Toxicity of bisphenol A to aerobic granular sludge in sequencing batch reactors. J Mol Liq. 2015, 209, 284–288.
[59] Wei, D., Li, M., Wang, X., Han, F., Li, L., Guo, J., Ai, L., Fang, L., Liu, L., Du, B., Wei, Q. Extracellular polymeric substances for Zn (II) binding during its sorption process onto aerobic granular sludge. J Hazard Mater. 2016, 301, 407–415.
[60] Gao, M., Diao, M., Yuan, S., Wang, Y., Xu, H., Wang, X. Effects of phenol on physicochemical properties and treatment performances of partial nitrifying granules in sequencing batch reactors. Biotechnol Rep. 2017, 13, 13–18.
[61] Liu, Q.S., Liu, Y., Tay, J.H. Toxicity effect of phenol on aerobic granules. Environ Technol. 2009, 30, 69–74.
[62] Ho, K.L., Chen, Y.Y., Lin, B., Lee, D.J. Degrading high-strength phenol using aerobic granular sludge. Appl Microbiol Biotechnol. 2010, 85, 2009–2015.
[63] Cajthaml, T., Křesinová, Z., Svobodová, K., Möder, M. Biodegradation of endocrine-disrupting compounds and suppression of estrogenic activity by ligninolytic fungi. Chemosphere. 2009, 75, 745–750.
[64] Libardi, N., Gern, R.M.M., Furlan, S.A., Schlosser, D. Laccase production by the aquatic ascomycete *Phoma* sp UHH 5-1-03 and the white rot basidiomycete *Pleurotus ostreatus* DSM 1833 during submerged cultivation on banana peels and enzyme applicability for the removal of endocrine-disrupting chemicals. Appl Biochem Biotechnol. 2012, 167, 1144–1156.
[65] Wang, S., Ma, X., Wang, Y., Duo, G., Tay, J., Li, J. Piggery wastewater treatment by aerobic granular sludge: Granulation process and antibiotics and antibiotic-resistant bacteria removal and transport. Bioresour Technol. 2019, 273, 350–357.
[66] Keskinkan, O., Balci, B. Biotic and abiotic bisphenol-A removal from wastewater by activated sludge: Effects of temperature, biomass, and bisphenol-A concentrations. Water Sci Technol. 2016, 73, 317–328.
[67] Wei, D., Wang, Y., Wang, X., Li, M., Han, F., Ju, L., Zhang, G., Li, K., Wang, B., Du, B., Wei, Q. Toxicity assessment of 4-chlorophenol to aerobic granular sludge an sots interaction with extracellular polymeric substances. J Hazard Mater. 2015, 30, 101–107.
[68] Cydzik-Kwiatkowska, A., Bernat, K., Zielinska, M., Bulkowska, K., Wojnowska-Baryla, I. Aerobic granular sludge for bisphenol A (BPA) removal from wastewater. Int Biodeterior Biodegradation. 2017, 122, 1–11.
[69] Yang, Y., Ok, Y.S., Kim, K.H., Kwon, E.E., Tsang, Y.F. Occurrences and removal of pharmaceuticals and personal care products (PPCPs) in drinking waster and water/sewage treatment plants: A review. Sci Total Environ. 2017, 15, 3-3-320.

[70] Liu, J., Wong, M. Pharmaceuticals and personal care products (PPCPs): A review on environmental contamination in China. Environ Int. 2013, 59, 208–224.

[71] Zhao, X., Chen, Z.L., Wang, X.C., Shen, J.M., Xu, H. PPCPs removal by aerobic granular sludge membrane bioreactor. Appl Microbiol Biotechnol. 2014, 98, 9843–9848.

[72] He, L., Ying, G., Liu, Y., Su, H., Chen, J., Liu, S., Zhao, J. Discharge of swine wastes risks water quality and food safety: Antibiotics and antibiotic resistance genes from swine sources to the receiving environments. Environ Int. 2016, 92, 210–219.

[73] He, L., Liu, Y., Su, H., Zhao, J., Liu, S., Chen, J., Liu, W., Ying, G. Dissemination of antibiotic resistance genes in representative broiler feedlots environments: Identification of indicator ARGs and correlations with environmental variables. Environ Sci Technol. 2014, 48, 13120–13129.

[74] Zhou, L., Ying, G., Liu, S., Zhang, R., Lai, H., Chen, Z., Pan, C. Excretion masses and environmental occurrence of antibiotics in typical swine and dairy cattle farms in China. Sci Total Environ. 2013, 444, 183–195.

[75] Liao, J., Liu, C., Liu, L., Li, J., Fan, H., Ye, J., Zeng, Z. Influence of hydraulic retention time on behavior of antibiotics and antibiotic resistance genes in aerobic granular reactor treating biogas slurry. Front Environ Sci Eng. 2019, 13, 31.

[76] Zhou, Y., Zhang, M., Wang, X., Huang, Q., Min, Y., Ma, T.J. Removal of crystal violet by a novel cellulose-based adsorbent: Comparison with native cellulose. Ind Eng Chem Res. 2014, 53, 5498–5506.

[77] Ladnorg, S., Libardi, N., Dall'Agnol, P., Domingos, D.G., Magnus, B.S., Wichern, M., Gehring, T., Da Costa, R.H.R. Alginate-like exopolysaccharide extracted from aerobic granular sludge as biosorbent for methylene blue: Thermodynamic, kinetic and isotherm studies. 2019, J Environ Chem Eng. 7, 1–9.

[78] Vijayaraghavan, K., Yun, Y.S. Bacterial biosorbents and biosorption. Biotechnol Adv. 2008, 26, 266–291.

[79] Saratale, R.G., Saratale, G.D., Chang, J.S., Govindwar, S.P. Bacterial decolorization and degradation of azo dyes: A review. J Taiwan Inst Chem Eng. 2011, 42, 138–157.

[80] Solis, M., Solis, A., Pérez, H.I., Manjarrez, N., Flores, M. Microbial decolourization of azo dyes: A review. Process Biochem. 2012, 47, 1723–1748.

[81] Sarvajith, M., Reddy, G.K.K., Nancharaiah, Y.V. Textile dye biodecolourization and ammonium removal over nitrite in aerobic granular sludge sequencing batch reactors. J Hazard Mater. 2018, 15, 536–543.

[82] Yan, L.K.Q., Fung, K.Y., Ng, K.M. Aerobic sludge granulation for simultaneous anaerobic decolorization and aerobic aromatic amines mineralization for azo dye wastewater treatment. Environ Technol. 2018, 39, 1368–1375.

[83] Muda, K., Aris, A., Salim, M.R., Ibrahim, Z., Yahya, A., Loosdrecht, M.C.M., Ahmad, M.A.B., Nawahwi, M.Z. Development of granular sludge for textile wastewater treatment. Water Res. 2010, 44, 4341–4350.

[84] Kolekar, Y.M., Nemade, H.N., Markad, V.L., Adav, S.S., Patole, M.S., Kodam, K.M. Decolorization and biodegradation of azo dye Reactive Blue 59 by aerobic granules. Bioresour Technol. 2012, 104, 818–282.

[85] Lotito, A.M., Sanctis, M., Iaconi, C., Bergna, G. Textile wastewater treatment: Aerobic granular sludge vs activated sludge systems. Water Res. 2014, 54, 337–346.

[86] Guimarães, L.B., Wagner, J., Akaboci, T.R.V., Daudt, G.C., Nielsen, P.H., Van Loosdrecht, M.C.M., Weissbrodt, D.G., Da Costa, R.H.R. Elucidating performance failures in use of granular sludge for nutrient removal from domestic wastewater in a warm coastal climate region. Environ Technol. 2018, 1551938, 1–16.

[87] Adav, S.S., Le, D.J., Show, K.Y., Tay, J.H. Aerobic granular sludge: Recent advances. Biotechnol Adv. 2008, 26, 411–423.
[88] Sguanci, S., Lubello, C., Caffaz, S., Lotti, T. Long-term stability of aerobic granular sludge for treatment of very low-strength real domestic wastewater. J Clean Prod. 2019, 222, 882–890.
[89] Rusanowska, P., Cydzik-Kwiatkowska, A., Świątczak, P., Wojnowska-Baryła, I. Changes in extracellular polymeric substances (EPS) content and composition in aerobic granule size-fractions during reactor cycles at different organic loads. Bioresour Technol. 2019, 272, 188–193.
[90] Zhu, L., Qi, H., Lv, M., Yu, Y., Xu, X. Component analysis of extracellular polymeric substances (EPS) during aerobic sludge granulation using FTIR and 3D-EEM technologies. Bioresour Technol. 2012, 124, 455–459.
[91] Hamza, R.A., Sheng, Z., Iorhemen, O.T., Zaghloul, M.S., Tay, J.H. Impact of food-to-microorganisms ratio on the stability of aerobic granular sludge treating high-strength organic wastewater. Water Res. 2019, 147, 287–298.
[92] Jiang, H.L., Tay, J.H., Maszenan, M., Tay, T.L. Enhanced phenol biodegradation and aerobic granulation by two co-aggregating bacterial strains. Environ Sci Technol. 2006, 40, 6137–6142.
[93] Morais, I., Silva, C., Zanuncio, J., Zanuncio, A. Structural stabilization of granular sludge by addition of calcium ions into aerobic bioreactors. Bioresources. 2018, 13, 176–191.
[94] Liu, L., Gao, D., Zhang, M., Fu, Y. Comparison of Ca^{2+} and Mg^{2+} enhancing aerobic granulation in SBR. J Hazard Mater. 2010, 181, 382–387.
[95] Pronk, M., Bassin, J.P., de Kreuk, M.K., Kleerebezem, M.C.M. Evaluating the main and side effects of high salinity on aerobic granular sludge. Appl Microbiol Biotechnol. 2014, 98, 1339–1348.
[96] Hou, M., Li, W., Li, H., Li, C., Wu, X., Liu, Y. Performance and bacterial characteristics of aerobic granular sludge in response to alternating salinity. Int Biodeterior Biodegradation. 2019, 142, 211–217.
[97] Tao, J., Qin, L., Liu, X., Li, B., Chen, J., You, J., Shen, Y., Chen, H. Effect of granular activated carbon on the aerobic granulation of sludge and its mechanism. Bioresour Technol. 2017, 236, 60–67.
[98] Liu, Z., Liu, Y., Kuschk, P., Wang, J., Chen, Y., Wang, X. Poly aluminum chloride (PAC) enhanced formation of aerobic granules: Coupling process between physicochemical-biochemical effects. Chem Eng J. 2016, 284, 1127–1135.
[99] Wang, X., Diao, M., Yang, Y., Shi, Y., Gao, M., Wang, S. Enhanced aerobic nitrifying granulation by static magnetic field. Bioresour Technol. 2012, 110, 105–110.
[100] Domingos, D.G., Henriques, R.O., Xavier, J.A., Libardi, N.da Costa, R.H.R. Increasing activated sludge aggregation by magnetite nanoparticles addition. Water Sci Technol. 2019, 79, 993–999.
[101] Pijuan, M., Werner, U., Yuan, Z. Reducing the start up time of aerobic granular sludge reactors through seeding floccular sludge with crushed aerobic granules. Water Res. 2011, 45, 5075–5083.
[102] Linlin, H., Jianlong, W., Xianghua, W., Yi, Q. The formation and characteristics of aerobic granules in sequencing batch reactor (SBR) by seeding anaerobic granules. Process Biochem. 2005, 40, 5–11.
[103] Liang, Z., Tu, Q., Su, X., Yang, X., Chen, J., Chen, Y., Liu, X., Li, H., He, Q. Formation, extracellular polymeric substances and microbial community of aerobic granules enhanced by microbial flocculant compared with poly-aluminum chloride. J Clean Prod. 2019, 220, 544–552.
[104] Yu, H.Q., Tay, J.H., Fang, H.H.P. The roles of calcium in sludge granulation during UASB reactor start-up. Water Res. 2001, 35, 1052–1060.

[105] Li, X., Liu, Q., Yang, Q., Guo, L., Zeng, G., Hu, J., Zheng, W. Enhanced aerobic sludge granulation in sequencing batch reactor by Mg^{2+} augmentation. Bioresour Technol. 2009, 100, 64–67.
[106] Ren, T., Liu, L., Sheng, H., Liu, X., Yu, H., Zhang, M., Zhu, J. Calcium spatial distribution in aerobic granules and its effects on granule structure, strength and bioactivity. Water Res. 2008, 42, 3343–3352.
[107] Li, X., Luo, J., Guo, G., Mackey, H.R., Hao, T., Chen, G. Seawater-based wastewater accelerates development of aerobic granular sludge: A laboratory proof-of-concept. Water Res. 2017, 115, 210–219.
[108] Jiang, H.L., Tay, J.H., Liu, Y., Tay, S.T.L. Ca^{2+}. Augmentation for enhancement of aerobically grown microbial granules in sludge blanket reactors. Biotechnol Lett. 2003, 25, 95–99.

Sreejita Ghosh, Dibyajit Lahiri, Sougata Ghosh,
Moupriya Nag and Rina Rani Ray

4 Biofilm-mediated industrial wastewater treatment

Abstract: Over the last few decades, the increase in urbanization and industrialization has resulted in the degradation of the quality of water. The contaminants present in water have toxic effects on living organisms and the surrounding environment. Hence, recycling is the best way to salvage this dwindling natural resource, and regulations concerning wastewater treatment have become more rigorous in recent years. A large amount of organic components discharged from the industrial effluents and household activities pose a serious matter of concern among the scientific communities. The use of biological treatment has been found to be an effective alternative strategy in controlling the water pollution. It involves disintegration of complex compounds into simpler forms by microorganisms before discharging into the natural environment. Biofilm is the consortia of sessile microbial communities that are being frequently used for the purpose of remediating wastewater. Being a complex community, it possesses the overall ability to enhance the rate of pollutant degradation through various mechanisms like biodegradation, bioaccumulation, biosorption, and biomineralization. The use of biofilm for wastewater treatment has several advantages, including operational plasticity, little space requirements, reduced time of hydraulic retention, flexibility to changes in the environment, increased biomass retention time, highly active biomass concentration, enhanced ability to manage and disintegrate recalcitrant compounds, as well as a sluggish microbial growth rate, resulting in lesser sludge production. This chapter focuses on the various types of microbial biofilms and their mechanisms of remediating wastewater, emphasizing the emerging technologies for wastewater treatment.

Keywords: wastewater treatment, biofilms, biodegradation, bioaccumulation, biosorption, biomineralization

Sreejita Ghosh, Rina Rani Ray, Department of Biotechnology, Maulana Abul Kalam Azad University of Technology, Haringhata, West Bengal, India, e-mail: raypumicro@gmail.com
Sougata Ghosh, Department of Microbiology, School of Science, RK University, Rajkot, Gujarat, India, e-mail: ghoshsibb@gmail.com
Dibyajit Lahiri, Moupriya Nag, Department of Biotechnology, University of Engineering and Management, Kolkata, West Bengal, India

https://doi.org/10.1515/9783110727227-004

4.1 Introduction

Advancements in biotechnology have led to the increased interest in scientific techniques for basic remodeling of science and its societal role [1]. Biotechnology makes use of the live microbial technologies to serve as tools by manipulating the cells to produce some innovative and alternative methods to more effectively maintain the natural environment by maintaining proper water quality and treatment of untreated industrial wastewater [2]. Various industrial activities like textile industries, chemical industries, tanneries, electroplating industries, ink industries, and other food and beverage industries deliver large amounts of wastewater contaminated with toxic chemicals, dyes, and heavy metals into the natural water sources, thereby polluting the environment in a significant way and harming the ecosystem. Discharge of heavy metals and radionuclides from metallurgical industries [3], wastewater/sewage sludge from industrial origin [4], and the rapid growth in industrialization have jeopardized the natural functioning of the ecosystem. Such situations require immediate measures in order to restore proper activities of the biogeochemical cycles, which is the main force in maintaining life on the Earth. Biogeochemical cycles are mediated by the metabolic activities of various microbes, which can prevent the pollutants from entering the biosphere [5].

Increased accumulation of these contaminants and their effects are detrimental to the biota system, environment, and humans as well as animals. Traditional methods for removal of heavy metals and radionuclides from contaminated waters involve immobilization/solidification/stabilization and physicochemical extractions such as precipitation, filtration, chelation, electrochemical treatment, oxidation and reduction, ion exchange, and reverse osmosis [6]. Such techniques require huge quantities of chemical reagents, huge capital cost, and production of toxic secondary byproducts, thereby making these conventional techniques ineffective and unsuccessful. Thus, it is necessary for an interdisciplinary approach by using biologically driven process and/or biologically derived products to enhance bioremediation processes via biogeochemical phenomenon.

Bioremediation is an environment-friendly and a cost-effective technique involving large numbers of remediation procedures like biofilm-mediated treatment, biodegradation, biosorption, bioaccumulation, and biomineralization [7]. These methods have so far proved to be promising enough to handle the current challenges for remediation of contaminated water. The technique of bioremediation is based on the principle of solubility reduction of the pollutant by changing pH, modification of redox reactions, and pollutant adsorption from contaminated environment.

This chapter highlights the various aforementioned techniques of bioremediation mediated by biofilm formation and the various biofilm reactors that are nowadays used in treatment of untreated industrial wastewater.

4.2 Biofilm-mediated treatment

Biofilms may be defined as some aggregated biological structures which are formed on various surfaces that are constantly coming in contact with water. They have complex structures possessing dynamic characteristics of primeval microbes attached together. Biofilms are composed of prokaryotic cells and cells of other microbes like fungi, protozoa, and yeasts. These microorganisms produce a mucilaginous defensive shield called exopolysaccharide or extra polymeric substance (EPS) on surfaces in which they aggregate. Biofilms actually provide the resistance of microbes against unfavorable conditions and also help them to survive under stress. In the biofilm system, the microbial cells remain embedded within the EPS matrix and consist of two layers: substratum (on which the biofilm formation occurs) and the bulk phase (phase floating over biofilm). Biofilms have some negative effects such as dental plaque formation leading to tooth decay and gum diseases [8], foodborne diseases, and also algal mat formation causing eutrophication in water bodies. In spite of some of the harmful effects, biofilm formation can be controlled and may be used to treat wastewater (in the present case, industrial wastewater).

4.2.1 Biofilm-supported reactors

Biofilm reactors primarily make use of biofilms to treat wastewater(s). Biofilms within these types of reactors act as the main mechanism to biologically transform nutrients, which are considered as potential environmental pollutants like nitrogen, phosphorus, and biodegradable organic matter. There are different types of biofilm reactors utilized for wastewater treatment but mainly integrated fixed-film-activated sludge (IFAS) system and membrane-supported biofilm reactors (MBfRs).

MBfRs and IFAS biofilm reactors are still evolving as the new technology for wastewater treatment. Modern IFAS and MBfRs systems utilize freely moving biofilm carriers that remain submerged and cause oxidation of carbon, nitrification, denitrification, and deammonification [9–11]. Over the recent few years, researches indicated a broader insight into the activities of the biofilm carriers on mass transfer and their effects on hydrodynamics to carry out various biochemical alteration procedures. MBfRs are mainly used to treat industrial wastewaters. These MBfRs are known to carry out partial nitrification and deammonification in aquacultures simultaneously (Figure 4.1).

4.2.1.1 Aerobic granular biomass

Another technique is the development of granular biomass in sequencing batch reactor (SBR) to treat contaminated water bodies. Aerobic granules which can be

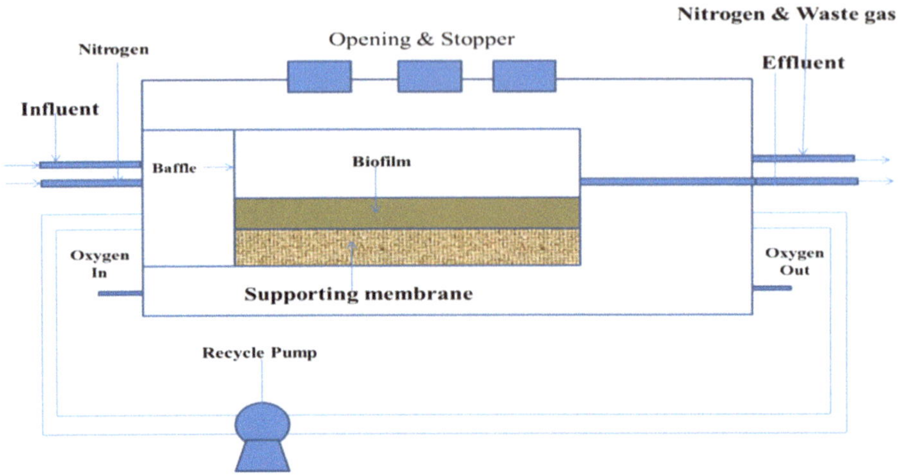

Figure 4.1: Membrane-supported biofilm reactor.

maintained within these SBRs are produced [12]. All these potentially stable aerobic granular sludge technologies are utilized to remove the biological nutrients from industrial wastewaters. Such processes conserve a fixed volume of liquid/biomass [13]. The filling, settling, and decanting processes take place simultaneously for almost 25–33% of the total period of operation. The rest of the operational time period is kept for the aeration process (reaction period). It is estimated that around 10–15 min is needed to reach reactor quiescence just before the following cycle is initiated by feed flow from the bottom of the reactor. These bioreactors can be explained as an empty tank with provision for aeration with fine bubbles and the feed flow of wastewater from the bottom of the tank. The effluent that has been already treated passes over an effluent barrage, which is present at the top of the reactor tank. These bioreactors contain no mixers for discharge of effluents through overspilled weirs; instead they contain a sludge waste collecting system, which is present around the top of the settling sludge bed to facilitate wastage of more slow settling sludge. Granular biomass can also be selectively retained on screens by means of cyclone separator. In PN/A systems, these granules are used to treat ammonia-nitrogen concentrated wastes released by digested sludge dewatering in anaerobic treatment of wastewater.

4.2.1.2 Membrane-supported biofilm reactors (MBfRs)

MBfRs exhibit a greater strength than any other available types of available biofilm bioreactor. In these reactors, gas is delivered to the liquid phase through a flat, tubular, or hollow-fiber membrane, which directly acts as a supporting material for biofilm growth. Electron acceptors and electron donors are being subjected to counter-current diffusion via biofilm from the lumen of the membrane as well as from the bulk fluid. The hydrogen-based MBfRs [14] transfer hydrogen (electron donor) to the biofilm while the air/oxygen-based MBfRs [15] transfers oxygen (electron acceptor) to the biofilm. This oxygen/air-based MBfRs are also known as membrane-aerated biofilm reactors [16]. Hydrogen-based MBfRs can be widely used to biochemically transform nitrite, nitrate, percholate, selenite/selenate, bromate, chromate, and arsenate. Since MBfRs allow to significantly control the electron donating/accepting systems, biofilms having increased redox potentialities are designed to simultaneously control oxic/anoxic systems.

4.2.1.3 Moving bed biofilm reactors (MBBRs)

Recently, increased demand to fulfill the current standards of effluent discharge essentially in terms of release of nutrients has led to the development of more advanced biotreatment processes. This is because nutrients in wastewaters possess harmful effects leading to water contamination [17]. Thus, nutrient removal from wastewaters is mandatory for preventing negative effects on the environment. Many such treatment strategies have been adopted so far but among them, the mostly used strategy is the biological nutrient removal (BNR). BNR takes place slowly as the microbes carrying out BNR grow at a slow pace to complete these removal reactions [18].

The mostly used BNR treatment process is the conventional activated sludge (CAS) process but this process is nowadays less preferred as it is not much energy efficient and possesses some problems in its operation. Moreover, this technique needs reactors with high volumes to perform nitrification to a greater level [19, 20]. So, moving bed biofilm reactors (MBBRs) can be efficiently used to incorporate biomass in the form of biofilms in place of using flocs. Gunnes et al. [19] performed a comparative study between nitrogen removal efficacies of CAS with that of an MBBR. It was found that MBBR decreased plant footprint by a level of 50% in comparison to that of CAS, thereby reducing the investment and cost of operation to a much lower level and making the wastewater treatment process a cost-effective method.

A study on the biotreatment of meat industry and slaughterhouse wastewater was done by Aziz et al. [21]. Slaughterhouse wastewater is one of the most challenging liquids for studying the effectiveness of various biotreatment processes as it contains a very high nutrient level. In this study, it was found that CAS treatment produced a maximum chemical oxygen demand (COD) removal of 98%, 65% phosphorus removal efficiency, and 99% ammonia removal efficiency while that with MBBR was 96% phosphorus removal and 81% nitrogen removal.

MBBR can operate with low reactor volumes with less time for sludge recycling. MBBR is capable of carrying out full nitrification in case of ammonia and carbon oxidation. During this nitrification, 75% dissolved oxygen (DO) is required for oxidation of ammonia and 25% is used for oxidation of nitrite to nitrate [22]. Lima et al. [23]

found that the suspended biomass plays an important role in the nitrification and denitrification simultaneously in high-volume MBBRs having low hydraulic retention time (HRT). Also, in order to operate an MBBR having low solid retention time at minimum temperatures, a minimum ammonium removal rate can be achieved because majority of the nitrification process will occur inside the biofilm (Figure 4.2).

Simultaneous nitrification and denitrification membrane bioreactor (SND-MBBR) has the capability to remove 80% of the total nitrogen (TN) and 48% of incoming TN via SND. Microbial populations loaded within the SND-MBBR was found to comprise *Nitrobacter*, *Nitrospira*, and *Paracoccus* [24].

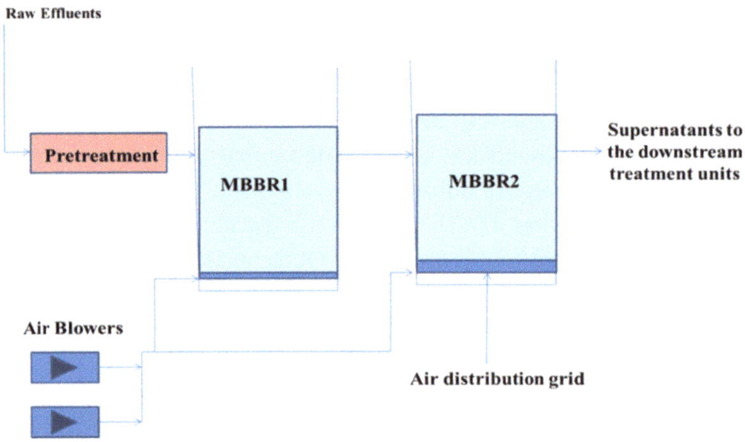

Figure 4.2: Moving bed biofilm reactor (MBBR).

4.2.2 Algal and prokaryotic interplay

The zone within which exudates of algae influence the activities of other coexisting microbes is called "phycosphere." The surfaces of algae are capable of providing a suitable microniche for opportunistic prokaryotes to grow [25]. Thus, the interaction between prokaryotes and algae has both biochemical and ecological significance at the level of microniche along with efficient cycling of nutrients and productivity of biomass at the systematic standards.

4.2.2.1 Carbon dioxide and oxygen exchange

The most important interactions between algae and prokaryotes are the exchange of carbon dioxide and oxygen. Oxygen produced due to algal photosynthesis is utilized by aerobic prokaryotes and carbon dioxide is used for the growth and development of algae. Balancing of the carbon dioxide and oxygen exchange will lead to the prevention of accumulating higher concentrations of DO, which is toxic to algae and also to the prokaryotes [26]. Nitrifying prokaryotes help reducing concentrations of free ammonia, thereby decreasing the pH inhibiting growth and development of algae. Interaction between algae and prokaryotes can even be antagonistic as both prokaryotes and algae will be competing for the available nutrients under condition, thereby limiting nutrient concentration [28]. Competition for other substrates such as HCO_3^-, CO_2, O_2, and NH_4^+ bring about shifting in the microbe communities and also nitrogen metabolism processes. Carbon dioxide sparging can increase nitrification.

4.2.2.2 Intensity of light

Light is one of the major factors in promoting growth of algae but intensity of light can affect prokaryotes and algae in different ways. Activities of algae normally increases with increased intensity of light till the saturation level of photosynthesis [29] but this increased activity with increased light intensity is species specific for algae. Nitrifiers like ammonia-oxidizing bacteria (AOB) and nitrite-oxidizing bacteria (NOB) have been found to be the most sensitive to light [30].

4.2.2.3 Algal-prokaryotic metabolites

Algae are benefitted from prokaryotes because of the release of prokaryotic growth promoters called phytohormones such as cytokinins and indole acetic acid and vitamin B_{12}. This is because most of the algal species need external vitamin B_{12} source for production of the enzyme methionine synthase. The biochemical mechanisms by which both prokaryotes and algae benefit from each other symbiotically is elucidated in Table 4.1. Prokaryotes like *Halomonas* sp. secrete vitamin B_{12} for supporting algal growth [31]. *Mesorhizobium* bacteria supported growth of green algae *Lobomonas rostrata* whose growth is dependent on vitamin B_{12} [32]. On the other hand, activities of prokaryotes are enhanced by the algal production of extracellular compounds such as transparent exopolymer particles, carbohydrates, and proteins. EPS produced by prokaryotes help in forming biofilms in algal-prokaryotic biofilm reactors, thereby facilitating retention of biomass.

Table 4.1: Biochemical mechanism for algal-prokaryotic intercations.

Metabolites	Producer organism	Biochemical mechanism of interaction	References
Phytohormones	Prokaryotes	Phytohormones like indole acetic acid, cytokinins, and other growth factors as well as metabolites initiates cell division in *Chlorella*	[31]
Primary metabolites	Algae	Prokaryotes are benefitted from production of carbohydrates, peptides, amino acids, proteins and transparent exopolymer particles. Heterotrophic prokaryotes release vitamin B_{12} supporting growth of algae but can also cause formation of floc, thereby decreasing available surface area for photosynthesis. Prokaryotes may even cause harm by entering algal tissues/membrane	[32]
Exotoxins/ bioactive metabolites	Both prokaryotes and algae	Secondary metabolites are produced by both of them for competitive advantage. Antimicrobial secondary metabolites are produced by algae to resist prokaryotic attack. Chlorellin (fatty acid mixture) produced by *Chlorella* inhibit gram-positive and gram-negative bacteria, and cyanobacteria secrete cyanotoxins inhibiting nitrifiers	[25]
Lytic enzymes	Prokaryotes	Lysis of algae is caused by cell-to-cell contact and bacterial production of lytic enzymes. Gram-negative myxobacteria can also cause algal lysis.	
Small molecules/auto-inducers	Prokaryotes/ algae	Algal surface of *Delisea pulchra* gets coated with halogenated furanone mixture, which are structural analogs to acyl homoserine lactone (AHL), and initiates community responses in algae and prokaryotes leading to biofilm/floc formation	[33]

4.2.3 Algal-prokaryotic bioreactors for treatment of wastewater

Types of reactors that are used for algal-prokaryotic-mediated treatment of wastewater are closed photobioreactors, open ponds (waste stabilization ponds (WSP) and high-rate algal ponds (HRAP)), and biofilm/attached growth reactors. Generally, WSPs and HRAPs require less investment cost but system control is not up to the mark. Closed reactors like flat plate or tubular photobioreactors require more investment cost and the conditions for growth can be easily controlled, thereby increasing biomass production in a comparatively smaller surface of area.

4.2.3.1 High-rate algal ponds (HRAPs)

HRAPs are generally shallow having a length of 0.2–1 m depending on the clarity of wastewater to penetration of light and a lining of PVC, asphalt, or clay in order to decrease infiltration into the neighboring groundwater and soil. They are provided with paddle wheels to produce turbulence and increase productivity of algal biomass, thereby enhancing efficiency rate of wastewater treatment. On a large-scale basis, HRAPs are actually used in the treatment of agricultural, industrial, and municipal wastewater. HRAPs contribute to improvised efficiency rates for phosphorus and nitrogen removal [34]. Deviller et al. 2004 designed an HRAP treatment for wastewater with very high concentration of nitrogen in a recirculating aquaculture system (RAS), which produces sea bass *Dicentrarchus labrax*. It was found that survival rates for fish in HRAP-integrated RAS increased with an increase in the photosynthesis rate. Water velocities for horizontal flow are maintained between 0.09 and 0.3 m/s for good mixing. HRT for HRAP is 3–15 days.

However, there are certain disadvantages of HRAPs such as reduced settling of biomass and higher evaporation rates although increased evaporation may be reduced by reducing turbulence. Chemical flocculation, dissolved air floatation (DAF), sedimentation, centrifugation, and filtration are some of the commercial technologies that are used in large-scale HRAPs. Sedimentation and coagulation are the most widely used and easiest method while DAF with additional coagulant happens to be more energy consuming. Separation efficiency of solid particles in HRAP can be enhanced by adding fine particles of clay, thereby reducing BOD.

4.2.3.2 Closed photobioreactors

These types of photobioreactors have increased surface area for biomass production and efficient photosynthesis. Closed photobioreactors possess least chance of volatilization of pollutants and reduced water loss via evaporation. They can function in any sort of open space as they are separated from external atmospheric conditions. Construction of closed photobioreactors requires transparent components such as acrylic or glass which make them very expensive. Again, the increased temperatures within these reactors have adverse effects on growth of biomass and so a cooling system is essential to maintain the culture temperature below 40 °C. Another disadvantage of closed photobioreactors is the accumulation of DO. In such closed systems, concentration of DO can be as high as 400% to that of saturation at the highest solar irradiations, which is harmful for algae as well as prokaryotes. Prokaryotic oxygen consumption in biomass of algae and prokaryotes reduce the harmful effects of higher DO on growth of algae [35]. Closed photobioreactors can be of horizontal, vertical, inclined, or spiral types. Tubular and flat plate photobioreactors are the most widely used because of their huge illuminated surface areas for increased productivity of algae. In comparison to tubular photobioreactors, flat plate photobioreactors were found to have decreased accumulation of DO [36].

4.2.3.3 Algal-prokaryotic biofilm reactors

Due to the difficulties in suspended algae harvesting from HRAP and closed photobioreactors, scientists have developed algal-prokaryotic biofilm reactors for effluent production with low total suspended solids concentration in comparison to systems involving suspended growth. In this system, wastewater is passed through the reactor and the biomass is attached to the mobile or stationary supporting medium, thereby reducing the residence time (mean cell residence time (MCRT)) than HRT for the prokaryotes and algae. Thus, the algal-prokaryotic biofilm reactors may be operated at high ammonium and organic rates of loading and reduced HRT than suspended systems of growth as communities with diminishing rates of growth (nitrifying bacteria) may be retained within the reactor. Besides, the attached system of growth has the capability to work with considerable depth, thereby, lowering the reactor footprint. In such systems, biomass can be harvested by scalding it from the supporting medium [37].

4.2.4 Nitrogen transformation by algal-prokaryotic interaction

Amalgamation of prokaryotes with microalgae to treat wastewater can significantly raise the nitrogen removal rate efficiency by various ways. Prokaryotes carry out the nitrification and denitrification steps essential for complete nitrogen removal as nitrogen gas (N_2). Algae in combination with prokaryotes can also remove nitrogen via assimilation since on illumination algae produce O_2 carrying out nitrification whereas reducing energy input required for aeration [38].

4.2.4.1 Process of nitrogen transformation

Conversion of NH_3 to N_2 consists of a sequence of oxidation followed by reduction steps, which are catalyzed by mainly prokaryotes, making use of these substances as electron donors and acceptors. Nitrification is the aerobic oxidation of the extremely reduced N/NH_3 to nitrite (NO_2^-) and then finally to nitrate (NO_3^-) while denitrification is the anaerobic process consisting of four major steps: reduction of nitrate to nitrite, nitrite to nitric oxide (NO), NO to nitrous oxide (N_2O), and subsequently N_2O to N_2. These steps of N_2 oxidation and reduction are catalyzed by various microbes. Ammonia oxidation to nitrate is carried out by two groups of bacteria: AOB that oxidizes NH_3 to NO_2^-; and NOB that oxidizes NO_2^- to NO_3^-. Denitrification steps are carried out by a diverse range of bacteria phylogenetically related in the presence of microaerophilic or anaerobic conditions.

These conventional nitrogen removal steps have been altered due to some recent discoveries in this respect. Daims et al. [39] discovered that some organisms can totally oxidize ammonia to nitrate. In short, removal of nitrogen involves NH_3 nitrification to NO_2^- and subsequently denitrification starting from NO_2^-, therefore bypassing the steps producing and consuming nitrates. Ammonia was even anaerobically oxidized by the use of anammox (anaerobic ammonia-oxidizing bacteria) producing nitrogen by combination of ammonia and nitrite if oxygen availability is very less followed by production of a very small quantity of nitrate) [40].

Besides, being used as electron donors/acceptors, nitrogen compounds may be ingested as biomass. Ammonia, nitrite, and nitrate get immediately assimilated by microbes. Algae as well as prokaryotes help in assimilation and reduction of nitrate, nitrite, and ammonia incorporating them into nucleotides, amino acids, and some other nitrogen-containing biomolecules. Within algae, nitrate gets transferred to the cytoplasm, in which nitrate reductase of cytoplasm will reduce nitrate to nitrite finally transporting it to the chloroplast, where it is subsequently reduced to ammonia and assimilated in the biomass.

Chlorella vulgaris can assimilate dissolved inorganic nitrogen in different forms. In the presence of multiple forms of nitrogen, ammonia will be assimilated first, then nitrite, and finally nitrate as ammonia is not required to get reduced for biosynthesis of amino acids. Since ammonia is partially soluble in lipids, it can easily diffuse through the membranes whereas ammonium will be utilized by energy-driven transport processes.

4.2.4.2 Nitrogen transformation process in wastewater treatment plants

Metabolic processes and final destination of nitrogen in algal-prokaryotic wastewater treatment plants are dependent on the characteristics of wastewater, regime of reactor operation and environmental conditions. Su et al. [41] reported that 40–53% and 17–20% of ammonia removal was because of assimilation followed by nitrification after treating wastewater in stirred tank photobioreactor. It was found that majority of the ammonium was eliminated by nitrification and consequently by assimilation and denitrification after anaerobic digestion of wastewater slurry treated in open ponds. However, some operational factors like MCRT and food-to-microbe ratio (F/M) can hamper nitrogen removal in algal-prokaryotic wastewater treatment systems. For example, HRAPs operating with less MCRT have high bacterial/algal ratios leading to high rates of nitrification [42]. It was further found that the settling flocs achieved higher efficiency of nutrient removal at a low F/M ratio of 0.15/day and HRT of 4 days. Seasonal changes in nitrogen removal pathways were also observed in outdoor use of HRAP. Ammonium removal rate by nitrification increased in summer season lowering ammonium concentrations. However, nitrogen loss via volatilization was comparatively high in winter season due to relatively higher concentration of ammonium within the reactor.

4.2.4.3 Enhanced algal-prokaryotic wastewater treatment plants

This method has been designed to further enhance the efficiency of nitrogen removal by supplying the appropriate environmental conditions essential for optimization of the functions of various microbial groups. Enhanced algal-prokaryotic wastewater treatment plant functions efficiently by taking the help of higher nitrogen and COD removal rates by bacteria and also production of DO by algae. Additionally, bacterial and algal EPS secretion initiates floc formation having well-settling capacity allowing HRT and MCRT separation [43].

Activated sludge system in combination with algae (also known as photoactivated sludge) has been found to enhance efficiency of nutrient removal essentially under very low aeration [44]. In this photoactivated sludge sequencing bioreactor (SBR), it was found that 81–85% ammonium was removed because of nitrification and due to the uptake of biomass. Oxygen production of algae was much higher than that of the HRAPs. Due to well-settling biomass, these SBR could be operated with long MCRT and short HRT, thereby allowing retention of the slow growing microbes within the bioreactor. In this photoactivated sludge process, activated sludge caused removal of COD while addition of algae enhanced the removal rate of nitrogen.

De Godos et al. [45] designed a two-stage anoxic-aerobic photobioreactor having a prokaryotic-algal consortium for simultaneous removal of nitrogen and carbon from industrial wastewater. Prolonged MCRT of about 20–30 days allowed growth and development of nitrifiers as well as denitrifiers. This system was found to remove 95% organic carbon and 90% nitrogen in treatment of wastewater containing 140 mg N/L of ammonium and 200 mg/L organic carbon. Even the biomass was found to have a good settling ability with a rate of sedimentation of about 0.28–0.42 m/h.

Since different microbes have different requirements for growth, they catalyze various steps in nitrogen transformation, and controlling the environmental conditions can help in selecting the exact population of algae-prokaryote consortium. For example, nitrogen can be readily removed by controlling concentrations of DO, thereby inhibiting NOB and favoring AOM [46]. Afterward, the wastewater gains entry into the anoxic phase, where the denitrifiers or anammox bacteria potentially reduce nitrite to nitrogen. If nitrite oxidation to nitrate is skipped, then the step of nitrite to nitrate reduction also gets skipped and hence less electron donors are required to carry out denitrification.

Nitrite has been found to be accumulated in algal-prokaryotic consortia to treat wastewater having a high content of ammonium [47]. On the other hand, removal of TN was very low in this case as the anoxic stage was not properly optimized. Another process is the algammox which is algal anaerobic ammonium oxidation and it is similar to the algal-prokaryotic nitrogen removal process but heterotrophic denitrification is replaced with anammox [48]. This system merged algae and AOM by means of anammox granules inside a photo-SBR with alternation of light and dark periods. Conversion of ammonia to nitrite was enhanced on adding anammox in a mixed culture of AOB and algae during the light period while nitrite was entirely removed by the activities of anammox in the dark period. No source of carbon needs to be provided in this system.

4.3 Biodegradation

Industrial wastewater containing a high amount of organic pollutants like phenols, bisphenol, azodyes, hydrocarbons, and various other aromatic compounds are degraded biologically. Halophilic microbes are of utmost importance for treatment of industrial wastewater because of their capacity to degrade harmful substances effectively [49]. The halophiles are capable of carrying out biodegradation which is the breakdown of organic matter by microbes. Various industrial activities such as petroleum industries, tanneries, and food industries release wastewater containing high amounts of organic wastes, and these wastewaters enter the normal aquatic system thereby disrupting water potability, aquatic life, and agriculture.

4.3.1 Biodegradation of biological organic matter

Wastewater generated from fish processing and pickled vegetable industries contain a high amount of organic matter expressed in terms of COD and high content of salt. Removal of COD can be done by activated sludge process but that is not effective at high salt concentrations and the biological degrading rates of organic matter get lowered. This problem can be overcome by pure culture consortia and a process called bioaugmentation. More than 95% of COD from pickling processing wastewater can be eliminated by bioaugmentation with *Halobacterium halobium* [50]. However, COD removal rates decrease with increased COD in feed, rate of COD loading, and salt content.

Similarly, *Staphylococcus xylosus* solely or in combination with activated sludge acting as inoculum in treating pickled vegetable industrial wastewater showed a COD removal efficiency of greater than 94% in the presence of 3% NaCl [51]. However, this COD removal efficiency will drop to 64% if treated with activated sludge only. This showed that halophiles can reduce COD to a significant level without requiring any prior treatment of the wastewater.

Efficiency of halophiles has also been tested under anaerobic conditions. The salt-tolerant anaerobic bacteria *Halanaerobium lacusrosei* was used in upflow anaerobic packed bed reactor for treatment of wastewater [52]. This showed a COD removal efficiency of 94% at 5% NaCl concentration.

4.3.2 Biodegradation of hydrocarbons

During crude oil and natural gas production, huge amount of reservoir water is released which is known as the produced water (PW). This PW mainly contains oil and greases but also consists of organic and inorganic compounds depending on their sites of extraction. PW released due to natural gas production contains benzene, toluene, ethyl benzene, and xylene, which are having very high toxicity as well as low molecular weights. PW released due to oil production contains polycyclic aromatic hydrocarbons (PAH), which can persist in the sediments and soils for a prolonged time period because of their hydrophobicity. PAHs are considered to be toxic, mutagenic, and carcinogenic.

Biodegradation of hydrocarbons and the organic pollutants by utilizing halophiles have been extensively researched by isolating and identifying the microbes from these hypersaline environments. In some experiments, it was found that the halophilic efficiency in hydrocarbon degradation significantly increased by adding some nutrients to enhance COD reduction and biodegradation of hydrocarbon [53].

Degradation of aromatic and aliphatic hydrocarbons has also been found to be carried out by both haloarchaeal and halobacterial strains. Members of genus *Haloferax* have been found to grow on a combination of PAH degradation intermediates acting as the sole source of carbon [54]. These intermediates were simultaneously broken down along with PAH mixture (anthracene, naphthalene, benzene, and pyrene) at 20% NaCl concentration. This indicates the catabolic adaptability of the haloarchaea in mineralizing aromatic compounds. Additionally, PAH degradation by members of *Haloferax* was enhanced on addition of 0.05% w/v yeast extract. However, capacity of halophiles in degrading PAHs diminishes with an increase in molecular weights, reduced solubility, and low bioavailability. The problem of decreased bioavailability can be overcome by using chemical dispersion or coculturing the bacterial consortium with biosurfactant-producing bacteria.

In order to improve pollutant biodegradation efficacy, different culturing conditions and media components used for biodegradation can be regulated. In industries with wastewater treatment plants, one-factor-at-a-time approach is used for removal of pollutants for carrying out bioaugmentation and biostimulation techniques. By using such approaches, it is possible to identify the contributing factors and their optimization in the biodegradation strategy. In addition to this, information about the various metabolic pathways is also important in foreseeing the fate of the degraded end products in the environment. Degradation pathways of PAH in halophiles have been found to employ oxidation of PAH to salicylate, which can be subsequently converted to gentisic acid or catechol [55]. The catechol ring was found to be cleaved by 1,2-dioxygenase (ortho conversion) or by 2,3-dioxygenase (meta conversion).

Halophile *Martelella* sp. performs PAH degradation via the gentisic acid pathway rather than utilizing the catechol pathway [55]. It was found that the enzyme catechol 1,2-dioxygenase was responsible for PAH biodegradation in *Halorubrum* sp., *Haloferax* sp., and *Halobacterium piscisalsi.* Catechol 2,3-dioxygenase unlike catechol 1,2-dioxygenase brought about degradation of toluene and benzene by *Arhodomonas* sp. at 29% NaCl concentrations [56].

4.3.3 Biodegradation of phenolic compounds

Phenolic compounds such as phenols and bisphenols are mainly present in the wastewater released from petrochemical, milling, and pulp industries along with some of the refineries. Treatment of these compounds with traditional processes does not prove to be effective because of their high concentration of salt, so this problem can be solved by using halophiles. It was reported in a study that *Halomonas* sp. can degrade almost 100% of phenol at a 5% NaCl concentration after treating the wastewater for 13 h [57]. Here, the function of catechol 1,2-dioxygenase was also identified in degrading crude extracts. At low concentrations of NaCl, phenol degradation was found to be followed by *cis*, *cis*-muconic acid accumulation. In comparison to this, haloalkaliphile *Halomonas campisalis* exhibited increased accumulation of *cis*, *cis*-muconate with increasing NaCl concentrations and increased pH. Bacterial production and accumulation of *cis*, *cis*-muconate in alkaline solutions also helps in producing many valuable raw materials.

Recently, in a phenol degradation study using various bioreactor configurations, found that 200 mg/L of phenol was degraded in aerobic granular sequencing batch reactor. Membrane biological reactors (MBRs) having submerged flat membrane are used to treat industrial wastewater containing 8–16 mg/L phenol. These reactors showed 96% phenol removal efficiency in SBRs. Microbial communities in such reactors consisted of *Halomonas* and *Marinobacter*, which help in phenol degradation [58]. It was also found that *Arhodomonas aquaeolei*, *Halomonas organivorans*, and *Modicisalibacter tunisiensis* contained a variety of genes encoding aromatic dioxygenase enzymes protocatechuate 3,4-dioxygenase and catechol 1,2-dioxygenase [59].

A combined mixture of phenolic compounds from different industries is released with differences in toxicity, biodegradability, and solubility. The available biological treating techniques on the basis of single microbial communities may prove insufficient in treating heterogeneous wastewater. Nowadays, microbial aggregations called periphyton technology is used in order to enhance in situ removal of pollutants from water surfaces [60]. Periphyton comprises phototrophic and heterotrophic microbes and showed improved removal of miscellaneous contaminants. This periphyton technology can be also incorporated or integrated within bioreactors on the basis of immobilization of cells in treating different pollutants.

4.3.4 Biodegradation of N-aliphatic and N-aromatic compounds

Nitro-substituted aromatic substances such as nitrophenol and nitrobenzene are used for manufacturing azodyes, pesticides, pharmaceuticals, and explosives. While degrading these nitro-substituted compounds biologically and chemically in water and soil, few of the nitroaromatic compounds or derivatives get released into the environment. Oren et al., 1991 found two strains of halophiles namely *Sporohalobacter marismortui* and *Haloanaerobium praevalens* capable of reducing *para*-nitrophenol and also some of the nitro-substituted aromatic compounds in order to release respective aromatic derivatives of amine.

Jin et al. [61] observed in an experimental study that on adding 10–70 mM sodium acetate enhances the rate of biodegradation of aniline and phenylamine by the halophilic strain *Dietzia natronolimnaea* JQ-AN. It was further found that at 3% concentration of NaCl, 60% aniline got degraded after incubating for 5 days at 30 °C and 150 rpm. Addition of 40 mM sodium acetate not only increased the growth of microbes but also increased the rate of aniline degradation to 87%. The researchers concluded that catechol was the first intermediate formed after metabolizing aniline by *D. natronolimnaea* in anaerobic conditions and finally catechol was biodegraded via tricarboxylic acid cycle yielding small organic compounds. Degradation of the remaining pollutants was even enhanced because of adding other cosubstrates to the existing medium. Campo et al. [62] carried out an experiment to study biodegradation of various amines in two different industrial wastewaters comprising 3% and 7% NaCl concentrations, respectively, utilizing consortia. However, salinity was not found to affect rates of biodegradation of amine (triethanolamine and *N,N*-bis 2-hyroxyethyl), tris(2-hyroxyethyl), and methylamine (methyldiethanolamine), whereas *N*-(2-aminoethyl) and *N,N*-diethylethanolamine were readily degraded at 3% NaCl concentration rather than at 7% NaCl concentration. Aniline was totally biodegraded after a period of 48 h at both concentrations of NaCl. In comparison to this, 100% of cyclohexylamine and 4,4′-methylenedianiline were biologically degraded at 3% NaCl concentrations after a period of 24 h while concentrations of these two compounds remained unchanged at 7% NaCl concentration. These outcomes indicate that the rate and extent of biodegradation are dependent on amine type and salt concentrations.

On the other hand, high levels of histamines in wastewaters released from fisheries can also have some detrimental effects on human health. So, three halophilic archaebacterial strains were *Natrinema* sp. HDS1-1 and HDS3-1 and *Halobacterium piscisalsi* HPC1-2 [63]. Among these, strain HDS3-1 showed a biodegrading efficiency of 80% in removing 5 nM histamine at 25% of NaCl concentrations and does not release any contaminants or harmful intermediates, thereby indicating the function of enzyme histamine dehydrogenase, which is a salt-tolerant enzyme.

4.3.5 Biodegradation of dyes

Azo dyes are the widely used dyes in paper printing and textile industries on a large scale. Biodegradation processes for azo dyes in culture broth with low salt concentrations are dependent on factors such as pH, temperature, dye concentration and structure, nitrogen and carbon sources, and concentration of oxygen and agitation.

Over the recent years, some halophilic/halotolerant microbes have been isolated and identified so far, which have the capacity to biodegrade the azo dyes generated from textile industries effectively. Discoloration of azo dye Acid Red B with a concentration of 50 mg/L was carried out by *Gracilibacillus* sp. GTY strain under 2–5% NaCl concentrations [64]. Low efficiency of discoloration was observed due to slower rate of growth of the particular strain. However, this strain was found to completely degrade the dye at 15% NaCl concentration and if incubated for a period of 96 h. At 25% concentration of NaCl, the discoloration was not significant because of bacterial growth inhibition brought about by high concentration of salt. The strain was found to express the enzyme azo reductase along with some other enzymes that lead to degradation of the entire dye.

Again, *Halomonas* sp. strains were found to decolorize azo dye Remazol Black B under microaerophilic as well as anaerobic conditions up to 20% w/v of NaCl concentrations [65]. Optimal conditions for complete degradation of dyes were a temperature of 35–40 °C and a pH of 9–11. High-performance liquid chromatographic chromatograms indicated the dye being reduced to aromatic amines, which are harmful intermediates. Another *Halomonas* sp. strain GTW could carry out anaerobic degradation of azo dye Reactive Brilliant Red K-2BP at a concentration of 10–15% NaCl and 100% dye concentration supplied with 30% yeast extract and at a temperature of 30 °C and a pH of 6.5–8.5 [66].

A consortium of *Pseudomonas aeruginosa* and *Bacillus circulans* also carried out anaerobic decolorization of 100 mg/L Reactive Black 5 with a degradation efficiency of 93% at 5% NaCl concentration [67]. Apart from this, the consortium can even cause decolorization of other azo dyes such as Reactive Orange 16, Reactive Violet 13, Reactive Red 141, Reactive Red 11, and Direct Yellow 12 with decolorization efficiencies of 80–90% while that of Acid Yellow 36 was 40–65%.

Some researchers further found that the halophiles can carry out biodegradation not only under anaerobic conditions but also under aerobic conditions. It was reported that the halophile *Bacillus lentus* BI377 could degrade Reactive Red 120 under aerobic and microaerophilic conditions [68]. Thus, they concluded that the biodegradation of azo dyes requires a sequential anaerobic–aerobic process, where the aromatic amines generated in the first step get oxidized in the second step under aerobic conditions. Utilization of periphyton technology may also be applied in degrading the wastewater contaminated with dyes produced from textile industries.

4.4 Bioaccumulation and biosorption

Due to some of the problems related to the traditional metal removal methods, biosorption and bioaccumulation methods have been developed over the recent years as economic, novel, eco-friendly, and efficient alternative procedures to remove heavy metals contaminating wastewaters released from various industries. However, the terms "biosorption" and "bioaccumulation" are not the same and must not be used interchangeably. Biosorption is the process of particle adsorption to a matrix of biological origin by means of either chemical interactions (proton/ion displacement) or physical interactions (electrostatic forces), chelation, or complexation. Microbial extracellular surfaces contain some anionic moieties at neutral pH for binding cationic heavy metal ions [69]. On the contrary, bioaccumulation is an active metabolic process, where microbes uptake heavy metals inside their cells intracellularly via importers, thereby creating a pathway for translocation by means of lipid bilayer or system of import [70].

4.4.1 Bioaccumulation

Bioaccumulation is defined as a metabolism-assisted active transport process, where the metallic ions get accumulated into the biosorbent in the living cells intracellularly. This process mainly consists of two steps, among which the first step is the metal ion adsorption on the cell surfaces and the second step is the metal ion transfer inside the cells via passive transport. The first step is similar to that of biosorption and occurs rapidly, whereas the second step is much slower than the first one. In contrast to biosorption, bioaccumulation is a complex and irreversible process, which is dependent on the metabolism of the living cells. Bioaccumulation takes place by biomass cultivation of a microbe in the surroundings of the metal ion to be accumulated. As the solution already contains the medium for growth, the microbe initiates its metabolism pathways, thereby activating the intracellular transport processes for sorbate accumulation. The major disadvantage of this process is that the nutrient medium for microbial growth consists of organic carbon sources. Bioaccumulation is an active transport system requiring living biosorbents and is dependent on the microbial metabolism. This process is being carried out by culturing the required microorganism in the presence of a metal ion to be accumulated/removed. Portion of the biosorbate gets accumulated within the cell, enabling the biomass to expand so that it can bind larger amounts of metal ions. The microorganisms that can resist higher loads of heavy metal ions are appropriate for accumulation of metal ions. These microbes do not exhibit any methods to hinder accumulation of metal ions in greater quantities. These microbes contain some specialized mechanisms to synthesize unique intracellular binding moieties, which are thiol-rich and act as a responsive mechanism to the metal ions present in their surrounding environment.

It has been found that the cell physiology and morphology get modified on increasing metal ion (to be accumulated) concentration [71]. Effective bioaccumulation can be attained by perfect microbial community selection, which is actually selected from wastewaters. Yeast *Pichia stipitis* can bioaccumulate trivalent chromium and cuprous ions with the highest uptake rate of 9.10 and 15.85 mg/g, respectively, from wastewaters containing a preliminary concentration of 100 ppm at a pH of 4.5. *Aspergillus niger* was also found to remove Pb(II) and Cu(II) with the highest uptake rate of 34.4 and 15.6 mg/g, respectively [72].

Since bioaccumulation is metabolically active, the host cell must always be alive and impose some critical challenges such as continuous nutrition for propagation and sustenance of biomass and continued aeration in order to meet up the anaerobic/aerobic requirements. This also means that the process of bioaccumulation includes cytosolic proteins and proteins that remain embedded within the lipid membrane, which in turn poses more critical challenges like reduced viability of cells because heterologous expression of storage imports proteins (although not required for growth), extensive aggregation of proteins, and phenotypic loss because of competition created by native microbes.

4.4.2 Biosorption

Biosorption can be defined as the simplest physicochemical and metabolically passive transport included in the metal ion binding (biosorbate) to the biosorbent surfaces, which are having a biological origin. Biological elimination involves using microbes, materials derived from plants, industrial or agricultural wastes, and biopolymers. This is a rapid and reversible process, which helps in binding ions over the functional groups present onto the biosorbent surfaces in aqueous solutions via different interactions but not oxidation by aerobic/anaerobic metabolism. The advantages of it are mainly easy operation, no extra nutrient requirements, sludge produced in low quantity, high effectivity, biosorbent regeneration, and reduced COD of water [73]. Biosorption can also eliminate contaminants from dilute solutions and has a special impact in removing heavy metals efficiently in terms of toxicity in ppb levels. Microbes (alive or dead) and various agricultural or industrial by-products can act as biosorbents for carrying out biosorption. The first step in biosorption process is suspension of the biosorbent in the biosorbate solution (mainly metal ions). Equilibrium is reached after being incubated for a particular interval of time. At this position, the biosorbent enriched with metals will be separated.

4.4.2.1 Biosorption mechanism

Biosorption mechanism is a complicated process involving sorbate binding to the biosorbent. Mostly natural materials are used as biosorbents which bind to the metal ions via chemical (bound metal cation/proton displacement or ion exchange) and physical (van der Waals or electrostatic interaction) binding, reduction, chelation, complexation, and precipitation. Biosorbents comprise chemical/functional groups such as amide, amine, thioether, imidazole, carbonyl, sulfonate, carboxyl, sulfhydryl, phenolic, phosphodiester, phosphate, and imino groups which help in attracting and sequestering metal ions.

4.4.2.1.1 Complexation

Two or more species associate together to form a complex. Ligands and metal ions form monodentate/mononuclear complexes, in which the metal ion takes up the central space. When more than a single metal ion takes up the central space, then the complexes are known as multidentate or polynuclear complexes, where the metal atom may carry a negative, neutral, or positive charge depending on how many ligands are involved in binding. Complexation by monodentate ligand is more preferred since the multidentate complexation will lead to the binding of multiple species. Attenuated total reflection infrared spectral analysis of *Cyanobacterium microcystis* after biosorbing antimony(III) indicated the participation of amine, hydroxyl, and carboxyl groups via surface complex formation [74]. Similar biosorption mechanisms were also observed in *Termitomyces clypeatus* and *Acidiphilum* in removing heavy metals like Cr(VI) and Cd(II) from sewage sludge, respectively [75, 76].

4.4.2.1.2 Chelation

Chelation is defined as the process of binding of chelating agents to a metal ion at multiple places simultaneously forming a ring-like structural complex known as chelate. Most of the times, multidentate ligands are involved in these reactions in order to form stable structural complexes by multiple bonding. On increasing ligand-binding sites, structural stability also increases. Since the chelating agents bind with the metal ion at multiple places, the chelates are more stable as compared to the complexes. Rice straw can be used as an effective biosorbent for Cd(II) removal from effluents. On biosorption of Cd(II), it forms chelates with functional groups such as C–O, C = O, O–H, and carboxylic acids present on the biosorbent surface. This biosorption mechanism was found in Cu(II) and Cr(III) removal by hydroxyl and carboxyl groups on soybean meal waste surfaces [77].

4.4.2.1.3 Coordination

Metal atom inside the complex can bind to its surrounding neighbors by means of a covalent coordinate bond on accepting one lone pair of electrons from a nonmetal atom. This nonmetal atom is called the coordinating or donor atom, and the metal atom accepting the lone pair of electrons is called the acceptor. Compounds with these types of structural bonds are called coordinate compounds. Examples of coordinating groups are $-NH_2$, = O, -N = , -NH, -S-, -OH, = NOH, and –O-R.

4.4.2.1.4 Ion exchange

Ion exchange is of great significance in biosorption in which binary metal ions are exchanged at the time of biosorption with counter-ions on the biosorbent surface. Almost all of the purification steps are dependent on the principle of ion exchange mechanism. Ion exchange occurs either by anion or by cation exchange. A good cation exchanger is a carboxyl group, whereas imidazole or amino groups are good anion exchangers. Biosorption process of *Spirulina* to biosorb Cu(II), Cd(III), or Cr(III) was observed [78]. Carboxyl, phosphate, and hydroxyl groups were found as cation exchangers on the surface. Ion exchange process of biosorption was observed in this study by the use of rice straw to remove cadmium by the exchange of Na^+, K^+, Ca^{2+}, and Mg^{2+} ions and watermelon rind was used to remove Zn(II), Pb(II), and Cu(II).

4.4.2.1.5 Precipitation

Functional groups existing on the microbial cell surfaces form precipitates with the metal ions so that either they enter the cell of the microbes or remain intact. Mostly, inorganic and insoluble metal precipitates are formed. By using microbial cells, organic metal precipitates are formed. Mostly, the EPS produced by the microorganisms get involved in forming organic precipitates. Cu(II) precipitation on *Mesorhizobium amorphae* leads to aggregation and deformation damaging the cell surface as obtained by scanning electron microscope (SEM) energy-dispersive X-ray (EDX) analysis [79]. This precipitation mechanism in biosorption of metal ions is done by utilization of soybean meal, husk of green tomato (*Physalis philadelphica* Lam), and watermelon rind to remove Cu(II) and Cr(III), iron and manganese, and Pb(II), Zn(II), and Cu(II), respectively.

4.4.2.1.6 Reduction

Process of reduction involves interaction of metals with the functional groups such as carboxyl and these on reduction lead to the formation of crystals. Elements such as palladium and gold are formed due to reduction. These metals are reduced on binding with the biosorbent at separate places. Toxic hexavalent chromium removal is mostly carried out by the process of reduction. There are many microbes that reduce Cr(VI) to Cr(III) by means of biosorption from aqueous solutions [80].

Biosorption mechanisms can be observed through various techniques. Basic and acidic characteristics of the functional groups on surfaces of materials and ion exchange capacities can be found out by the potentiometric titration called the Boehm method [81]. Fourier transform infrared spectroscopy indicates the types of functional groups such as amino, carboxyl, amide, sulfate, hydroxyl, ester, carbonyl, and ether that are present on biosorbent surfaces and also the type of bonding causing biosorption. SEM is used for qualitative evaluation of the morphological and structural changes of the biosorbent previously and after the biosorption of metals. EDX reveals information about bioavailability of the different elements on the biosorbent surface. X-ray photoelectron spectroscopy is used for quantitative analysis of the surface chemistry of biosorbent, that is, electronic state and empirical formula of the elements that are present and oxidation state of the biosorbed metal ion [82].

4.5 Biomineralization

Biomineralization is a process in which living organisms can form minerals resulting in the removal of metals from aqueous solutions, thereby supplying ways of biorecovery as well as detoxification. Mineral and metal transformations are the main foundations in geomicrobiological processes. The various mechanisms by which microbes change the mobility and speciation of metals and their mineral dissolution properties and formation of minerals are of utmost importance. These processes underline the natural biogeochemical cycles for metals and other related elements such as metal radionuclides and metalloids and also the elements being associated with minerals and metallic compounds like sulfur (S) and phosphorus (P). These processes help in deciding the fate of radionuclides and contaminating toxic metals when redistributed or introduced in the environment by means of any anthropogenic/natural activities. In addition to this, derived products and dead biota can provide a platform for deposition of minerals, their reversibility, and some of the physicochemical reactions. The commonly precipitated biominerals are phosphates, oxides, oxalates, and sulfides as they have some specialized chemical properties like redox catalysis and increased metal sorption capabilities. Moreover, some of the biominerals as well gets deposited in nanoscale proportions, giving more important information on the biological, chemical, and physical properties.

4.5.1 Bioleaching

Bioleaching is one of the main ways of metal recovery, and metal-associated bioleachates may serve as suitable substrates for bioprecipitation of metals via biomineralization reactions. Gadd and Pan [83] screened mechanisms of bioleaching to recover nickel (Ni) from the sludge released by electroplating industries and discovered that the sulfur reduction system of *Acidithiobacillus thiooxidans* is the most effective and depends on the direct contact between the sludge and the cells. Bioprecipitation of sulfide is carried out by sulfate-reducing bacteria, which is one of the efficient methods in metal removal process from bioleachates.

Redox reactions are also important for metal and mineral transformations, mainly in the subsurface regions, where reduction of immobilized metals restrict the mobility of radionuclides and toxic metals. Reduction of Fe(III) is one of the common properties of anaerobic bacteria that can affect geochemistry of many metals. Brookshaw et al. [84] showed that the mineral-associated reduction of Fe(III) led to the enhanced dissolution of chlorite as well as biotite, thereby reducing the available sites for cesium (Cs) and strontium (Sr) biosorption. This study also indicated that microbial reduction of Fe(III) can increase Sr and Cs migration via phyllosilicate-dominated strata.

4.5.2 Radionuclides and heavy metal removal

Radionuclides and heavy metals are some of the major challenges in waste disposal. Not only are they harmful to the humans but also to the entire environment. Conventional methods of using pumps for wastewater treatment have proved to be ineffective and thus, nowadays, ureolytic bacteria, which can cause calcium carbonate precipitation, are widely used. In this process, biomineralization of toxic heavy metals and radionuclides into calcite takes place due to competitive coprecipitation reaction, where appropriate divalent positively charged ions get incorporated within the calcium lattice [85]:

$$^{90}Sr^{2+} + OH^{-} + HCO_3^{-} \rightarrow SrCO_3(s) + H_2O \tag{4.1}$$

These radionuclides and cations get integrated within the calcite lattice by calcium ion substitution in the mineral precipitate microenvironment to form minerals of low-strontium carbonate having very less solubility. Warren et al., 2005 showed capturing of 95% total strontium in solid phase via microbially induced calcite precipitation (MICCP) by the use of *Sporosarcina pasteurii*. MICCP can also aid in removing arsenic contamination from the industrial wastewaters. Positive potential of *Sporosarcina ginsengisoli* strain CR5 was reported to be used for arsenic removal and *Kocuria flava* strain CR1 was found in removal of copper-contaminated wastewaters [86].

Uranium is also biomineralized over a vast range of pH by means of a metal-tolerant *Serratia* sp., strain OT II7, which was isolated from soil subsurface in an uranium ore deposit of Domiasiat in India [87]. The cells of *Serratia* were found to be involved in active expression of alkaline and acid phosphatase enzymes that can hydrolyze differential quantities of phosphate from organophosphate substrate when uranium is present at a pH of 5–9. These cells were shown to precipitate 91% uranium from the aqueous solutions containing 1 mM uranyl nitrate within a time period of 120 h and at a pH of 5. Increased and rapid precipitation was found within a pH of 7–9, where the cells are capable of removing almost 93–94% of uranium from solutions supplemented with 1 mM uranyl carbonate within a period of 24 h. The aqueous speciation of uranyl that prevailed under the given conditions of pH affected precipitates of uranyl phosphate crystallization and localization, thereby influencing the cell viability to a much larger extent. Thus, this experimental setup showed that the natural *Serratia* strain could ecologically adapt in a uranium-rich environment and aid in immobilization of uranium by forming stable minerals of uranyl phosphate over a wide range of pH in contaminated subsurfaces.

4.5.3 Polychlorinated biphenyls (PCBs) and calcium ion removal

Existence of calcium ions in high concentration (500–1,500 mg/L) in the wastewaters can lead to acute scaling in the reactors and the pipelines because of the formation of different calcium compounds like calcium carbonate, calcium phosphate, and/or gypsum. Hammes et al. [88] examined the calcium ion removal potential from industrial wastewaters by means of microbiologically induced calcium carbonate precipitation (MICCP). Addition of urea in a lower concentration (16 g/L) could remove almost 90% of the total calcium ion concentration from the given wastewater.

Recently, there has been an increase in the release of a new type of pollutant that is polychlorinated biphenyls (PCBs). Microbe-driven processes are in a developing stage for cleaning up of such recalcitrant contaminants. Traditional methods for PCB removal involve sandblasting or hydroblasting and solvent washing followed by epoxy-coating encapsulation. However, all these treatment procedures have proved to be ineffective because of oil resurfacing with due course of time. Thus, carbonate biomineralizing along with bacterial isolates is nowadays used for PCB removal. Over the recent years, Okwadha and Li [89] studied the positive potential of the ureolytic cultures of *S. pasteurii* along with calcium/urea to treat PCB-laden wastewater through superficial PCB-containing oil encapsulation.

4.5.4 Phosphorus removal

Increased industrial and demographical growth rate over the recent years has also increased water pollution. Particularly, discharge of nutrients in natural water bodies will lead to increased problems of eutrophication having severe consequences for aquatic life and also degrade the water quality for drinking and domestic purposes. Food industries that produce pre-fried potatoes and frozen vegetables led to the accumulation of large amounts of phosphates in the wastewater released by them. Struvite

($NH_4MgPO_4·6H_2O$) is one of the major contaminants released. In wastewater treatment plants, deposition of struvite mainly causes increased turbulence in aerators, pipe curves, and pumps.

The traditional way for phosphate removal is through addition of flocculants like aluminum and ferric chloride [90], but the main disadvantage of this technique is the high cost of flocculant chemicals; hence, this process is replaced by phosphorus biomineralization and biological crystallization.

The main focus of wastewater treatment is prevention of eutrophication as well as phosphorus recovery so that it can be used as a fertilizer by ureolytic bacterial mechanism. Phosphate can be removed from anaerobic effluents via the method of ureolytic precipitation. This study used the method of phosphate precipitation by ureolytic bacteria where phosphate concentration in the wastewater generated from vegetable processing industry was <30 mg/L. $Mg(OH)_2$ or MgO along with urea was supplied for precipitation to take place, thereby reducing the phosphate concentration to almost 5–7 mg/L. In another study, it was found that phosphate can be precipitated from anaerobic industrial effluents through magnesium ammonium phosphate (MAP)/ struvite precipitation [91]. They showed in their study that a pH more than 8.5 and using $MgCl_2$ gave better results in comparison to MgO-mediated phosphate precipitation from anaerobic industrial effluent. They also used water from a potato-processing industry to compare MAP precipitation with ureolytic crystallization in combination with an autotrophic removal process for nitrogen, where the pH is increased through NaOH to recover orthophosphate and ammoniacal nitrogen in order to form pure struvite crystals. They even determined various factors influencing precipitation of struvite like concentration of calcium, concentration of magnesium and phosphate, their ionic strength, and their molar ratio.

4.5.5 Ion precipitation from wastewater

Calcium ion removal from a paper recycling factory through biomineralization by using a semicontinuous reactor (BCR) was found to remove about 90% of the total calcium concentration [88]. However, they did not mention any detail about the type of microbes they used but they described the method of calcium precipitation for ureolytic metabolism. They found that the calcium carbonate is chemically precipitated by the influence of four different factors which are concentration of calcium ions, concentration of carbonates, pH, and existence of sites for nucleation. In the process of calcium carbonate precipitation by microbes, the first three factors are the most important while the fourth factor can be replaced by the bacterial presence since they can themselves serve as sites for nucleation. Effects of concentration of urea in the BCR bioreactors showed organic matter removal over a prolonged period of time, and also the ammonia toxicity was evaluated in the mixed ureolytic culture of *S. pasteurii* proving that this microbe isolated from alkaline conditions (pH of 9–13) can initiate biomineralization of calcite [92]. Moreover, this microbe can be mostly used under controlled conditions in the laboratory and also in the existence of solid industrial wastes such as lime kiln dust and cement kiln dust (CKD). The microbe has a unique potential to consolidate grain microbes for a mixture of CKD and granulated blast furnace slag and can be applied for biocementation and bioclogging. The experimental results indicated that stable biocalcite was formed in the existence of CKD giving a yield depending on the pH value and free calcium ion content.

In another study, calcium carbonate biomineralization was mediated by bacteria *Virgibacillus* and *Bacillus* genus that were isolated from saline conditions when cultured within a culture media provided with organic matter [93].

Magnesium precipitation is a part of struvite crystal formation and is associated with phosphate recovery from wastewater via chemical crystallization. In case of biological struvite precipitation, certain bacteria isolated from soils and waters like *Arthrobacter* sp., *Myxococcus* sp., and *Pseudomonas* sp. can bring about struvite precipitation in laboratory conditions [94]. Among these microbes, *Myxococcus xanthus* can mediate struvite biomineralization [25] and microbes such as *Halobacterium salinarium*, *Halorubrum distributum*, and *Brevibacterium antiquum* have also been found to aid in the formation of crystals of magnesium phosphate [95]. Ureolytic bacteria found in the Atacama desert can serve as biomineralizers of magnesium and calcium ions for desalination of wastewater. In this study, *Rhodococcus erythropolis* was found to precipitate 8% magnesium and 95% soluble calcium. The crystals were then analyzed, and it was found that crystal components consist of 30.72% struvite, 12.69% monohydrate, and 56.59% halite showing that this bacterium has immense potential to be used to pretreat industrial wastewaters.

Fungi also have important geoactive properties because they grow in filamentous branching system having a chemoorganotrophic metabolism that determines their biomineralization activities [96]. Biosorption by fungi is followed by the phenomenon of microbial biomineralization so that the branched mycelium of the fungi can become an efficient platform for removing metals and formation of biominerals. Both dead and living biomass of fungi can help in biosorption of nickel and cadmium ions with respect to the different models of sorption that can characterize this process. Oxalate is a significant metabolite in most of the fungi, and the most harmful metal oxalates remain insoluble, thereby providing a way for immobilization of metals and also a mechanism for metal tolerance.

4.6 Conclusion

Over the recent years, there has been a significant scientific progress in successful application of biotechnological approaches or tools for treatment and management of industrial wastewater, thereby mitigating the amounts of water pollution all over the world. The problems of water pollution have so far led to some serious environmental and health hazards and so the solution for this problem is the integrative and multipronged application of microbial procedures in combination with the physical and chemical ones. The use of biofilm reactors to treat industrial wastewaters has emerged as a reliable technique developed on the basis of the fundamental mechanism of biofilm formation. On the other hand, biosorption is capable of removing all the minute or residual contaminants after the wastewater is treated by biofilm reactors and biodegradation of the contaminants. Thus, biosorption can be implemented as an aftermath of other traditional treatments. On the other hand, biomineralization becomes the platform for a number of technologies that would allow ion removal from industrial discharge of wastewater. Various studies using ureolytic microorganisms have been successfully carried out for removal of heavy metals like cadmium, nickel, strontium, and chromium and radionuclides like uranium via calcite coprecipitation. Even though nowadays many of these biological tools are employed to treat the untreated industrial wastewater, more detailed knowledge about the biofilm ecology and further studies to overcome certain limitations of implementation of these technologies are still required.

References

[1] Kastenhofer, K. Converging epistemic cultures?. Innovation. 2007, 30(4), 359–373. doi: https://doi.org/10.1080/13511610701767908.

[2] Mani, D., Kumar, C. Biotechnological advances in bioremediation of heavy metals contaminated ecosystems: An overview with special reference to phytoremediation. Int J Environ Sci Technol. 2014, 11, 843–872. doi: https://doi.org/10.1007/s13762-013-0299-8.

[3] Govarthanan, M., Lee, K.J., Cho, M., Kim, J.S., Kamala-Kanna, S., Oh, B.T. Significance of autochthonous Bacillus sp. KK1 on biomineralization of lead in mine tailings. Chemosphere. 2013, 90(8), 2267–2272. doi: 10.1016/j.chemosphere.2012.10.038.

[4] Deng, L., Su, Y., Su, H., Wang, X., Zhu, X. Sorption and desorption of lead (II) from wastewater by green algae Cladophora fascicularis. J Hazard Mater. 2007, 143(1–2), 220–225. doi: 10.1016/j.hazmat.2006.09.009.

[5] Li, M., Cheng, X., Guo, H. Heavy metal removal by biomineralization of urease producing bacteria isolated from soil. Int Biodeterioration Biodegradation. 2013, 76, 81–85. doi: https://doi.org/10.1016/j.ibiod.2012.06.016.

[6] Abdel-Sabour, M.F. Remediation and bioremediation of uranium contaminated soils. Electron J Environ Agric Food Chem. 2007, 6(5),2009–2023. URL: https://www.researchgate.net/publication/279624136_Remediation_and_bioremediation_of_uranium_contaminated_soils

[7] Jain, S., Arnepalli, D.N. Biomineralisation as a Remediation Technique: A Critical Review. *Indian Geotechnical Conference*, 2016, (pp. 1–4). Chennai. URL:https://link.springer.com/chapter/10.1007/978-981-13-0899-4_19

[8] Kolenbrander, P.E., Palmer, R. Jr, Periasamy, J., Jakubovics, N.S. Oral multispecies biofilm development and the key role of cell-cell distance. Nat Rev Microbiol. 2010, 8(7), 471–480. doi: 10.1038/nrmicro2381.

[9] McQuarrie, J.P., Boltz, J.P. Moving bed biofilm reactor technology: Process applications, design and performance. Water Environ Res. 2011, 83(6), 560–575. doi: 10.2175/106143010x12851009156286.
[10] Odegaard, H., Christensson, M., Sorensen, K.K. Hybrid systems. In: Jenkins, D., Wanner, J., eds., Activated sludge – 100 years and counting, London, IWA Publishing, 2014, URL: https://www.iwapublishing.com/books/9781780404936/activated-sludge-100-years-and-counting.
[11] Rusten, B., Eikebrokk, B., Ulgenes, Y., Lygren, E. Design and operations of the Kaldnes moving bed biofilm reactors. Aquacult Eng. 2006, 34(3), 322–331. doi: https://doi.org/10.1016/j.aquaeng.2005.04.002.
[12] Kreuk, M.K., Kishida, N., Loosdrecht, M. Aerobic granular sludge – State of the art. Water Sci Technol. 2007, 55(8–9), 75–81. doi: 10.2166/wst.2007.244.
[13] Pronk, M., Kreuk, M.K., Bruin, B.D., Kamminga, P., Kleerebezem, R., Loosdrecht, M.C. Full scale performance of the aerobic granular sludge process for sewage treatment. Water Res. 2015, 84, 207–217. doi: 10.1016/j.watres.2015.07.011.
[14] Rittmann, B.E. The membrane biofilm reactor: The natural partnership of membranes and biofilm. Water Sci Technol. 2006, 53(3), 219–225. doi: 10.2166/wst.2006.096.
[15] Syron, E., Casey, E. Membrane aerated biofilms for high rate of biotreatment: Performance, appraisal, engineering principles, scale-up and development requirement. Environ Sci Technol. 2008, 42(6), 1833–1844. doi: https://doi.org/10.1021/es0719428.
[16] Martin, K.J., Nerenberg, R. The membrane biofilm reactor (MBfR) for water and wastewater treatment: Principles applications, and recent developments. Bioresour Technol. 2012, 122, 83–94. doi: 10.1016/j.biortech.2012.02.110.
[17] Wang, X., Daigger, G., Vries, W.D., Kroeze, C., Yang, M., Ren, N.Q., et al. Impact hotspots of reduced nutrient discharge shift across the globe with population and dietary changes. Nat Commun. 2019, 10(1), 2627. doi: 10.1038/s41467-019-10445-0.
[18] Chaali, M., Naghdi, M., Brar, S.K., Avalos-Ramirez, A. A review on the advances in nitrifying biofilm reactors and their removal rates in wastewater treatment. Chem Technol Biotechnol. 2018, 93, 3113–3124. doi: https://doi.org/10.1002/jctb.5692.
[19] Gunnes, G., Hallac, E., Ozgan, M., Erturk, A., Tas, D.O., Cokgor, E., et al. Enhancement of nutrient removal performance of activated sludge with a novel hybrid biofilm process. Bioprocess Biosyst Eng. 2019, 42, 379–390. doi: 10.1007/s00449-018-2042-9.
[20] Shah, M.P. Removal of Emerging Contaminants through Microbial Processes, Springer, 2021.
[21] Aziz, A., Basheer, F., Sengar, A., Irfanullah Khan, S.U., Farooqi, I.H. Biological wastewater treatment (anaerobic-aerobic) technologies for safe discharge of treated slaughterhouse and meat processing wastewater. Sci Total Environ. 2019, 686. 681–708.
[22] Biase, A.D., Kowalski, M.S., Devlin, T.R., Oleszkiewicz, J.A. Moving bed biofilm reactor technology municipal wastewater treatment: A review. J Environ Manage. 2019, 247. 849–866.
[23] Lima, P.S., Dezotti, M., Bassin, J.P. Interpreting the effect of increasing COD loading rates on the performance of a pre-anoxic MBBR system: Implications on the attached and suspended biomass dynamics and nitrification-denitrification activity. Bioprocess Biosyst Eng. 2016, 39(6), 945–957. doi: 10.1007/s00449-016-1574-0.
[24] Paetkau, M., Cicek, N. Comparison of nitrogen removal and sludge characteristics between a conventional and a simultaneous nitrification–denitrification membrane bioreactor. Desalination. 2011, 283, 165–168. doi: https://doi.org/10.1016/j.desal.2011.02.018.
[25] Goecke, F., Imhoff, J.F. Chemical interactions between marine macroalgae and bacteria. Mar Ecol Prog Ser. 2010, 409, 267–300. doi: 10.3354/meps08607.

[26] Wang, M., Keeley, R., Zalivina, N., Halfhide, T., Scott, K., Zhang, Q., et al. Advances in algal-prokaryotic wastewater treatment: A review of nitrogen transformations, reactor configurations and molecular tools. J Environ Manage. 2018, 217, 845–857. doi: 10.1016/j.jenvman.2018.04.021.
[27] Shah, M.P. Advanced Oxidation Processes for Effluent Treatment Plants, Elsevier, 2020.
[28] Choi, O., Das, A., Yu, C.P., Hu, Z. Nitrifying bacterial growth inhibition in the presence of algae and cyanobacteria. Biotechnol Bioeng. 2010, 107(6), 1004–1011. doi: 10.1002/bit.22860.
[29] Ogbonna, J.C., Tanaka, H. Light requirement and photosynthetic cell cultivation development of processes for efficient light utilization in photobioreactors. J Appl Phycol. 2000, 12, 207–218. doi: 10.1023/A:.
[30] Vergara, C., Munoz, R., Campos, J.L., Seeger, M., Jeison, D. Influence of light intensity on bacterial nitrifying activity in algal-bacterial photobioreactors and its implications for microalgae based wastewater treatment. Int Biodeterioration Biodegradation. 2016, 114, 116–121. doi: https://doi.org/10.1016/j.ibiod.2016.06.006.
[31] Croft, M.T., Lawrence, A.D., Raux-Deery, E., Warren, M.J., Smith, A.G. Algae acquire vitamin B12 through a symbiotic relationship with bacteria. Nature. 2005, 438(7064), 90–93. doi: 10.1038/nature04056.
[32] Kazamia, E., Czesnick, H., Nguyen, T.V., Croft, M.T., Sherwood, E., Sasso, S., et al. Mutualistic interactions between vitamin B12-dependent algae and heterotrophic bacteria exhibit regulation. Environ Microbiol. 2012, 14(6), 1466–1476. doi: 10.1111/j.1462-2920.2012.02733.x.
[33] Camili, A., Bassler, B.L. Bacterial small molecule signaling pathways. Science. 2006, 311 (5764), 1113–1116. doi: 10.1126/science.1121357.
[34] Garcia, J., Mujeriego, R., Hernandez-Marine, M. High rate algal pond operating strategies for urban wastewater nitrogen removal. J Appl Phycol. 2000, 12(3), 331–339. doi: 10.1023/A:.
[35] Bilanovic, D., Holland, M., Starosvetsky, J., Armon, R. Co cultivation of microalgae and nitrifiers for higher biomass production and better carbon capture. Bioresour Technol. 2016, 220. 282–288.
[36] Ugwu, C.U., Aoyagi, H., Uchiyama, H. Photobioreactors for mass cultivation of algae. Bioresour Technol. 2008, 99(10), 4021–4028. doi: 10.1016/j.biortech.2007.01.046.
[37] Kesaano, M., Sims, R.C. Algal biofilm based technology for wastewater treatment. Algal Res. 2014, 5, 231–240. doi: https://doi.org/10.1016/j.algal.2014.02.003.
[38] Chae, K.J., Kang, J. Estimating the energy independence of a municipal wastewater treatment plant incorporating green energy resources. Energy Convers Manage. 2013, 75, 664–672. https://doi.org/101.1016/j.enconman.2013.08.028.
[39] Daims, H., Lebedeva, E.V., Pjevac, P., Han, P., Herbold, C., Albertsen, M. Complete nitrification by Nitrospira bacteria. Nature. 2015, 528(7583), 504–509. doi: 10.1038/nature16461.
[40] Sonthiphand, P., Hall, M.W., Neufeld, J.D. Biogeography of anaerobic ammonia oxidizing (Anammox) bacteria. Front Microbiol. 2014, 5, 399. doi: 10.3389/fmicb.2014.00399.
[41] Su, Y., Mennerich, A., Urban, B. Municipal wastewater treatment and biomass accumulation with a wastewater born and settlable algal-bacterial culture. Water Res. 2011, 45(11), 3351–3358. doi: 10.1016/j.watres.2011.03.046.
[42] Park, J.B., Craggs, R.J. Nutrient removal in wastewater treatment high rate algal ponds with carbon dioxide addition. Water Sci Technol. 2011, 63(8), 1758–1764. doi: 10.2166/wst.2011.114.
[43] Hende, S.V., Beelen, V., Julien, L., Lefoulon, A., Vanhoucke, T., Coolsaet, C., et al. Technical potential of microalgal bacterial floc raceway ponds treating food-industry effluents while producing microalgal bacterial biomass: An outdoor pilot scale study. Bioresour Technol. 2016, 218, 969–979. doi: http://dx.doi.org/10.1016/j.biortech.2016.07.065.

[44] Steen, P.V., Rahsilawati, K., Rada-Ariza, A.M., Lopez-Vazquez, C.M., Lens, P.N. A new photoactivated sludge system for nitrification by an algal-bacterial consortium in a photobioreactor with biomass recycle. Water Sci Technol. 2015, 72(3), 443–450. doi: 10.2166/wst.2015.205.
[45] Godos, I.D., Vargas, V.A., Guzman, H.O., Soto, R., Garcia, B., Garcia, P.A., et al. Assessing carbon and nitrogen removal in a novel anoxic-aerobic cyanobacterial-bacterial photobioreactor configuration with enhanced biomass sedimentation. Water Res. 2014, 61, 77–85. doi: 10.1016/j.watres.2014.04.050.
[46] Peng, Y., Zhu, G. Biological nitrogen removal with nitrification and denitrification via nitrite pathway. Appl Microbiol Biotechnol. 2006, 73(1), 15–26. doi: 10.1007/s00253-006-0534-z.
[47] Gonzalez, C., Marciniak, J., Villaverde, S., Garcia-Encina, P.A., Munoz, R. Micro algae based processes for the biodegradation of pretreated piggery wastewater. Appl Microbiol Biotechnol. 2008, 80(5), 891–898. doi: 10.1007/s00253-008-1571-6.
[48] Manser, N.D., Wang, M., Ergas, S.J., Mihelcic, J.R. Biological nitrogen removal in a photosequencing batch reactor with an algal-nitrifying bacterial consortium and anammox granules. Environ Sci Technol Lett. 2016, 3(4), 175–179. doi: 10.1021/acs.estlett.6b00034.
[49] Donati, E.R., Sani, R.K., Goh, K.M., Chan, K.G. Recent advances in bioremediation/biodegradation by extreme microorganisms. Front Microbiol. 2019, 10, 1–2. doi: https://doi.org/10.3389/fmicb.2019.01851.
[50] Kargi, F., Dincer, A.R., Pala, A. Characterization and biological treatment of pickling industry wastewater. Bioprocess Eng. 2000, 23(4), 371–374. doi: 10.1007/s004499900174.
[51] Abou-Elele, S.I., Kamel, M.M., Fawzy, M.E. Biological treatment of saline wastewater using salt tolerant microorganisms. Desalination. 2010, 250(1), 1–5.
[52] Kapdan, I.K., Erten, B. Anaerobic treatment of saline wastewater by Halanaerobium lacusrosei. Process Biochem. 2007, 42(3), 449–453. doi: https://doi.org/10.1016/j.procbio.2006.09.001.
[53] Piubeli, F., Grossman, M.J., Fantinatti-Garboggini, F., Durrant, L.R. Enhanced reduction of COD and aromatics in petroleum-produced water using indigenous microorganisms and nutrient addition. Int Biodeterioration Biodegradation. 2012, 68, 78–84. doi: https://doi.org/10.1016/j.ibiod.2011.11.012.
[54] Bonfa, M.R., Grossman, M.J., Mellado, E., Durrant, L.R. Biodegradation of aromatic hydrocarbons by haloarchaea and their use for the reduction of the chemical oxygen demand of the hypersaline petroleum produced water. Chemosphere. 2011, 84(11), 1671–1676.
[55] Feng, T.C., Cui, C.Z., Dong, F., Feng, Y.Y. Phenanthrene biodegradation by halophilic Martelella sp. AD-3. J Appl Microbiol. 2012, 113(4), 779–789. doi: 10.1111/j.1365-2672.2012.05386.x.
[56] Borgne, S.L., Paniagua, D., Vazquez-Duhalt, R. Biodegradation of organic pollutants by halophilic bacteria and archaea. J Mol Microbiol Biotechnol. 2008, 15(2-3), 74–92.
[57] Garcia, M.T., Ventosa, A., Mellado, E. Catabolic versatility of aromatic compound-degrading halophilic bacteria. FEMS Microbiol Ecol. 2005, 54(1), 97–109. doi: 10.1016/j.femsec.2005.03.009.
[58] Dosta, J., Nieto, J.M., Vila, J., Grifoll, M., Mata-Alvarez, J. Phenol removal from hypersaline wastewaters in a membrane biological reactor (MBR): Operation and microbiological characterization. Bioresour Technol. 2011, 102(5), 4013–4020. doi: 10.1016/j.biortech.2010.11.123.
[59] Bonfa, M.R., Grossman, M.T., Piubeli, F., Mellada, E., Durrant, L.R. Phenol degradation by halophilic bacteria isolated from hypersaline environments. Biodegradation. 2013, 24(5), 699–709.
[60] Wu, Y., Xia, L., Yu, Z., Shabbir, S., Kerr, P.G. In situ bioremediation of surface waters by periphytons. Bioresour Technol. 2014, 151, 367–372. doi: 10.1016/j.biortech.2013.10.088.

[61] Jin, Q., Hu, Z., Jin, Z., Qiu, L., Zhong, W., Pan, Z. Biodegradation of aniline in an alkaline environment by a novel strain of the halophilic bacterium, Dietzia natronolimnaea JQ-AN. Bioresour Technol. 2012, 117, 148–154. doi: 10.1016/j.biortech.2012.04.068.
[62] Campo, P., Platten III, W., Suidan, M.T., Chai, Y., Davis, J.W. Aerobic biodegradation of amines in industrial saline wastewater. Chemosphere. 2011, 85(7), 1199–1203. doi: 10.1016/j.chemosphere.2011.08.039.
[63] Tapingkae, W., Tanasupawat, S., Parkin, K.L., Benjakul, S., Visessanguan, W. Degradation of histamine by extremely halophilic archaea isolated from high salt-fermented fishery products. Enzyme Microb Technol. 2010, 46(2), 92–99. doi: https://doi.org/10.1016/j.enzmictec.2009.10.011.
[64] Uddin, M.S., Zhou, J., Qu, Y., Guo, J., Wang, P., Zhao, L.H. Biodecolorization of azo dye acid red B under high salinity condition. Bull Environ Contam Toxicol. 2007, 79(4), 440–444. doi: 10.1007/s00128-007-9260-1.
[65] Asad, S., Amoozegar, M.A., Pourbabaee, A.A., Sarbolouki, M.N., Dastgheib, S.M.M. Decolorization of textile azo dyes by newly isolated halophilic and halotolerant bacteria. Bioresour Technol. 2007, 98(11), 2082–2088. doi: 10.1016/j.biortech.2006.08.020.
[66] Guo, J., Zhou, J., Wang, D., Tian, C., Wang, P., Uddin, M.S. A novel moderately halophilic bacterium for decolorizing azo dye under high salt condition. Biodegradation. 2008, 19(1), 15–19. doi: 10.1007/s10532-007-9110-1.
[67] Dafale, N., Rao, N.N., Meshram, S.U., Wate, S.R. Decolorization of azo dyes and simulated dye bath wastewater using acclimatized microbial consortium – Biostimulation and halo tolerance. Bioresour Technol. 2008, 99(7), 2552–2558. doi: 10.1016/j.biortech.2007.04.044.
[68] Oturkar, C.C., Nemade, H.N., Mulik, P.M., Patole, M.S., Hawaldar, R.R., Gawai, K.R. Mechanistic investigation of decolorization and degradation of reactive red 120 by Bacillus lentus BI377. Bioresour Technol. 2011, 102(2), 758–764. doi: 10.1016/j.biortech.2010.08.094.
[69] Diep, P., Mahadevan, R., Yakunin, A.F. Heavy metal removal by bioaccumulation using genetically engineered microorganisms. Front Bioeng Biotechnol. 2018, 6, 157. doi: 10.3389/fbioe.2018.00157.
[70] Lagenbach, T. Persistence and bioaccumulation of persistent organic pollutants (POPs). In: Patil, Y.B., Rao, P., eds., Applied bioremediation-active and passive approaches, 3rd ed., Rio de Janeiro, Brazil, 2013, 3659. doi: DOI: 10.5772/56418.
[71] de Silóniz, M.-I., Balsalobre, L., Alba, C., Valderrama, M.J., Peinado, J.M. Feasibility of copper uptake by the yeast Pichia guilliermondii isolated from sewage sludge. Res Microbiol. 2002, 153(3), 173–180. doi: 10.1016/s0923-2508(02.
[72] Dursun, A., Uslu, G., Cuci, Y., Aksu, Z. Bioaccumulation of copper(II), lead(II) and chromium (VI) by growing Aspergillus niger. Process Biochem. 2003, 38(12), 1647–1651. doi: https://doi.org/10.1016/S0032-9592(02.
[73] Chojnacka, K. Biosorption and bioaccumulation–the prospects for practical applications. Environ Int. 2010, 36(3), 299–307. doi: 10.1016/j.envint.2009.12.001.
[74] Wu, F., Sun, F., Wu, S., Yan, Y., Xing, B. Removal of antimony(III) from aqueous solution by freshwater cyanobacteria Microcystis biomass. Chem Eng J. 2012, 183, 172–179. doi: https://doi.org/10.1016/j.cej.2011.12.050.
[75] Chakravarty, R., Banerjee, P.C. Mechanism of cadmium binding on the cell wall of an acidophilic bacterium. Bioresour Technol. 2012, 108, 176–183. doi: 10.1016/j.biortech.2011.12.100.
[76] Ramrakhiani, L., Majumder, R., Khowala, S. Removal of hexavalent chromium by heat inactivated fungal biomass of Termitomyces clypeatus: Surface characterization and mechanism of biosorption. Chem Eng J. 2011, 171(3), 1060–1068. doi: https://doi.org/10.1016/j.cej.2011.05.002.

[77] Witek-Krowiak, A., Reddy, D.H.K. Removal of microelemental Cr(III) and Cu(II) by using soybean meal waste – Unusual isotherms and insights of binding mechanism. Bioresour Technol. 2013, 127, 350–357. doi: https://doi.org/10.1016/j.biortech.2012.09.072.
[78] Chojnacka, K., Chojnacki, A., Gorecka, H. Biosorption of Cr3+, Cd2+ and Cu2+ ions by blue–green algae Spirulina sp.: Kinetics, equilibrium and the mechanism of the process. Chemosphere. 2005, 59(1), 75–84. doi: 10.1016/j.chemosphere.2004.10.005.
[79] Mohamad, O.A., Hao, X., Xie, P., Hatab, S., Lin, Y., Wei, G. Biosorption of copper (II) from aqueous solution using non-living Mesorhizobium amorphae strain CCNWGS0123. Microbes Environ. 2012, 27(3), 234–241. doi: 10.1264/jsme2.me11331.
[80] Park, D., Yun, Y.S., Jo, J.H., Park, J.M. Mechanism of hexavalent chromium removal by dead fungal biomass of Aspergillus niger. Water Res. 2005, 39(4), 533–540. doi: 10.1016/j.watres.2004.11.002.
[81] Michalak, I., Chojnacka, K Interactions of metal cations with anionic groups on the cell wall of the macroalga Vaucheria sp. Eng Life Sci. 2010, 10, 209–217. doi: https://doi.org/10.1002/elsc.200900039.
[82] Michalak, I., Chojnacka, K., Witek-Krowiak, A State of the art for the biosorption process–a review. Appl Biochem Biotechnol. 2013, 170(6), 1389–1416. doi: 10.1007/s12010-013-0269-0.
[83] Gadd, G.M., Pan, X Biomineralization, bioremediation and biorecovery of toxic metals and radionuclides. Geomicrobiol J. 2016, 33(3–4), 175–178. doi: https://doi.org/10.1080/01490451.2015.1087603.
[84] Brookshaw, D.R., Lloyd, J.R., Vaughan, D.J., Patrick, R.A.D Effects of microbial Fe(III) reduction on the sorption of Cs and Sr on biotite and chlorite. Geomicrobiol J. 2016, 33(3–4), 206–215. doi: https://doi.org/10.1080/01490451.2015.1076543.
[85] Dhami, N.K., Reddy, M.S., Mukherjee, A Biomineralization of calcium carbonates and their engineered applications: A review. Front Microbiol. 2013, 4, 314. doi: 10.3389/fmicb.2013.00314.
[86] Achal, V., Pan, X., Fu, Q., Zhang, D Biomineralization based remediation of As(III) contaminated soil by Sporosarcina ginsengisoli. J Hazard Mater. 2012, 201. 178–184.
[87] Chandwadkar, P., Misra, H.S., Acharya, S Uranium biomineralization induced by a metal tolerant Serratia strain under acid, alkaline and irradiated conditions. Metallomics. 2018, 10(8), 1078–1088. doi: 10.1039/c8mt00061.a.
[88] Hammes, F., Boon, N., Villiers, J.D., Verstraete, W., Siciliano, S.D Strain-specific ureolytic microbial calcium carbonate precipitation. Appl Environ Microbiol. 2003, 69(8), 4901–4909. doi: 10.1128/aem.69.8.4901-4909.2003.
[89] Okwadha, G.D., Li, J. Optimum conditions for microbial carbonate precipitation. Chemosphere. 2010, 81(9), 1143–1148. doi: 10.1128/AEM.69.8.4901-4909.2003.
[90] Alamdari, A., Rohani, S. Phosphate recovery from municipal wastewater through crystallization of calcium phosphate. Pac J Sci Technol. 2007, 8(1), 27–31.
[91] Desmidt, E., Verstratete, W., Dick, J., Meesschaert, B.D., Carballa, M. Ureolytic phosphate precipitation from anaerobic effluents. Water Sci Technol. 2009, 59(10), 1983–1988. doi: https://doi.org/10.2166/wst.2009.193.
[92] Isk, M., Altas, L., Ozcan, S., Simsek, I., Agdag, O.N., Alas, A. Effect of urea concentration on microbial Ca precipitation. J Ind Eng Chem. 2012, 18(6), 1908–1911. doi: https://doi.org/10.1016/j.jiec.2012.05.002.
[93] Uad, I., Gonzalez-Lopez, J., Silva-Castro, A.G., Vilchez, J.I. Precipitation of carbonates crystals by bacteria isolated from a submerged fixed-film bioreactor used for the treatment of urban wastewater. Int J Environ Res. 2014, 8, 435–446. doi: 10.22059/IJER.2014.735.

[94] Pratt, C., Parsons, S.A., Soares, A., Martin, B.D. Biologically and chemically mediated adsorption and precipitation of phosphorus from wastewater. Curr Opin Biotechnol. 2012, 23(6), 890–896. doi: https://doi.org/10.1016/j.copbio.2012.07.003.
[95] Smirnov, A., Suzina, A., Chudinova, N., Kulakovskaya, T., Kulaev, I. Formation of insoluble magnesium phosphates during growth of the archaea Halorubrum distributum and Halobacterium salinarium and the Brevibacterium antiquum. FEMS Microbiol Ecol. 2005, 52(1), 129–1373. doi: 10.1016/j.femsec.2004.10.012.
[96] Crusberg, T.C. Final Report: Biomineralization of Heavy Metals Within Fungal Mycelia A New Technology for Bioremediation of Hazardous Wastes. Worcester Polytechnique Institute. United States Environmental Protection Agency. 1998, URL:https://cfpub.epa.gov/ncer_abstracts/index.cfm/fuseaction/display.highlight/abstract/590/report/F

Getu Kassegn Weldegebrieal and Assefu Kassegn Sibhatu

5 Photocatalytic degradation of organic contaminants in wastewater treatment

Abstract: The expansion of urbanization and industrialization have primarily given rise to the contamination of the water body via direct or indirect routes. Wastewater treatment is pivotal to avoid its potential deleterious impact on animals and the environment and minimize clean water shortage. Effective removal of pollutants in wastewater using conventional treatment approaches such as adsorption, membrane filtration, flocculation, and flotation is often difficult to achieve due to the complex nature of wastewater matrix. These approaches have drawbacks such as incomplete pollutant removal and sludge formation, and does not involve degradation. Advanced oxidation processes (AOPs) are promising technologies for degrading refractory organic contaminants, removing toxic heavy metal ions and microorganisms from water and wastewater. Heterogeneous photocatalysis is one of the AOPs involving the acceleration of the rate of chemical reactions in the presence of semiconductor catalyst, radiation, and pollutant. It has such merits as high removal efficiency, low cost, no sludge formation, environmental benignity, and easiness to integrate it with other treatment methods. However, its performance is influenced by factors such as nature and concentration of pollutants, catalyst loading, pH, presence of interferents, and presence of oxidants, which are discussed in this chapter one by one based on recent reports in the literature.

Keywords: organic contaminants, photocatalysis, semiconductor catalysts, wastewater treatment

5.1 Introduction

Although about 70% of the surface of the Earth is covered with water, only a small fraction of it is available in fresh form. Many people across the globe are exposed to water-related diseases, including typhoid, cholera, and dysentery, owing to lack of clean and safe drinking water. Clean water is an invaluable asset for ensuring

Acknowledgment: Authors are thankful to their editor, Prof. Maulin Shah, for inviting them to write this book chapter.

Getu Kassegn Weldegebrieal, Department of Chemistry, College of Natural and Computational Sciences, Debre Berhan University, Ethiopia, e-mail: gk21@dbu.edu.et
Assefu Kassegn Sibhatu, Department of Physics, College of Natural and Computational Sciences, Debre Berhan University, Ethiopia

https://doi.org/10.1515/9783110727227-005

sustainable quality standard of living in society. It is always in grave need, not just for drinking and cooking but also for running businesses such as food processing and pharmaceutical industries.

Thus, wastewater reclamation remains a crucial step for alleviating the shortage of clean water and its sustainable utilization. Wastewater may be defined as water contaminated by external foreign substances named pollutants as a result of natural or anthropogenic activities. Proportionally, it is predominantly water which is unfit for many applications because of its composition including pollutants of various nature such as suspended solid particles or floating oil droplets, dissolved organic or inorganic substances, and microorganisms such as bacteria, fungi, viruses, and protozoa. Organic pollutants in wastewater consist of a complex array of compounds including dyes, humic substances, phenolics, oils and fats, surfactants, pesticides and herbicides, pharmaceuticals and personal care products discharged from domestic, municipal, industrial, agricultural, food processing, textile, leather industries, mining, and oil refineries. Some of these pollutants could have deleterious effects on the life of aquatic flora and fauna via creating oxygen deficiency and precluding sunlight penetration, inducing different health problems in humans and also in agriculture, for example, by hampering proper growth of cereals such as wheat [1]. Treatment of wastewater prior to discharge to the environment and leakage to the water stream is, therefore, immensely important to avoid the potential havoc it could cause to humans and animals' lives.

Generally, wastewater can be treated using physical, chemical, and biological methods. In the physical method, physical pollutants of the wastewater such as suspended solid particles and floating oil or foam are removed by settling and skimming from the top surface, respectively. While degradation of pollutants is carried out by the action of microorganisms in the biological method, chemical substances are deliberately added as flocculants to enhance precipitation of dissolved substances or as disinfecting agents such as chlorine, ozone, or chlorine dioxide to neutralize microorganisms in the chemical method. Conventional wastewater treatment approaches including adsorption on activated carbons, chemical oxidation and reduction, flocculation, and membrane filtration do not involve destruction of pollutant substances and have inefficient removal efficiency. Due to the complex nature of the wastewater matrix, adequate decontamination of wastewater is difficult to achieve and in practice a combination of these methods is employed to get more quality water in a more economical manner [2].

Advanced oxidation processes (AOPs) are promising technologies for wastewater decontamination. In these processes, water and wastewater pollutants are removed via oxidation and reduction reactions by in situ generated reactive radical species, primarily the $HO^{\cdot}$. The reactive radicals produced inside the reaction system can not only remove toxic heavy metal ions such as Hg^{2+}, Pb^{2+}, and Cr^{6+} by reducing to their lower valency or elemental form but also can nonselectively destruct organic compounds including the toxic refractory organic pollutants. Such systems include the addition of ozone (O_3), UV/O_3, H_2O_2, UV/H_2O_2, UV/O_3/H_2O_2, and Fe^{2+}/H_2O_2 (Fenton's reagent). AOPs have advantages such as fast reaction kinetics, small footprint in the treated water, no sludge production involved, and hence incurs no additional cost of further waste processing and disposal, can occur at mild temperature and pressure conditions, complete degradation can be achieved under appropriate experimental condition, is simpler to operate, and easily integrable with other treatment approaches [3–5]. However, one of the main drawbacks of this approach is the high cost involved as a continuous input of expensive chemicals is needed to maintain the operation of the system, and also pretreatment might be necessary so as to improve the performance of the system.

Heterogeneous photocatalysis, which is one among the AOPs, is radiation-induced acceleration of the rate of chemical reactions in the presence of a semiconductor catalysts. For these reactions to feasibly proceed, not only should the catalyst, radiation, and pollutant molecules come in physical contact or close proximity to each other but also an appropriate semiconductor catalyst be employed and that the radiation incident on the surface of catalyst should be energetic enough to induce electronic excitation from the filled valence band (VB) to the empty conduction band (CB) of the semiconductor, leaving holes behind. After that, reactive oxidizing radical species such as superoxide radical ion ($O_2^{-\cdot}$) and $HO^{\cdot}$ are generated by the reduction of dissolved oxygen by the photoexcited electrons and oxidation of H_2O or hydroxyl ion (HO^-) by the holes in the VB, respectively. These reactive radicals once produced nonselectively decompose the organic contaminants by setting in a chain of reactions resulting in harmless end products of CO_2, H_2O, and mineral acids at the end of complete degradation process [6].

Semiconductor photocatalysts could be generally metal oxides, sulfides, nitrides, organic substances, or doped or composites of two or more of them. Specific commonly used metal oxide catalysts include CuO, ZnO, NiO, TiO_2, Co_3O_4, WO_3, Fe_2O_3, $BiVO_4$, and doped forms and coupled composites forming heterojunction structures of one or more of them.

As in other AOPs, heterogeneous photocatalysis could be used to treat organic and inorganic substances or even disinfect wastewater from recalcitrant microorganisms [7]. The advantages of this method in wastewater treatment include complete mineralization or formation of easily biodegradable intermediates, no sludge formation, no need of adding costly chemicals, low cost, environmentally friendliness, and easy maintenance and operation of operating system.

In general, heterogeneous photocatalysis has several other multiple potential applications in addition to wastewater remediation including air cleaning, fuel generation such as hydrogen gas via photocatalytic water splitting reaction, as an antimicrobial agent for inhibiting microbial growth, synthesis of chemicals, cancer therapy, anticorrosion to prevent rusting, and in material design such as in self-cleaning paints and tiles.

In this chapter, semiconductor's photocatalytic degradation of organic contaminants in water and wastewater is reviewed. In doing so, the different types and classification of organic contaminants in wastewater and factors affecting the photocatalytic degradation efficiency of the organic contaminants such as type and initial concentration of organic pollutant compounds, type and initial catalyst dose, pH, energy and intensity of the radiation source, as well as the effect of the addition of oxidants and the presence of interfering substances will be overviewed individually based on latest reports in the literature.

5.2 Classification of organic contaminants in wastewater

A typical wastewater effluent discharged from its source may constitute a cocktail of organic and inorganic chemicals, suspended matter, and microorganisms. Organic pollutants in such wastes are very common and can often have predominant percentage in the overall composition of contaminants. These organic molecules could be of various types in structure and property and available dissolved or undissolved form in water.

The extent of organic pollution in wastewater can be grossly quantified by carrying out chemical oxygen demand (COD), biological oxygen demand (BOD), or total organic carbon (TOC) tests. While COD is the amount of oxygen required for organic compounds to be oxidized into CO_2 with strong oxidizing chemicals such as $KMnO_4$ or $K_2C_2O_7$ in an acidic environment, BOD refers to the amount of oxygen required to decompose organic compounds with biological aerobic microorganisms in a specified time at a given temperature. TOC gives direct measure of the total organic content of a waste via a combustion technique. All of the three testing methods cited give information on the total organic compounds in a sample but not the types of individual compounds in it, and separation and analysis of such compounds may be achieved with chromatography–mass spectrometry methods.

Organic pollutants in wastewater could be highly varied and complex. Nonetheless, they can possibly be classified on the basis of source, toxicity, and biodegradability (Figure 5.1). Based on the source, they can be further classified into point and nonpoint sources, depending on whether the waste is discharged from a single source such as domestic or industrial effluents or from multiple sources as in urban or agricultural runoff [2]. So, wastewater released from different sources could bear various classes of organic compounds. For example, agricultural wastewater could carry pesticides such as dichlorodiphenyltrichloroethane (DDT), hexachlorobenzene (HCB), γ-hexachlorocyclohexane (lindane), and organophosphorus pesticides such as malathion, herbicides including atrazine and 2,4-dichlorophenoxyacetic acid, and humic substances such as humic acid and fulvic acid. Domestic wastes mostly carry compounds such as surfactants mainly from soaps and detergents, disinfectants such as ethanol, dichlorobenzene, and acetone in vinegar, and pharmaceutical compounds like ampicillin, amoxicillin, and cloxacillin excreted after consumption along with urine.

Dyes are colored organic compounds used for coloring substrates and are common in the effluents from textile, pulp and paper, and leather industries. Moreover, various classes of organic compounds such as aldehydes, ketones, polycyclic aromatic hydrocarbons, pesticides, and alcohols enter to the water and wastewater stream from the atmosphere by precipitation in the form of rain [8].

Based on toxicity, organic contaminants in wastewater could be grouped as toxic and nontoxic to humans, plants, animals, and/or the environment. While the environmental damage or potential influence of many contaminants of emerging concern on living beings is not yet monitored and unknown, some organic compounds are known to have induced various health problems including genotoxicity, teratogenicity, endocrine disruption, and neurotoxicity on exposed humans and animals. For instance, phenol is a known human carcinogen and has the potential to inhibit the growth and reproductive capacity of aquatic organisms [9]. It is one of the lists of priority pollutants designated by the US environmental protection agency for which it has set a water purity standard of less than 1 ppb in surface water. Benzidine is reported to be carcinogenic to different organs of the human body such as urinary and gallbladder, pancreas, liver, lung, stomach, and large intestine and the dyes that have benzidine moiety in their molecular structure and could metabolically be reduced to benzidine in the body such as direct brown 95, direct black 38, direct blue 6, and direct red 28 (Congo red) are potentially carcinogenic [6]. As a result of their high absorption coefficient even in small concentrations, dyes are vivid in color and can pose threat to aquatic biota through preventing sunlight penetration and reducing the amount of dissolved oxygen in water. Another class of toxic organic pollutants is persistent organic pollutants (POPs). These refer to a broad class of chemical compounds including DDT, HCB, Mirex, lindane, polychlorinated biphenyls (PCBs), polychlorinated naphthalenes, atrazine, dioxins, and bisphenol A which are hazardous to animals and humans and are, therefore, regulated under the Stockholm convention adopted in 2001. Structural formula and common application of some POPs is given in Table 5.1. What makes such pollutants even worse is their physical and chemical properties that enable them to bioaccumulate in lipid tissues and biomagnify across the food chain, resist chemical, photochemical, and biological degradation, and migrate long-range distances by wind and water currents. POPs, even at low concentration levels, are the causes of many such health problems as increased cancer risk, reproductive dysfunction, genotoxicity, endocrine disruption, increased birth defects, alteration of the immune system, and neurobehavioral impairment [10].

In general, wastewater with high degree of organic pollution has lower capability of supporting life in it and that the exposure to some organic compounds could pose a threat to the health of living beings.

On the basis of biodegradability, organic pollutants can be biodegradable and nonbiodegradable. Those contaminant compounds with good hydrophilicity and simpler molecular structures such as alcohols, acids, esters, proteins, and carbohydrates are easily degradable by microorganisms in the environment and their influence to the environment does not last long. Whereas pollutants such as *N*, *N*-dimethylformamide, 4-methyloctane, *N*,*N*-diethylbenzamide, and POPs mentioned above, chemical dyes based on aromatic and heterocyclic ring structures produced from different chemical industries, lignin and lignin derivatives commonly discharged from paper mill factories are recalcitrant to biodegradation [11, 12].

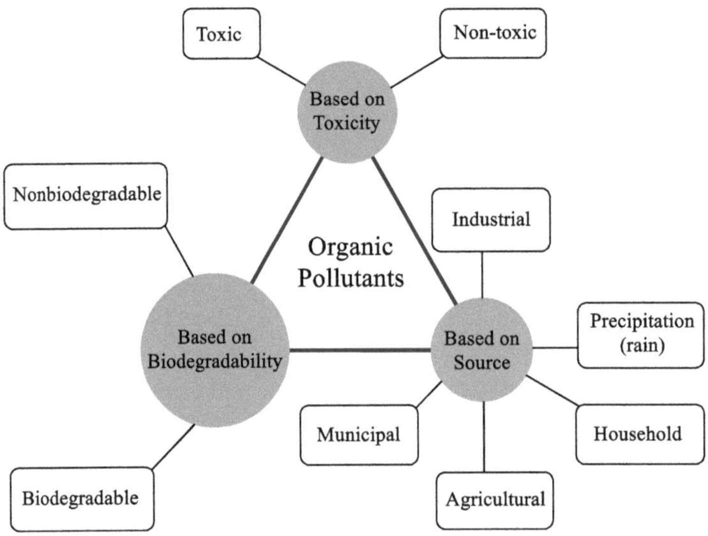

Figure 5.1: Possible major classification of organic pollutants in wastewater.

The extent of biodegradable organics may be estimated with BOD test, which gives the amount of oxygen needed for oxidation by aerobic bacteria, usually expressed as milligrams of oxygen consumed per liter of sample for specified number of days at a fixed temperature. Environmental conditions such as temperature, exposure to radiation, and pH of media could also affect not only the toxicity and long-term stability of the organic contaminants but also the degrading efficiency of the microorganisms to them in the wastewater.

Table 5.1: Structural formula and common application of some POP compounds.

S. no.	Name	Structural formula	Common application
1	Lindane	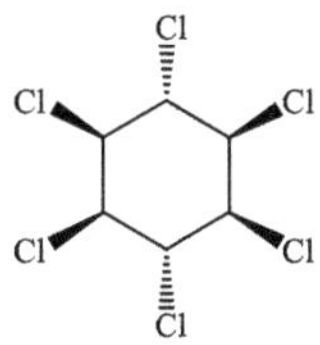	Used to treat scabies and lice in the form of lotion
2	DDT	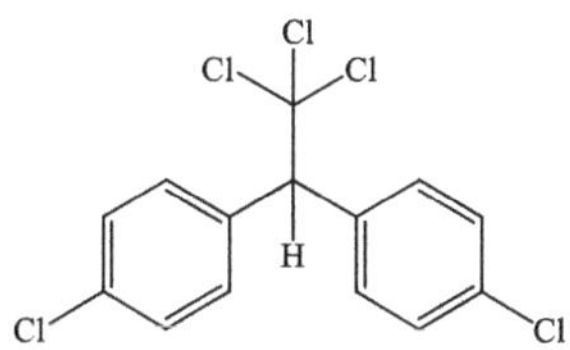	It was initially used to treat malaria, typhus, and other insect-borne human diseases
3	Bisphenol A	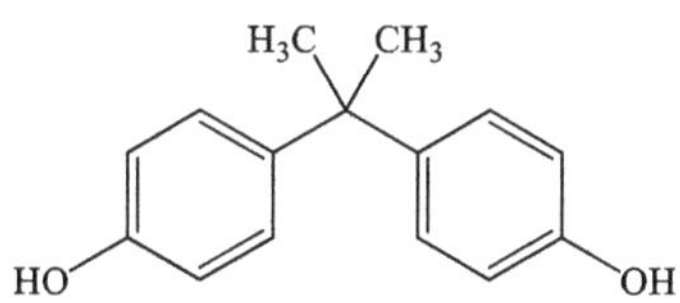	Used in the synthesis of hard and clear polycarbonate plastic. It is also used in epoxy resins as a protective lining in some metal-based food and beverage cans
4	HCB	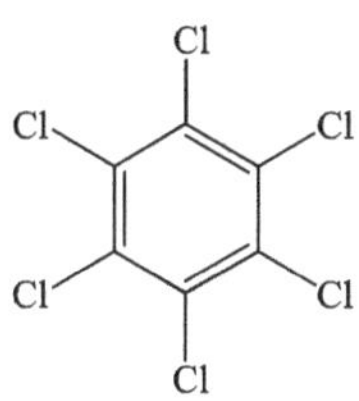	Used as a seed dressing to protect them against fungi
5	Atrazine	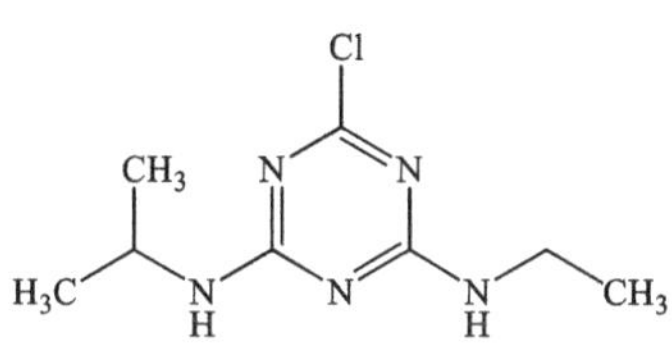	Used as a herbicide for selectively killing weeds

5.3 Photocatalytic degradation of organic contaminants in wastewater

Heterogeneous photocatalysis is one of the AOPs with a promising application in the degradation of organic pollutants in water and wastewater. It may be defined as light-induced acceleration of the rate of chemical reactions in the presence of semiconductor photocatalyst. It is based on the in situ generation of reactive oxidizing radical species on the catalyst surface when irradiated with electromagnetic radiation of enough energy and their subsequent chain of reactions with the organic pollutants. In such processes, the reactions could also be initiated via the absorption of photons by coexisting molecules, named photosensitizers, and transferring the energy to the photocatalyst. Employing a photocatalyst possessing features such as high photonic efficiency, large specific surface area, full utilization of sunlight, and good recyclability enhances the removal efficiency. The advantages of heterogeneous photocatalysis include environmental benignity, low cost since small amount of catalyst and sunlight can be used in the reaction, can proceed at room temperature and pressure, can completely degrade the pollutants into harmless CO_2 and H_2O products, and in due process no formation of sludge is involved [6].

5.3.1 Mechanism of photocatalytic degradation of organic pollutants

In a photocatalytic degradation process, the first step involves the formation of electron–hole (e^-/h^+) pairs (eq. (5.1)), where electromagnetic radiation of energy ($h\nu$) greater than or equal to the band gap energy (E_g) of the semiconductor photocatalyst promotes electron from the VB to the CB, leaving holes behind. Then these photogenerated charge carriers diffuse onto the catalyst surface and undergo redox reactions with adsorbed species. While the holes (h^+) in the VB oxidize H_2O or HO^- into hydroxyl radical ($HO^{\cdot}$) (eqs. (5.3) and (5.4)), the excited electrons in the CB reduce the dissolved oxygen into superoxide radical ion ($O_2^{-\cdot}$) (eq. (5.2)). This reduction reaction is an important step in two ways: firstly, it ensures the separation of e^-/h^+ pairs, which otherwise would lead to their recombination and no formation of

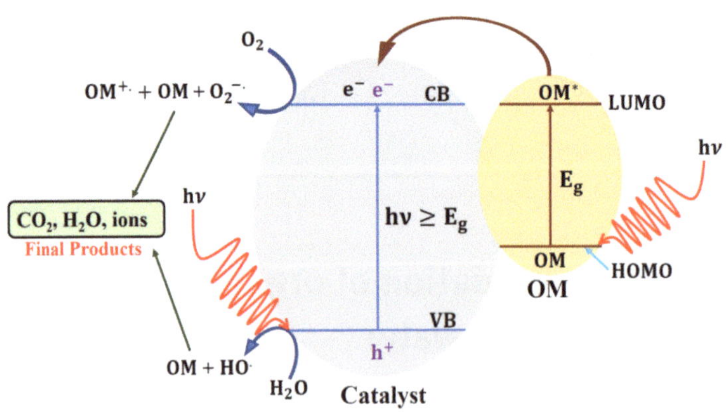

Figure 5.2: Illustration of the general photocatalytic degradation mechanism of pollutant organic molecule in the presence of semiconductor photocatalyst (reprinted with permission from [6]. Copyright 2020. Elsevier).

radicals and no degradation reaction. Secondly, the produced superoxide ion can execute further reactions, such as its protonation to give hydroperoxyl radical ($HOO^{\cdot}$) followed by the formation of H_2O_2 and the most reactive $HO^{\cdot}$ radical (eqs. (5.5)–(5.7)). In addition, the incident radiation may also cause the organic molecules to be excited from the highest occupied molecular orbital (HOMO) to the lowest unoccupied molecular orbital (LUMO) and inject electron into the CB of the catalyst, thereby reducing dissolved oxygen by changing itself to cationic radical ($OM^{+\cdot}$) (eqs. (5.8) and (5.9)). After that, these reactive radicals once formed commence the reaction to decompose the pollutants nonselectively into smaller intermediate species and finally into CO_2 and H_2O (eq. (5.10)). The whole process is illustrated in Figure 5.2.

The various possible reactions that occur are shown in eqs. (5.1)–(5.10) using hypothetical catalyst (C) and pollutant organic molecule (OM):

$$C + h\nu \rightarrow C(e^- + h^+) \tag{5.1}$$

$$O_2 + e^- \rightarrow O_2^{-\cdot} \tag{5.2}$$

$$H_2O + h^+ \rightarrow HO^{\cdot} + H^+ \tag{5.3}$$

$$HO^- + h^+ \rightarrow HO^{\cdot} \tag{5.4}$$

$$O_2^{-\cdot} + H^+ \rightarrow HOO^{\cdot} \tag{5.5}$$

$$2HOO^{\cdot} \rightarrow H_2O_2 + O_2 \tag{5.6}$$

$$H_2O_2 \rightarrow 2HO^{\cdot} \tag{5.7}$$

$$OM + h\nu \rightarrow OM^* \tag{5.8}$$

$$OM^* + O_2 \rightarrow O_2^{-\cdot} + OM^{+\cdot} \tag{5.9}$$

$$OM + OM^{+\cdot} + HO^{\cdot}/O_2^{-\cdot} \rightarrow \text{Intermediate products} \rightarrow CO_2 + H_2O \tag{5.10}$$

The mechanism of organic pollutant degradation can be divided into two: direct and indirect mechanisms. The degradation by $HO^{\cdot}$ radicals produced from the injection of electrons to the CB of the semiconductor, resulting in the reduction of dissolved O_2 to $O_2^{-\cdot}$ and then subsequently to $HO^{\cdot}$ as indicated by eqs. (5.8) and (5.9) and eqs. (5.5)–(5.7) is called direct mechanism, whereas decomposition due to the formation of charge carriers in the semiconductor and then production of radicals is called indirect mechanism, and it has been found that this later route is kinetically faster and plays a predominant role in the degradation of dyes [13].

For all these reactions to be feasible, the generation and efficient separation of e^-/h^+ pairs are indispensable. The photoproduced VB holes are able to oxidize H_2O or HO^- species: one requirement that need to be fulfilled is that the VB edge potential of the semiconductor used should be more positive than the oxidation potential of H_2O or HO^- into hydroxyl radical. Similarly, the bottom of the CB potential should be more negative than the reduction potential of O_2 molecules into oxygen radical anion, $O_2^{-\cdot}$. The same idea also works for the reduction of toxic metal ions in wastewater, that is, the CB edge potential of the semiconductor should be more negative than the reduction potential of the metal ions.

Overall, the photocatalytic degradation process can be categorized into five independent steps [6]: (i) diffusion of pollutant molecules onto the catalyst surface, (ii) adsorption of the molecules on the catalyst surface, (iii) generation of reactive radicals followed by the degradation of the molecules, (iv) desorption of intermediates or final products from the surface, and finally (v) diffusion of intermediates or final products into the bulk of the liquid, which is outlined in Figure 5.3.

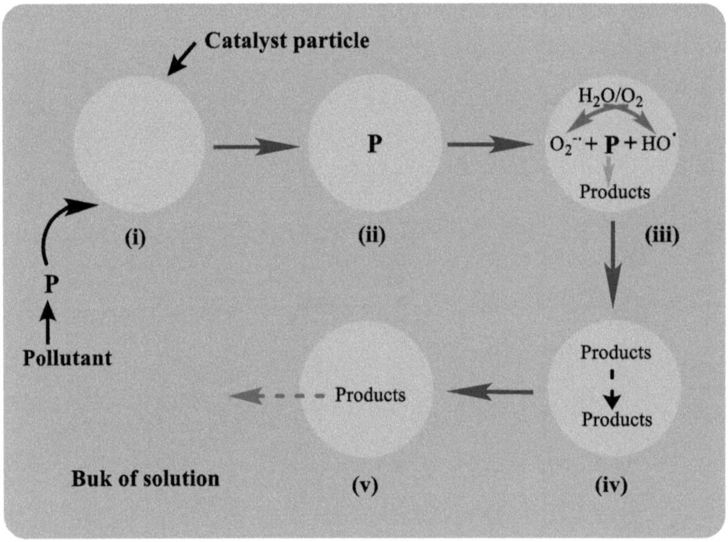

Figure 5.3: Steps of reaction in photocatalytic degradation of organic pollutant (P) molecules on the surface of photocatalyst particles in an aqueous media.

The degree of organic pollutants degradation achieved during photocatalytic degradation process may be quantitatively determined by measuring COD test at the beginning and end of the processes, as given by Eq. (5.11):

$$\text{COD removal} \left(\%\right) = \frac{\text{COD}_\text{i} - \text{COD}_\text{f}}{\text{COD}_\text{i}} \times 100 \tag{5.11}$$

where COD_i and COD_f refer to the initial and final COD values measured, respectively.

Moreover, if we are studying the degradation of a particular organic pollutant, for example, a dye, the progress of degradation can be observed visually by change in color and also quantitatively through measuring the absorbance and/or concentration of the dye at its maximum absorbance at regular time intervals, Eq. (5.12):

$$\text{Degradation}\,(\%) = \frac{A_0 - A_t}{A_0} \times 100 = \frac{C_0 - C_t}{C_0} \times 100 \tag{5.12}$$

where A_0, C_0, A_t, C_t refer to the absorbance and concentration values measured initially and after any time t of the reaction, respectively.

5.3.2 Factors affecting the photocatalytic degradation of organic pollutants in wastewater

Heterogeneous photocatalytic reactions may be experimentally carried out in slurry or fixed-bed configurations. In the slurry form, the powdered catalyst is suspended in solution, whereas in the fixed-bed form the catalyst is immobilized on inert substrates such as glass, metallic supports. Each configuration has its own pros and cons. While the merits of the slurry configuration include simplicity of configuration and high surface area contact between catalyst and pollutant molecules, the drawback is associated with the difficulty of recovering spent catalyst particles as well as the tendency of agglomeration of particles when used in excess which prevent radiation penetration deep into the solution and lower degradation. However, in the fixed-bed form, difficulty of recovering the spent catalyst does not exist but its disadvantage is low contact surface of catalyst and pollutant molecules, resulting in lower degradation efficiency.

For photocatalytic reactions to proceed effectively, the catalyst, pollutant, and radiation should be in close proximity or in physical contact to each other, and the efficiency of degradation depends on the physicochemical properties of the catalyst and pollutants and also on the intensity and wavelength of the radiation source employed. At the heart of these reactions is the photocatalyst. For possible large-scale application of heterogeneous photocatalysis, the catalyst used should be of low cost, easily recoverable after use so that it may be repeatedly reused, and visible light active so that solar radiation, about 50% of which that arrives on the surface of the Earth is composed of visible light wavelengths (400–750 nm), may be used. To be visible light active, its bandgap should be generally less than 3.0 eV but not very small so as to minimize the chance of rapid recombination of photogenerated e^-/h^+ pairs, in which case the removal efficiency is hindered. Moreover, the visible light activity of the catalyst can be enhanced by either improving its surface features such as increasing specific surface area as well as porosity or by chemically modifying it via doping with metallic or nonmetallic elements or coupling with other semiconductors. Coupling semiconductors and forming heterojunctions are the most common forms of reducing the recombination rate of e^-/h^+ pairs and enhancing the photocatalytic activity.

In addition to the reactor design, bandgap, and physicochemical features of the particular photocatalyst employed, several other factors and operational parameters influence the degree of photocatalytic degradation of the target organic pollutants in wastewater treatment. The most important ones that we are going to see here individually include the type and initial concentration of the organic pollutant, type and initial catalyst dose, pH, the energy and intensity of the radiation used, the addition of oxidizing substances, and the presence of interfering substances.

5.3.2.1 Effect of type and initial pollutant concentration

The photocatalytic degradation efficiency of organic compounds in wastewater is affected by the initial concentration of organic compounds in wastewater. The degradation efficiency is reduced with more concentration of organic compounds. This is because as the pollutant concentration is higher, the path length of light is reduced and this in turn reduces the number of available active surface sites on the catalyst and hence the number of hydroxyl radicals responsible for the degradation is immensely reduced [14]. So, photocatalytic degradation is most efficient generally when the concentration of organic compounds is low using optimal catalyst dosage. For example, Rajamanickam and Santi [5] studied the effect of varying the concentration of organic pollutant 4-nitrophenol from 0.2 to 0.8 mM, which decreased by less than half of the initial 97.7% efficiency at a fixed amount of ZnO catalyst and pH media. Similarly, the degradation efficiency of 2,4-dichlorophenol was reduced when increasing its initial concentration from 10 to 50 ppm over a period of 2 h at a fixed ZnO supported on activated carbon catalyst loading of 75 mg and at the solution pH [15]. Song et al. [16] also found the degradation of aqueous solution of CI Direct red 23 azo dye to decrease when dye concentration was increased from 100 to 500 ppm at pH 12 using 1.5 g/L of $SrTiO_3/CeO_2$ composite as a catalyst. Similar decrement of the rate of degradation was reported by Daneshvar et al. [17] for the aqueous solution of the organophosphorus pesticide Diazinon by varying its concentration from 16 to 30 ppm using ZnO catalyst. The authors attributed the observed reduced degradation to the fewer adsorption and generation of $HO^{\cdot}$ radicals owing to fewer active sites left by the adsorption of more organic molecules and also due to the less amount of radical generation as a result of the interception of photons by the pollutant molecules before they reach the catalyst surface.

The nature of the organic pollutant also has its own impact on the rate of photocatalytic degradation efficiency. While some simpler organic pollutants get directly converted into CO_2 and H_2O, others pass through a series of intermediates, and complete degradation takes longer time. For example, while 4-chlorophenol passes through degradation intermediates of benzoquinone and hydroquinone, oxalic acid undergoes direct conversion into CO_2 and H_2O [18].

5.3.2.2 Effect of type and initial catalyst dose

The type of catalyst and its initial dosage used affect the rate of photocatalytic degradation of an organic pollutant keeping other experimental variables fixed. Using catalyst loading of more than optimum may result in lowering of degradation percentage due to the formation of colloids by the excess catalyst particles, preventing the penetration of radiation deep into the solution by scattering the radiation and reducing the number of available active sites [19]. Moreover, more catalyst particles than optimum could diminish the efficiency owing to particle agglomeration thereby reducing active sites, and deactivation of activated molecules could result via particle collision, resulting in lowering efficiency [14]. By contrast, lower dose of catalyst than optimum results in diminished degradation for only smaller fraction of radiation will be utilized for affecting the degradation and the rest passes through the reactor without being used. An ideal photocatalyst should have high photocatalytic efficiency, high specific surface area, full utilization of sunlight, and good recyclability [20]. Although no single catalyst can fulfill all these criteria, the catalytic efficiency of a photocatalyst can be enhanced via improving particle size and surface morphologies, doping, and forming composites with other substances. For example, the degradation percentage of Rhodamine B and acid green 25 was enhanced when 10 wt% GO/TiO_2 composite was used compared to that achieved by pristine TiO_2 under other similar experimental parameters [21]. This enhancement was ascribed to the increased adsorption capacity and electron transfer capability of graphene oxide (GO) sheets present in the composite. The photocatalytic degradation percentage of methylene blue (MB) dye was observed to increase when reduced GO (rGO) was introduced to pure TiO_2 and increased in the following order: $TiO_2 < rGO/TiO_2 < rGO/TiO_2/ZnO$ [22]. Similar explanation as above was offered as to why the degradation in the presence of composites was higher than using just pure TiO_2. Nathir et al. [23] reported higher photocatalytic degradation efficiency of the composite GO/ZnO/Ag than GO/ZnO/Cu toward MB and reasoned that the higher activity of the former is due to reduced recombination of e^-/h^+ pairs by serving as a carrier trapper and also higher surface area. In addition to the catalyst type, the crystal structure of a specific catalyst also affects its photocatalytic activity toward target organic pollutants. For example, TiO_2 exists in three different crystal structures such as anatase, rutile, and brookite, where the first two have tetragonal structure and the last one has orthorhombic structure. Anatase polymorph is usually a more photoactive phase than the other two and this is probably because of its indirect bandgap and lightest average effective mass of photogenerated charge carriers which result in longer lifetime e^-/h^+ pairs and their fastest migration onto the surface from the bulk, resulting in lower recombination rate and faster reaction kinetics [24].

Rajamanickam and Shanthi [5] studied the effect of increasing the ZnO catalyst dose from 25 to 150 mg per 50 mL of solution on the degradation efficiency of

4-nitrophenol. When they raised the amount of catalyst from 25 to 75 mg, the degradation efficiency increased from 56.8% to 79.9% in a period of 1 h irradiation time. However, any further increase of catalyst dose reduced the efficiency. The authors reasoned that while the increase in degradation with the increase up to 75 mg could be attributed to the increase in the number of ZnO particles which increased photon absorption and adsorption of pollutant molecules, the reduction in degradation beyond 75 mg per 50 mL of solution might be due to the screening effect of particles to the incident radiation and also to particle aggregation which reduces its activity. Jamal et al. [25] studied the effect of raising the amount of ZnO/nanocellulose composite on the removal efficiency of Congo red and found that on increasing the loadings from 0.025 to 0.25 g, the efficiency went decreasing after reaching a peak at 0.075 g. The authors elaborated the increased activity up to that optimal value to be owing to the increased number of surface-active sites which increased the number of hydroxyl radicals, whereas the reduction after the optimal value was due to the shielding effect of the light by the particles that reduce the generation of reactive radical species despite the increase in the number of particles. Similarly, Ghasemi et al. [26] investigated the degradation of furfural with TiO_2 on granulated activated carbon composite catalyst by varying its dosage from 1 to 3.5 g and found the degradation increasing and peaked at 2.5 g. However, further increase of catalyst beyond that decreased the removal rate and was reasoned as stated in the above case.

However, a composite of two oxides does not necessarily result in higher degradation performance for a pollutant molecule. For example, TiO_2/ZnO was better at removing methyl orange than TiO_2 alone but not as efficient as ZnO under visible light irradiation and explained this to be due to the particle size difference between the two oxides mixed [27].

5.3.2.3 Effect of pH

The pH of the media plays a prominent role as the degradation efficiency of pollutants by photocatalysts. This is because it affects not only the chemical stability of the employed catalyst but also its surface charge properties which in turn affects the adsorption capacity of the catalyst surface for organic pollutant molecules in the media. In addition, the solution pH also affects the VB and CB positions of the employed photocatalyst itself, which in turn affects the range of the electromagnetic spectrum to be absorbed by it and hence the degradation efficiency.

Various studies reported the influence of pH on the photocatalytic degradation of sample organic contaminants in water and wastewater. For instance, Chen et al. [28] studied the influence of changing pH on the photocatalytic removal efficiency of three different azo dyes, namely, methyl orange, Congo red, and direct black 38, using the same ZnO photocatalyst. They found that the efficiency of degradation was higher at acidic pH than at alkaline pH for all of the dyes and explained this to be mainly because of two reasons. Firstly, at acidic pH, the surface charge of ZnO particles is positive which makes it convenient for anionic dye molecules to get easily adsorbed and undergo redox reactions. Secondly, at such media, the dye molecules prominently exist in their quinonoid structure, making them unstable and easier for breakage than at alkaline media which occurs in their azo molecular forms and more difficult for degradation. Paul et al. [29] found that the removal efficiency of MB using graphitic carbon nitride (g-C_3N_4) catalyst particles, prepared by calcination of urea at 550 °C for 3 h, at an alkaline pH of 11 was greater by 1.57 times to that at normal pH of 6.6. According to the authors, the higher catalytic activity at the higher pH is because MB is negatively charged at pH greater than 6.8 which results in strong adsorption of dye molecules on the positively charged g-C_3N_4 particles via electrostatic interaction. Moreover, the formation of reactive radicals is also increased in alkaline medium resulting in the enhancement of the rate of catalysis. Gupta et al. [30] studied the influence of changing media pH from 3 to 13 on the removal efficiency of a mixture of crystal violet and methyl orange dyes using Ag/ZnO catalyst. The degradation was monitored at their respective λ_{max} of 590 and 430 nm and found that while no influence of pH for crystal violet was observed in the entire pH range, for methyl red, however, it first increased and peaked to more than 99% at 10, then declined in the rest range. This decrement was explained to be due to the electrostatic repulsion between anionic dye molecules and negatively charged particles of the catalyst as the pH is greater than the zero point charge pH_{zpc} of the catalyst, which is 6.6. Whereas, Sahoo et al. [31] studied the influence of the same pH range (3–13) utilizing Ag^+/TiO_2 catalyst for the degradation of MB and found the degradation increasing with pH which peaked to about complete degradation at pH 13. This was ascribed to the suitable electrostatic attraction between the cationic dye and surface charge of the catalyst, which is negative at pH higher than pH_{zpc} (6.6). Table 5.2 gives additional examples of optimum pH where the catalytic efficiency for different organic pollutants in wastewater has been found to be maximum at the specified experimental conditions.

5.3.2.4 Effect of the addition of oxidants

An oxidant is a substance which oxidizes other substances by accepting their electrons. An optimum amount of it is sometimes added deliberately to photocatalytic reaction systems to fasten the rate of reactions. As photocatalytic reactions are initiated first by the formation of electron-hole pairs, the recombination of these pairs severely limits the degradation process and such oxidants are added to lessen this phenomenon. Oxidants differ in terms of their oxidizing power. The hydroxyl radical ($HO^·$) with an oxidation potential of +2.8 V versus standard hydrogen electrode (SHE) is the second most powerful oxidant after fluorine in aqueous media which oxidizes

Table 5.2: Removal efficiency (D (%)) of pollutant molecules using a specific catalyst and at given optimum pH value.

Catalyst	Pollutant	Optimal pH	Other conditions	*D* (%)	Ref.
ZnO	Penicillin G	5	Catalyst load = 0.1 g/L; pollutant conc. = 20 ppm; irradiation time = 2 h	47.88	[32]
$Fe_3O_4/SiO_2/TiO_2$/rGO	2,4-Dinitrophenol	5	Catalyst load = 0.1 g/L; pollutant conc. = 10 ppm; irradiation time = 30 min.	100	[33]
ZnO	*m*-Cresol purple	8	Catalyst load = 1.5 g/L; pollutant conc. = 20 ppm; irradiation time = 150 min.	97	[34]
ZnO/ nanocellulose	Congo red	5.5	Catalyst load = 0.1 g; pollutant conc. = 25 ppm; irradiation time = 2 h	98	[25]
$BiVO_4$	Rhodamine B	10	Catalyst load = . . . ; pollutant conc. = 10 ppm; irradiation time = 3 h	90	[35]
MWCNT/TiO_2	Tetracycline	5	Catalyst load = 0.1 g/L; pollutant conc. = 5 g/L; irradiation time = 2 h	74.4	[36]
Pt/TiO_2	Phenol	3	Catalyst load = 1 g/L; pollutant conc. = 50 ppm; irradiation time = 3 h	87.7	[37]
	2-Chlorophenol	3	Catalyst load = 1 g/L; pollutant conc. = 50 ppm; irradiation time = 3 h	93.8	[37]
ZnO-ZnS/ Ag_2O-Ag_2S	2-Chlorophenol	5	Catalyst load = 0.25 g/L; pollutant conc. = 10 ppm; irradiation time = 270 min.	96	[38]
TiO_2/GO	Methylene blue	10	Catalyst load = 0.2 g/L; pollutant conc. = 5 ppm; irradiation time = 4 h	~100	[39]

many organic compounds nonselectively [32]. Depending on the nature of the organic pollutant to be degraded, it can attack the pollutant via four mechanisms of hydrogen abstraction, electron transfer, radical addition, and radical combination [33]. Other common oxidants used in photocatalytic reactions include sulfate radical ion ($SO_4^{-\cdot}$), generated from the addition of persulfate ($S_2O_8^{2-}$) salts or peroxymonosulfate (HSO_5^-) or oxone, H_2O_2, permanganate (MnO_4^-), and chlorine dioxide (ClO_2) with their respective oxidation potentials of +2.5–3.1, 1.8, 1.7, and 1.5 V versus SHE [34].

These radical species may be produced in the reaction media via either direct bond breakage by the action of radiant energy, for example, hydroxyl radical from hydrogen peroxide (H_2O_2, eq. (5.13)), hydroxyl and sulfate radical ions from oxone ($H_3K_5O_{18}S_4$, eq. (5.14)), or indirectly via oxidation with the photogenerated e^-/h^+ carriers (eqs. (5.15)–(5.17):

$$H_2O_2 + h\nu \rightarrow 2HO^{\cdot} \tag{5.13}$$

$$HSO_5^- + h\nu \rightarrow HO^{\cdot} + SO_4^{-\cdot} \tag{5.14}$$

$$Cl^- + h^+ \rightarrow Cl^{\cdot} \tag{5.15}$$

$$H_2O_2 + h^+ \rightarrow HOO^{\cdot} + H^+ \tag{5.16}$$

$$H_2O_2 + e^- \rightarrow HO^{\cdot} + HO^- \tag{5.17}$$

Various studies have shown the effect of adding oxidant species on the photocatalytic efficiency of catalyst toward target pollutant molecules. Aber et al. [35] reported enhancement in the removal efficiency of Acid Blue 9 using ZnS catalyst on adding persulfate ion ($S_2O_8^{2-}$) and periodate ion (IO_4^-) separately by varying the amount of $(NH_4)_2S_2O_8$ and $NaIO_4$ added, respectively, from 0 to 100 ppm. They proposed the resulting pronounced catalytic efficiency to be due to effective separation of charge carriers by scavenging the photoexcited electrons in the CB of the catalyst while producing sulfate anion and iodate radicals (eqs. (5.18)–(5.20)), and other reactive radical species which can participate in the degradation process of the dye:

$$S_2O_8^{2-} + e^- \rightarrow SO_4^{2-} + SO_4^{-\cdot} \tag{5.18}$$

$$IO_4^- + 8e^- + 8H^+ \rightarrow 4H_2O + I^- \tag{5.19}$$

$$IO_4^- + h\nu \rightarrow IO_3^{\cdot} + O^{-\cdot} \tag{5.20}$$

In another study, Yao and Wang [36] reported increased degradation of MB dye in the presence of H_2O_2 using TiO_2 catalyst. They found that adding 0.03 M H_2O_2 has resulted in an increase in the degradation of the dye and proposed it to be due to the photoelectron capturing role played by it (eq. (5.17)), thereby facilitating the effective separation of e^-/h^+ pairs and also its reaction with superoxide radical ion ($O_2^{-\cdot}$) to produce hydroxyl radical, Eq. (5.21):

$$H_2O_2 + O_2^{-\cdot} \rightarrow HO^{\cdot} + HO^- + O_2^{\cdot} \tag{5.21}$$

Similar enhancement in the sun-light-assisted degradation of the same dye using CuO catalyst was observed [37]. Abdelsamad et al. [38] investigated the effect of adding 0.005–0.16 mL of 30% H_2O_2 to three dyes (Reactive Red 195, Reactive Yellow 145, and Reactive Orange 122) individually using Ag/ZnO thin-film catalyst over an irradiation time of 2 h. They found that while the degradation was increased on increasing the added volume up to 0.01 mL, it resulted in the opposite effect on further increase of the volume beyond that amount and explained this to be due to the hole scavenging (eq. (5.16)), and hydroxyl radical scavenging (eq. (5.22)) effects producing the weaker hydroperoxyl radial ($HOO^{\cdot}$) than the hydroxyl radical when it is added in excess:

$$H_2O_2 + HO^{\cdot} \rightarrow HOO^{\cdot} + H_2O \tag{5.22}$$

Moreover, addition of bromate ion (BrO_3^-) in the degradation percentage of Reactive Blue 19 dye using a fixed dose of TiO_2 catalyst increased the removal efficiency when 10–30 mM concentration was added but any further addition from 30 to 100 mM resulted in no further significant change [39]. This increase in the catalytic activity has been proposed to be because of the electron capturing capability of the anion, thereby ensuring the separation of charge carriers (eq. (5.23)) and increased rate of degradation. However, the insignificant effect of the rate of degradation in the concentration range from 30 to 100 mM concentration is attributed to the adsorption effect of in situ generated Br^- ions on the surface of TiO_2, which inhibits its catalytic activity:

$$BrO_3^- + e^- + 6H^+ \rightarrow Br^- + 3H_2O \tag{5.23}$$

Thus, the optimal amount of oxidant utilization is crucial when necessary to effect better removal efficiency of target organic pollutant molecules during water or wastewater treatment.

5.3.2.5 Effect of energy and intensity of radiation source

The energy and intensity of radiation employed also plays a significant role on the degradation performance of catalysts toward organic pollutants. The energy of electromagnetic radiation is related to its frequency, and that higher the frequency, higher is the energy of the radiation, whereas the intensity of the radiation refers to the number of photons that pass per unit area per unit time and higher the intensity, the brighter it gets. Both the energy and intensity of the radiation may affect the degree of degradation of pollutants. The higher the intensity, the higher the likelihood of photogeneration of e^-/h^+ pairs and reactive radicals and rate of degradation. The rate of reaction generally increases with an increase in intensity up to a certain threshold value. However, after that value, the rate becomes independent of the light intensity because the number of activation sites remains the same irrespective of further increase in intensity. In contrast, at lower light intensity, e^-/h^+ pair separation competes with recombination and decreases the number of reactive radicals and hence the removal efficiency [14].

5.3.2.6 Effect of the presence of interfering substances

During photocatalytic degradation study of a catalyst toward a specific pollutant, the presence of other interfering substances significantly affects its removal efficiency if the catalyst is used in slurry form or in fixed-bed form. As typical industrial wastewater is a cocktail of chemicals, studying the effect of interfering substances, which can be neutral or ionic substances, on the photocatalytic activity of a catalyst in the degradation of target pollutant is highly essential. The interfering substances which influence the rate of photodegradation of pollutant molecules could also come from the intermediate products formed in situ during the photodegradation process which could foul the employed catalyst and hence affect degradation efficiency [40]. The most common inorganic ions available in wastewater include Cl^-, NO_3^-, PO_4^{3-}, SO_4^{2-}, CO_3^{2-}, and HCO_3^-. In the dying process in textile industry, NaCl is added to facilitate diffusion and its adsorption onto fabrics, whereas CO_3^{2-} and HCO_3^- are used to enhance the fixation ability of the dye molecules onto the fibers of the fabrics [41]. These ions acting individually may fasten, retard, or have no or little effect on the rate of photocatalytic degradation of organic pollutants in wastewater. On the one hand, they can retard the rate of degradation by competing for catalyst surface adsorption with the contaminant molecules and poisoning or coagulating it or serving as radical scavengers, thereby decreasing the concentration of active radical species and rate of degradation. On the other hand, they can also fasten the rate of reaction via forming additional reactive radicals in the reaction medium. Hedayat et al. [42] studied the effect of Cl^-, SO_4^{2-}, and CO_3^{2-} ions on the photocatalytic degradation of 2-4-dinitrophenol using $Fe_3O_4/SiO_2/TiO_2$/rGO catalyst both individually and in combined form. They observed that the removal efficiency was reduced in the order $(Cl^- + SO_4^{2-} + CO_3^{2-}) > SO_4^{2-} > Cl^- > CO_3^{2-}$. According to the authors, the reduced catalytic efficiency of the catalyst toward the target pollutant in the presence of these interferents is attributed to the surface deactivation by their adsorption onto the catalyst surface as well as serving as hydroxyl radical and hole scavengers. In an earlier study, Neppolian et al. [43] reported that the photodegradation of Reactive Blue 4 using TiO_2 catalyst decreased consistently up on addition of 0.25–2 g/L of Cl^- from NaCl and ascribed this to the blockage of surface-active sites and hole scavenging effects of the ions (eq. (5.15)). However, it was observed that the presence of interfering ions such as SO_4^{2-}, NO_3^-, Cl^-, Mg^{2+}, and Ca^{2+} on the degradation of methyl orange using Ag/ZnO catalyst even at high concentration did have little impact on the removal efficiency [44].

In addition to inorganic ions, the presence of other organic pollutant molecules than the target organic molecules could hamper the degradation process by either competing for adsorption onto the active surface sites of the catalyst with the target organic molecules or by their co-degradation. For example, during the photodegradation study of Basic Blue 41 dye using TiO_2/ZnO catalyst, it was reported that the presence of the two most common additives in the dyeing process, surfactant 1227 (dodecyl dimethyl benzyl ammonium chloride) and antistatic agent SN (octadecyl dimethyl hydroxyethyl quaternary ammonium nitrate), hindered the degradation of the dye [45]. This was explained to be due to the competition for adsorption of these interferents onto the catalyst surface as well as their simultaneous degradation along with the target dye, thereby lowering the efficiency of its degradation. Similarly, the addition of anionic surfactant sodium dodecyl sulfate and cationic surfactant cetyltrimethylammonium bromide decreased the photodegradation of acridine orange dye using ZnO catalyst and that this was attributed to their preferential adsorption onto the catalyst surface [31]. In another study, a significant reduction in the degradation of methyl red using ZnO in the presence of malic acid, maleic acid, and tartaric acid was observed while humic acid, oxalic acid, and citric acid added separately exhibited no reduction in the degradation of the dye [46].

5.4 Conclusion

Clean water is increasingly in need these days as naturally occurring water is continuously contaminated mainly with anthropogenic activities, owing to ever-rising expansion of urbanization and industrialization. Wastewater is water that has been used and is therefore inappropriate for many applications and needs treatment. It could contain microorganisms as well as a mixture of different chemicals which could be organic, inorganic, ionic, or nonionic substances. Among these, organic pollutants are most common and often highly varied and complex in composition. One possible classification of these pollutants in wastewater is based on their source of origin, biodegradability, and toxicity. Among organic pollutants, POPs have received a lot of attention. These refer to a broad class of chemical compounds, including DDT, HCB, Mirex, lindane, PCBs, polychlorinated naphthalenes, atrazine, dioxins, and bisphenol A, which are hazardous to animals. What makes such pollutants even worse is their physical and chemical properties that enable them to bioaccumulate in lipid tissues and biomagnify across the food chain, resist chemical, photochemical, and biological degradation, and migrate long-range distances by wind and water currents. POPs, even at low concentration levels, are the causes of many health problems such as increased cancer risk, reproductive dysfunction, genotoxicity, endocrine disruption, increased birth defects, alteration of the immune system, and neurobehavioral impairment.

Wastewater could be treated using physical, chemical, and biological methods or hybrid of these methods. Heterogeneous photocatalytic degradation of organic pollutants is a promising technology which involves the acceleration of the rate of chemical reactions in the presence of semiconductor catalysts. It is one class of AOPs which proceeds by the action of oxidizing species mainly the $HO^{\cdot}$ radical formed either directly through the oxidation of adsorbed H_2O or HO^- species with photogenerated holes in the VB of the semiconductor or directly by the injection of elections from the LUMO of the pollutant into the CB of the semiconductor forming superoxide radical ion and finally $HO^{\cdot}$. For e^-/h^+ pairs to be formed in the semiconductor, radiation of energy greater than or equal to the band gap energy of the semiconductor should irradiate it. Then, electrons get promoted from the filled VB to the empty CB, leaving holes behind. Next, the holes produced would oxidize adsorbed H_2O into $HO^{\cdot}$ radicals while the photoexcited electrons in the CB reduce oxygen into $O_2^{-\cdot}$ radical anion. Once these radicals are created, the reaction progresses by a series of chain reactions which ends in the destruction of organic pollutants in the wastewater into environmentally benign outputs of CO_2 and water.

Heterogeneous photocatalysis, however, has its own pros and cons. It has advantages such as low cost, environmentally friendly, easy to handle, and under appropriate conditions can virtually completely degrade the pollutants into harmless end products. In contrast, its effectiveness only at low concentration of pollutant molecules and its dependence on experimental conditions such as pH, catalyst type and concentration, and intensity and energy of radiation, and the influence of interfering substances make it disadvantageous. As such, practical real wastewater treatment approaches employ a combination of different methods according to the cost incurred and the level of purification desired to be achieved for a specified application.

References

[1] Gunture, A.S., Bhati, A., Khare, P., Tripathi, K.M., Sonkar, S.K. Soluble graphene nanosheets for the sunlight-induced photodegradation of the mixture of dyes and its environmental assessment. Sci Rep. 9, 2019, 2522.
[2] Crini, G., Lichtfouse, E. Advantages and disadvantages of techniques used for wastewater treatment. Environ Chem Lett. 17, 2019, 145–155.
[3] Linden, K.G., Mohseni, M. 2.8 – advanced oxidation processes: applications in drinking water treatment. In: Ahuja, S., ed. Comprehensive Water Quality and Purification, Waltham, Elsevier, 2014, 148–172.
[4] Garrido-Cardenas, J.A., Esteban-García, B., Agüera, A., Sánchez-Pérez, J.A., Manzano-Agugliaro, F. Wastewater treatment by advanced oxidation process and their worldwide research trends. Int J Environ Res Public Health. 17, 2020, 170.
[5] Rajamanickam, D., Shanthi, M. Photocatalytic degradation of an organic pollutant by zinc oxide – Solar process. Arabian J Chem. 9, 2016, S1858–S1868.
[6] Weldegebrieal, G.K. Synthesis method, antibacterial and photocatalytic activity of ZnO nanoparticles for azo dyes in wastewater treatment: A review. Inorg Chem Commun. 120, 2020, 108140.
[7] Jiang, L., Wang, Y., Feng, C. Application of photocatalytic technology in environmental safety. Procedia Eng. 45, 2012, 993–997.
[8] Polkowska, Ż., Kot, A., Wiergowski, M., Wolska, L., Wołowska, K., Namieśnik, J. Organic pollutants in precipitation: Determination of pesticides and polycyclic aromatic hydrocarbons in Gdańsk. Pol Atmos Environ 34, 2000, 1233–1245.
[9] Zheng, C., Zhao, L., Zhou, X., Fu, Z., Li, A. Treatment technologies for organic wastewater. Water Treat 11, 2013, 250–286.
[10] WHO Team. Food safety: Persistent organic pollutants (POPs). In: WHO, 2020. https://www.who.int/news-room/questions-and-answers/item/food-safety-persistent-organic-pollutants-(pops)

[11] Zheng, C., Zhao, L., Zhou, X., Fu, Z., Li, A. Treatment technologies for organic wastewater, water treatment. In: Walid Elshorbagy, Rezaul Chowdhury, eds., Water Treatment. Norderstedt, Germany, IntechOpen, 2016, 249–286.

[12] Luo, Q., Wang, J., Wang, J., Shen, Y., Yan, P., Chen, Y., Zhang, C. Fate and occurrence of pharmaceutically active organic compounds during typical pharmaceutical wastewater treatment. J Chem. 2019, 2019, 2674852.

[13] Ajmal, A., Majeed, I., Malik, R.N., Idriss, H., Nadeem, M.A. Principles and mechanisms of photocatalytic dye degradation on TiO_2 based photocatalysts: A comparative overview. RSC Adv 4, 2014, 37003–37026.

[14] Gnanaprakasam, A., Sivakumar, V.M., Thirumarimurugan, M. Influencing parameters in the photocatalytic degradation of organic effluent via nanometal oxide catalyst: A review. Indian J Mater Sci. 2015, 2015, 601827.

[15] Gopinath, A., Krishna, K. Photocatalytic degradation of a chlorinated organic chemical using activated carbon fiber coupled with semiconductor. Photochem Photobiol. 95, 2019, 1311–1319.

[16] Song, S., Xu, L., He, Z., Ying, H., Chen, J., Xiao, X., Yan, B. Photocatalytic degradation of C.I. Direct Red 23 in aqueous solutions under UV irradiation using $SrTiO_3/CeO_2$ composite as the catalyst. J Hazard Mater. 152, 2008, 1301–1308.

[17] Daneshvar, N., Aber, S., Seyed Dorraji, M.S., Khataee, A.R., Rasoulifard, M.H. Photocatalytic degradation of the insecticide diazinon in the presence of prepared nanocrystalline ZnO powders under irradiation of UV-C light. Sep Purif Technol. 58, 2007, 91–98.

[18] Chong, M.N., Jin, B., Chow, C.W.K., Saint, C. Recent developments in photocatalytic water treatment technology: A review. Water Res. 44, 2010, 2997–3027.

[19] Bajorowicz, B., Kobylański, M., Malankowska, A., Mazierski, P., Nadolna, J., Pieczyńska, A., Zaleska-Medynska, A. Application of metal oxide-based photocatalysis, 2018.

[20] Zhang, F., Wang, X., Liu, H., Liu, C., Wan, Y., Long, Y., Cai, Z. Recent advances and applications of semiconductor photocatalytic technology. Appl Sci. 9, 2019, 2489.

[21] Adly, M.S., El-Dafrawy, S.M., El-Hakam, S.A. Application of nanostructured graphene oxide/titanium dioxide composites for photocatalytic degradation of rhodamine B and acid green 25 dyes. J Mater Res Technol. 8, 2019, 5610–5622.

[22] Raghavan, N., Thangavel, S., Venugopal, G. Enhanced photocatalytic degradation of methylene blue by reduced graphene-oxide/titanium dioxide/zinc oxide ternary nanocomposites. Mater Sci Semicond Process. 30, 2015, 321–329.

[23] Al-Rawashdeh, N.A.F., Allabadi, O., Aljarrah, M.T. Photocatalytic activity of graphene oxide/zinc oxide nanocomposites with embedded metal nanoparticles for the degradation of organic dyes. ACS Omega. 5, 2020, 28046–28055.

[24] Zhang, J., Zhou, P., Liu, J., Yu, J. New understanding of the difference of photocatalytic activity among anatase, rutile and brookite TiO_2. Phys Chem Chem Phys. 16, 2014, 20382–20386.

[25] Jamal, N., Radhakrishnan, A., Raghavan, R., Bhaskaran, B. Efficient photocatalytic degradation of organic dye from aqueous solutions over zinc oxide incorporated nanocellulose under visible light irradiation. Main Group Metal Chem. 43, 2020, 84–91.

[26] Ghasemi, B., Anvaripour, B., Jorfi, S., Jaafarzadeh, N. Enhanced photocatalytic degradation and mineralization of furfural using $UVC/TiO_2/GAC$ composite in aqueous solution. Int J Photoenergy. 2016, 2016, 1–10.

[27] Raliya, R., Avery, C., Chakrabarti, S., Biswas, P. Photocatalytic degradation of methyl orange dye by pristine titanium dioxide, zinc oxide, and graphene oxide nanostructures and their composites under visible light irradiation. Appl Nanosci. 7, 2017, 253–259.

[28] Chen, X., Wu, Z., Liu, D., Gao, Z. Preparation of ZnO photocatalyst for the efficient and rapid photocatalytic degradation of azo dyes. Nanoscale Res Lett. 12, 2017, 143.

[29] Paul, D.R., Sharma, R., Nehra, S.P., Sharma, A. Effect of calcination temperature, pH and catalyst loading on photodegradation efficiency of urea derived graphitic carbon nitride towards methylene blue dye solution. RSC Adv. 9, 2019, 15381–15391.
[30] Gupta, A.K., Pal, A., Sahoo, C. Photocatalytic degradation of a mixture of crystal violet (Basic Violet 3) and methyl red dye in aqueous suspensions using Ag^+ doped TiO_2. Dyes Pigm 69, 2006, 224–232.
[31] Pare, B., Jonnalagadda, S., Tomar, H., Singh, P., Bhagwat, V. ZnO assisted photocatalytic degradation of acridine orange in aqueous solution using visible irradiation. Desalination. 232, 2008, 80–90.
[32] Mousset, E., Oturan, N., Oturan, M., Mousset, E., Oturan, N., Oturan, M. An unprecedented route of OH radical reactivity: Ipso-substitution with perhalogenocarbon compounds. Appl Catal B: Environ. 226, 2018, 135–146.
[33] Pirilä, M., Saouabe, M., Ojala, S., Rathnayake, B., Drault, F., Valtanen, A., Huuhtanen, M., Brahmi, R., Keiski, R.L. Photocatalytic degradation of organic pollutants in wastewater. Top Catal 58, 2015, 1085–1099.
[34] Guerra-Rodríguez, S., Rodríguez, E., Singh, D.N., Rodríguez-Chueca, J. Assessment of sulfate radical-based advanced oxidation processes for water and wastewater treatment: A review. Water. 10, 2018, 1828.
[35] Aber, S., Mehrizade, H., Khataee, A.R. Preparation of ZnS nanocrystal and investigation of its photocatalytic activity in removal of CI acid blue 9 from contaminated water. Desalin Water Treat. 28, 2011, 92–96.
[36] Yao, J., Wang, C. Decolorization of methylene blue with sol via UV irradiation photocatalytic degradation. Int J Photoenergy. 2010, 2010, 643182.
[37] Weldegebrieal, G.K. Photocatalytic and antibacterial activityof CuO nanoparticles biosynthesized using Verbascum thapsus leaves extract. Opt. 204, 2020, 164230.
[38] Abdelsamad, A.M., Gad-Allah, T.A., Mahmoud, F.A., Badawy, M.I. Enhanced photocatalytic degradation of textile wastewater using Ag/ZnO thin films. J Water Process Eng. 25, 2018, 88–95.
[39] Vučić, M.D.R., Mitrović, J.Z., Kostić, M.M., Velinov, N.D., Najdanović, S.M., Bojić, D.V., Bojić, A.L. Heterogeneous photocatalytic degradation of anthraquinone dye Reactive Blue 19: Optimization, comparison between processes and identification of intermediate products. Water SA. 46, 2020, 291–299.
[40] Saggioro, E.M., Oliveira, A.S., Moreira, J.C. Heterogeneous photocatalysis remediation of wastewater polluted by indigoid dyes. Text Wastewater Treat. 2016, 93, 93–114.
[41] Kabir, S.M.M., Koh, J. Dyeing chemicals. In: Ashis Kumar Samanta, Nasser Awwad, Hamed Majdooa Algarni eds. Chemistry and technology of natural and synthetic dyes and pigments. Norderstedt, Germany, IntechOpen, 2018.
[42] MirzaHedayat, B., Noorisepehr, M., Dehghanifard, E., Esrafili, A., Norozi, R. Evaluation of photocatalytic degradation of 2,4-Dinitrophenol from synthetic wastewater using Fe_3O_4@SiO_2@TiO_2/rGO magnetic nanoparticles. J Mol Liq. 264, 2018, 571–578.
[43] Neppolian, B., Choi, H., Sakthivel, S., Arabindoo, B., Murugesan, V. Solar light induced and TiO_2 assisted degradation of textile dye reactive blue 4. Chemosphere. 46, 2002, 1173–1181.
[44] Chamjangali, M.A., Boroumand, S. Synthesis of flower-like Ag-ZnO nanostructure and its application in the photodegradation of methyl orange. J Braz Chem Soc. 24, 2013, 1329–1338.
[45] Jiang, Y., Sun, Y., Liu, H., Zhu, F., Yin, H. Solar photocatalytic decolorization of CI basic blue 41 in an aqueous suspension of TiO_2–ZnO. Dyes Pigm 78, 2008, 77–83.
[46] Singh, N.K., Saha, S., Pal, A. Solar light-induced photocatalytic degradation of methyl red in an aqueoussuspension of commercial ZnO: A green approach. Desalin Water Treat. 53, 2015, 501–514.

Sarangi N. V. and Rajkumar R.

6 Biodegradation of organic pollutants by microbial process

Abstract: Expeditious growth in population and industrial revolution has resulted in direct and indirect contamination of air, water, and soil with toxic chemicals. Organic pollutants are the toxic chemicals of worldwide concern as a consequence of their durability, biomagnification, and bioaccumulation ability in the ecosystem. With the exception of physical and chemical decontamination treatment of these pollutants, biological approaches are cost-effective and environmental-friendly. Diverseness of microorganisms in our environment provides the possibility to transfer toxic compounds to less toxic by-products. Many of the microorganisms are capable of biodegradation and biotransformation of organic pollutants from the environment. Therefore, microorganisms can be used in bioremediation of toxic substances. This chapter referenced and briefly described about the various microorganisms and their capability to bioremediation and the major factors responsible for microbial degradation.

Keywords: biodegradation, bioremediation, organic pollutants, microorganisms, factors

6.1 Introduction

Increased human population and their needs are the major reason for the industrialization during the 1900s. This leads to the growth of various industries of plastic, petroleum, chemicals, herbicides, pesticides, and so on. Effluents discharged from these industries without any treatment caused the introduction of various toxic chemicals into the air, water, and soil. The extensive use of organic pollutants (OPs) during the past decades led to deep-rooted environmental problems.

Persistent organic pollutants (POPs) are noxious materials which cause inauspicious effects on the health of environment and human being. POPs are halogenated organic compounds, which include hexachlorobenzene (HCB), polychlorinated

Acknowledgment: This study was financially supported by Rashtriya Uchcha Shiksha Abhiyan (RUSA – phase 2.0) under grant number BU/RUSA/BEICH/2019/299-31, Bharathiar University, Tamil Nadu.

Sarangi N. V., Department of Environmental Sciences, Bharathiar University, Coimbatore 641 046, Tamil Nadu
Rajkumar R., Department of Environmental Sciences, Bharathiar University, Coimbatore 641 046, Tamil Nadu, e-mail: micro_rajkumar@yahoo.co.in

https://doi.org/10.1515/9783110727227-006

biphenyls (PCBs), dichlorodiphenyltrichloroethane (DDT), dioxins, and polychlorinated dibenzofurans (PCDF). Industrialization, direct use of pesticides, insecticides, herbicides, and atmospheric transportation of persistent pollutants are the major reason for global environmental pollution. Each pollutant introduced into the environment will undergo a life cycle based on its characteristics, persistence, biodegradability, medium, and so on. Degradation of introduced toxic substances from the environment is an important area of concern. Plants and microorganisms take part in the degradation of harmful chemical substances.

Bioremediation is better than the usual process of degradation or transformation of pollutants into less toxic substances by microorganisms. There are various studies about the potential of biodegradation by bacteria, fungus, algae, and some plants. The bioremediation done by plants is called phytoremediation. It may have some limitations like time consumption. So the bioremediation can be modified with the help of certain microorganisms in which the microbial enzyme will degrade the particular pollutants based on the medium, microbial community, and toxicity to pollutants.

6.2 Major pollutants in the environment and its sources

Pollution is expounded as the unpleasant replace in chemical, physical, and biological characteristics of water, land, and air. Environmental pollution is the introduction of harmful pollutants toward the environment. Pollutants are the impure elements, molecules, and particles which pollute something, especially environment, and affect the well-being of human and animals when exposed to these substances. Contaminants can go into the habitats naturally or through anthropogenic activities. Natural sources of pollutants such as volcanic eruptions and hazardous pollutants with diverse chemical structures are originating due to anthropogenic activities. A huge amount of organic-based chemicals are used by various industries like paper and pulp, textile, petroleum, pesticide, and paint. Final effluents from such industries comprise enormous quantities of those chemicals that are harmful to the environment.

Pollutants can be characterized as primary and secondary pollutants. Contaminants that directly enter into the nature are called as primary pollutants and secondary pollutants are the substances formed from primary pollutants and other external factors. Oxides of nitrogen, sulfur, volatile organic compounds, particulate matter, OPs, and heavy metals are some of the major pollutants entering into the environment in various ways. There are five types of environmental pollution such as air, water, soil, light, and noise pollution. Air pollution may be caused by primary sources such as smoke, ash, fumes, and dust. Nitric oxide and sulfur dioxide also come under the category of primary sources. Secondary pollutants are smog, ozone, sulfur trioxide, and nitrogen dioxide.

Water pollution is the decaying of quality of water bodies and occurs when pollutants are directly or indirectly sent out to the water body without any processing. If the pollutant source is single and identifiable, it is a point source and includes nitrogen, phosphorus, POPs, and hydrocarbons. Nonpoint sources of pollution are not originating from a single discrete source including urban run-off and agricultural run-off. Soil pollution is the adulteration of soil with hazardous chemical substances that affect the quality of soil and life of organisms living on it. Soil pollution occurs with two types of pollutants: agricultural pollutants and chemical pollutants. Agricultural soil pollution is caused by the use of harmful pesticides, insecticides, and fungicides. Chemical soil pollutants are the toxic substances that enter into the environment from various industries without any treatment.

Considerable OPs such as chlorophenols, PAHs, biopolymers, heavy metals, dyes, PCBs, chloroanilines, and dioxins are generated by industries, agricultural practices, and military activities. For industrial purposes, large amount of hazardous chemical substances are producing yearly and are entering into the environment directly or indirectly. Some of the compounds are pesticides, fuels, polycyclic aromatic hydrocarbons (PAHs), PCBs, and chlorophenols. Most dangerous properties of these substances are resistance to degradation, toxicity, persistence, and accumulation in the body. Synthetic chemicals with complex chemical structures are highly resistant to biodegradation. For that reason, hazardous wastes and chemicals in the environment have become one of the substantial threats of the worldwide modern society.

6.3 Organic pollutants and its type

During the 1800s and 1900s, the Earth's population had grown to 6 billion. Industrialization was the only way to fulfill the needs of humans. As a result, the world looks on at an abrupt increase of various chemical-based products which are useful to us. The combined effect of those chemicals made dreadful impacts on the environment. So the residence of organic contaminants in the environment is due to the continuous industrial activity, farming, and other human activities.

OPs are the substances which live long in the environment. These are organic compounds that contain carbon–hydrogen bonding, which are resistant to environmental degradation. It can enter into the environment either unintentionally through combustion process or intentionally by the industrial chemicals [1]. The major properties of OPs are persistence to degradation, bioaccumulation in the fat tissues of organisms, high toxicity, and the ability to move through any kind of medium such as air, water, and soil. Activities like transportation, combustion, construction, and industrial activities are the major sources of OPs in urban areas. Long-range transportation of OPs causes the remote areas to get affected. OPs mainly originate from nonpoint sources in the urban areas, and the pollutants will enter into the flowing storm water and finally reach the remote areas also. This will affect humans and other organisms when they come in contact with contaminated water and air. For aquatic ecosystem, urban run-off is the main reason for organic pollution [2].

POPs are the primary and by-products of industrial processes. They are present in almost all areas, including food, soil, water, and air. Humans and other organisms come in contact with POPs through these media and cause harmful effects such as nervous system destruction, immune system damage, developmental and reproductive disorders, and cancers. There are numerous POPs in the world due to their anthropogenic action and 12 of them are most persistent and bioaccumulative chemicals, including pesticides, industrial chemicals, and combustion of some chemicals. Aldrin, chlordane, DDT, dieldrin, endrin, heptachlor, mirex, and toxaphene are included in the category of pesticides. PCBs and HCB are the industrial chemicals having persistent nature. Dioxins and furans are the toxic substances released to the environment during the combustion of chlorine and chlorine-containing materials.

6.3.1 Types of organic pollutants

OPs are broadly classified into three classes [3]:
- Hydrocarbons
- Oxygen, nitrogen, and phosphorus compounds
- Organometallic compounds

6.3.1.1 Hydrocarbons

Hydrocarbons are the combination of hydrogen and carbon atoms. It is present in the nature in various forms. Some of the compounds that contain hydrocarbons are DDT, dioxins, and PAHs. C–H, C–C, C–Cl, C=C, and C=C (aromatic) are the chemical bonds present in the group of hydrocarbons, where they contain less types of chemical bonds. Chemical bonds in hydrocarbons have the peculiarity such as stability and comparatively less polarity. The character of low polarity gives the lipophilic, persistent, and less soluble nature to hydrocarbons.

This class of chemicals includes deadly organic compounds such as 2, 3, 7, 8-tetrachlorodibenzo (1, 4) dioxin (TCDD). Hydrocarbons occur naturally in crude oils like petroleum and natural gas and are the widely used organic compounds in the world. Hydrocarbons are generally colorless and hydrophobic with weak odors. Because of their complex chemical structures, they are classified as saturated hydrocarbons (alkanes) and simplest hydrocarbon species consists of single bonds and saturated with hydrogen (C_nH_{2n+2}). Saturated hydrocarbons are the basis of petroleum fuel. Unsaturated hydrocarbons have one or more double or triple bonds between carbon atoms. Double-bonded compounds are called alkenes (C_nH_{2n}) and those compounds containing triple bonds are called alkyne (C_nH_{2n-2}). Aromatic hydrocarbons or arenes are the compounds having at least one aromatic ring.

These are the major components of volatile organic compounds, fossil fuels, hydrocarbon lighters, grills, and fuel. Most of the hydrocarbons are introduced into the environment due to the processes like industrial processing of crude oil, ethanol, and so on. Hydrocarbons are compounds having low risk unless they come in contact with solar radiation or nitrogen oxides and undergo chemical reaction. These are the major compounds of crude oil, natural gas, and most of the pesticides. The chemical reaction to these compounds will result in various hazardous pollutants and lead to environmental pollution.

6.3.1.2 Oxygen, nitrogen, and phosphorus compounds

The group containing these compounds has high water solubility and low fat solubility so that the persistence of these compounds is low. Phosphorus and nitrogen are the primary nutrients that in excessive amounts pollute our water bodies.

6.3.1.3 Organometallic compounds

These are the compounds with least one metal-to-carbon bond. The carbon attached to this is a part of the organic group. The complex chemical structure of organometallic compounds gives various physical and chemical properties for each substance. Most of these compounds are in soil state, especially for the group with ring-shaped or aromatic hydrocarbons. Some of them are liquids and gases. Carbon–metal bonds determine the nature of organometallic compounds. These compounds are persistent in the environment. It can easily concentrate and are highly toxic [4]. When the compound enters into the environment, it will undergo reactions with other chemicals and biota. Once the organometallic compounds enter into the environment it can affect the functioning of ecosystem and organism and also contaminate the food chain. Natural occurrence of organometallic compounds are very less in the environment but anthropogenic activities have greatly increased their input into the environment.

6.3.2 Persistent organic pollutants

POP are the organic compounds which are a global challenge because of its persistence in the environment for a very long time as they resist biological, chemical, and phytolytic degradation. There are many factors that give lethal effect to POPs such as persistency, immutable nature during transportation, bioaccumulation, and toxicity. Normal plants or microbes cannot degrade the POPs present in the environment. It persists in the environment as solid, liquid, or gaseous form. While considering the origin of POPs, the main sources are human-induced. It can reach the environment through various ways such as urban run-off, deposition from the atmosphere, and industrial effluents. In urban areas, agricultural practices with various organochlorine insecticides lead to a very common source for POP deposition in the water bodies, soil, and atmosphere. As compared to nonagricultural areas, agricultural areas are having high organochlorine concentrations [5]. Other insecticides, substances used for vector control, and so on are other sources of POPs such as polychlorinated dibenzodioxins (PCDD) and PCDF, in urban areas.

Open burning of waste material, improper land filling, vegetation fire, and volcanic eruptions are the other major sources of POPs in developing countries. Because of the threat due to POPs, in 2001, an international environmental treaty signed that to eliminate the manufacture and utilization of POPs: Stockholm Convention (SC). The major objective of the global treaty was to protect the environment and humans from the adverse effects of POPs through controlling over the production and elimination of POPs. As per the guidelines of treaty, there are 22 POPs listed in SC. Table 6.1 shows the POPs listed in SC amendment.

SC in 2004 listed 12 deadly POPs in a class known as "dirty dozen." This group includes aldrin, dieldrin, endrin, chlordane, heptachlor, HCB, mirex, toxaphene, DDT, PCBs, PCDDs, and PCDFs. Again in 2009, a group of POPs was added to the category such as chlordecone, lindane, α-HCH, β-HCH, hexabromobiphenyl, tetra-BDE, penta-BDE, hexa-BDE, hepta-BDE, PFOs and its salts, PFOSF, and pentachlorobenzene. During 2011, endosulfan became the 22nd POP. From the 22 POPs listed in SC, 15 are pesticides. The listed substances are in Table 6.1.

Table 6.1: POPs listed in SC amendment (source [6]).

Chemical	Annex	Year of listing decision
Aldrin	A: Elimination	2001
A-hexachlorocyclohexane	A: Elimination	2009
B-hexachlorocyclohexane	A: Elimination	2009
Chlordane	A: Elimination	2001
Chlordecone	A: Elimination	2009
Decarbromodiphenyl ether	A: Elimination	2017
DDT	B: Restriction	2001

Table 6.1 (continued)

Chemical	Annex	Year of listing decision
Dicofol	A: Elimination	2019
Dieldrin	A: Elimination	2001
Endosulfan	A: Elimination	2011
Endrin	A: Elimination	2001
Heptachlor	A: Elimination	2001
Hexabromobiphenyl	A: Elimination	2009
Hexabromocyclododecane	A: Elimination	2013
Hexabromodiphenylether	A: Elimination	2009
Hexachlorobenzene	A: Elimination	2001
Hexachlorobutadiene	A: Elimination	2015
Lindane	A: Elimination	2009
Mirex	A: Elimination	2001
Pentachlorobenzene	A: Elimination	2009
Pentachlorophenol	A: Elimination	2015
Perfluorooctanoic acid (PFOA)	A: Elimination	2019
Perfluorooctane sulfonic acid (PFOS)	B: Restriction	2009
Polychlorinated biphenyls (PCBs)	A: Elimination	2001
Polychlorinated dibenzodioxins	A: Elimination	2001
Polychlorinated naphthalenes	A: Elimination	2015
Tetrabromodiphenylether and pentabromodiphenylether	A: Elimination	2009
Chlorinated parafffins	A: Elimination	2017
Toxaphene	A: Elimination	2001

According to the Ministry of Environment, forests, and climate change (MoEF&CC), "POPs are the chemical substances that persist in the environment, bio-accumulate in living organisms, adversely affect human health/environment and have the property of long-range environmental transport." The Union Cabinet meeting held in India on 2020 October sanctioned the ban of seven hazardous POPs such as chlordecone, hexabromobiphenyl, hexabromodiphenylether and heptabromodiphenylether, tetrabromodiphenylether and pentabromodiphenylether, pentachlorobenzene, hexabromocyclododecane, and hexachlorobutadiene. Before the ban in India, these chemicals were widely used as pesticides, fungicides, fire retardants, insecticides, and so on. These chemical exposures will cause cancer, nervous system damages, immune system diseases, and other diseases. Because of the harmful effects of POPs to the humans and environment, legislative discussions were initiated and lead to the notification of the Regulation of Persistent Organic Pollutants Rules, 2018, prohibiting the manufacture, trade, use, import, and export of these seven chemicals.

6.4 Physical, chemical, and biological degradation of organic pollutants

POPs entering into the environment involve various pathways. Degradation is the collapse of xenobiotic compounds through physical, chemical, or biological mechanism. POPs remains as such in the environment for a long period of time and the environment system will influence the overall fate of the chemical compound. POPs will not degrade completely, but some molecular alterations can be done to the compounds to form simple and less toxic compounds through various processes. Major processes are given in the following sections:

6.4.1 Physical degradation of POPs

Physical degradation of POPs has been introduced for long time and is performed by using various processes like Ag-modified Zn_2GeO_4 nanorods, TiO_2/graphene oxide (GO) nanocomposite hydrogels, and biosilica coated with amorphous manganese oxide [3]. Catalytic or photocatalytic oxidation is the green method to treat organic wastes by a catalyst or photocatalyst which are the substances activated by absorbing a photon and are able to accelerate a reaction without being devoured. Semiconducting oxide photocatalysts are used to treat organic contaminants in air and water, and this procedure is the modified oxidation process (AOP). Environmental purification with heterogeneous photocatalysts is developed in the 1970s, with an application to the mineralization of recalcitrant organic compounds followed by decomposition. There are many catalysts such as TiO_2, ZnO, Fe_2O_3, CdS, and ZnS used in the photodegradation or physical degradation of POPs. TiO_2 is a catalyst with wide acceptance because of its cost-effectiveness, less toxicity, and high chemical stability. There are a lot of solids that can be used for photocatalysis. Photocorrosion resistances of metal oxide semiconductors are very less. This peculiarity of metal oxide semiconductors is utilized as photocatalysts.

The photocatalytic task of TiO_2 depends on the surface and structural properties like crystal composition, surface area, porosity, particle size, and bandgap energy [7]. Photocatalytic degradation promises the degradation of toxic components, rather than transformation from one form to another. This process can remove an extensive range of POPs from the environment.

6.4.2 Chemical degradation of POPs

Electrochemical dehalogenation is the method used for the chemical degradation. For example, chlorinated benzenes can electrochemically dehalogenated to form less toxic benzene. Rhodamine B, methylene blue, and benzyl alcohol are also chemically degraded.

6.4.3 Biological degradation of POPs

Treatment of polluted soil, contaminated with OPs, is costly and the production of other toxic substances will also be an issue. Biological degradation of POPs is the depletion of chemical toxicity of toxic/chemical substances on biologically catalyzed reactions in the environment. It is also defined as the change in the chemical nature of toxic substances with biological activity.

Biodegradation of complex chemical compounds is done by the living microbial organisms and they break down the organic substances into smaller particles. Microorganisms comprise fungi, bacteria, yeast, algae, and protozoans (in a small range) are responsible for the biodegradation of POPs. The final product of microbial degradation is carbon dioxide. Bioremediation is the action that involves plants and microorganisms. The active organism can deplete or nullify the toxicity of pollutants through two ways: in situ or ex situ bioremediation. Onsite treatment of contaminated channel is called in situ bioremediation. When the pollutants are moved from the contaminated site for treatment is called ex situ bioremediation. Bioremediation can be performed by using the indigenous organisms or introducing exogenous species. Bioremediation process can be divided into three stages: bioattenuation, biostimulation, and bioaugmentation (see Figure 6.1).

6.4.3.1 Bioattenuation

Bioattenuation or natural attenuation is a process of degradation of pollutants naturally with the help of microorganisms present in that area. This process undergoes the transformation of highly toxic pollutants into less toxic by-products. Microorganisms have a major role in the stimulation of chemical reaction in the process of bioattenuation. Degree of pollution and site setup will determine the requisite time for bioattenuation. It is a cost-effective and natural process of restoration of polluted area with the help of appropriate microbes.

6.4.3.2 Biostimulation

It is the process of decontaminating polluted soil, and the growth of microbes is facilitated by the modifying environment. Fertilizers such as water-soluble nitrate and phosphate are used for the enhancement of this process. The amount of nutrients present also affect the activity of bacterial community [8].

6.4.3.3 Bioaugmentation

It is the method to improve the efficiency of bioremediation. For this purpose, pre-adapted microbes are used as a catalyst. Modification will be done with toxic pollutants to increase the productivity. Pretreatment with modified microorganisms will be introduced to the contaminated site to modify the biodegradation efficiency and persistence in the soil. But in some cases, the lab performance during testing will be high and on site will be comparatively low.

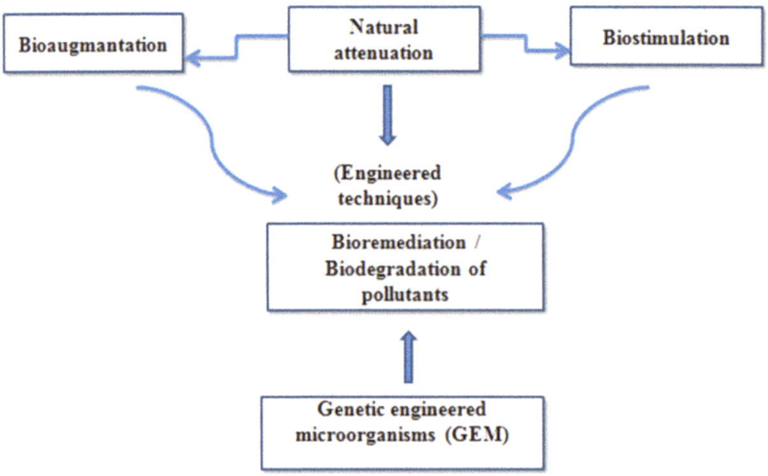

Figure 6.1: Biodegradation of pollutants.

Bioremediation can be done in two ways;

- In situ: This is the method of bioremediation feasible to be done at the contaminated site itself. Site factors such as water, CO_2, and temperature will influence the biological degradation of toxic pollutants. Vital advantages of onsite bioremediation are cost-effectiveness, eco-friendly, and sustainability. Maintenance is not required during this process. Displacement of toxic contaminants during the transportation can be avoided in this method. So the acceptance of this technique is increasing.

- Ex situ: It is the treatment of polluted area by removing the waste material or polluted media to another site. Extreme climatic conditions such as very low temperature, nutrient scarcity or stress conditions to microbes, and the activity of microbes that carry out the bioremediation will be limited. For the enhancement of microbial activity and chemical degradation, adequate conditions are required. Ex situ bioremediation is the way to resolve the flaw of in situ method. In case of contaminated soil, the surface soil will be removed from the site and transported to the area where the bioremediation process happens. The major limitation of this method is the higher cost, requirement of man power, and so on (Table 6.2).

Table 6.2: In situ and ex situ bioremediation [3–8].

Bioremediation	Examples	Benefits	Limitations
In situ	– In situ bioremediation – Biosparging – Bioventing – Bioaugmentation	– Cost-effective method – Natural treatment of soil and water – No need of site for process – Less contaminants to be released	– Time consumption – Monitoring difficulties
Ex situ	– Land farming – Composting – Biopiles	– Cost-efficient – Easy to control – Uniformity of treatment	– Site requirement – Extended treatment time – Risk of handling

Biological processes have more advantages in comparison to physicochemical processes. During the biological processes, there is no formation of harmful intermediate products. Complete destruction of the xenobiotic is possible in these methods. Formation of harmful by-products will be also low while using these methods. On-site bioremediation process will not muddle the natural activities on the site. This can reduce the risk of spreading of pollutants from one site to another. Cost, man power, and threat to human health will be less or negligible during on-site treatment of contaminated area. Natural methods for the restoration will produce less harmful or no harmful products through biotransformation.

6.5 Role of microorganisms in biodegradation

Microorganisms are diffused everywhere in the world. Because of its metabolic activity, they can divide easily in a broad environment. This advantage can be used in the degradation of pollutants. For several decades, the studies regarding the biodegradation of OPs have been going on. In the process of microbial bioremediation, some of the microbial communities are suitable for the conversion of pollutant with high toxicity into the energy required for their growth. Instead of storing the toxic compounds, microbial-assisted biodegradation is the convincing procedure of transformation of highly toxic OPs into less toxic or nontoxic compounds. Microorganisms such as bacteria and fungi can reduce the toxicity of pollutants via deterioration, transformation, or accumulation through their metabolic activity.

Some of the persistent pollutants in the environment are biodegradable;

- Hydrocarbons: These are the organic compounds with hydrogen and carbon in its structure.
- PAHs: It is a kind of hydrophobic organic contaminants. Because of its toxicity, prevalence, and persistence, PAHs are very dangerous in the environment.
- PCBs: Due to the nonflammability and chemical stability, it has been used for a long time, and the amount of PCBs present in the environment produces various endocrine diseases and cancers.
- Pesticides: These are the compounds which are very important in the field of agriculture. Pesticides present in the soil will degrade by soil microorganisms such as fungi and bacteria.
- Radionuclides: These are the atoms that release radiation during radioactive decay. The emitted radiation will be alpha particles (α), beta particles (β), or gamma rays (γ).

6.5.1 Microbial degradation of organic pollutants

Various microorganisms are capable of degrading various pollutants in the environment. Biodegradability is the property of plant, animal, or plant- and animal-derived substances to break down into a final product with the help of microorganisms. The by-product of degradation will be humus or CO_2. Microbial degradation can be defined as the destruction or transformation of xenobiotics present in the environment with the use of oxygen, nutrients, temperature, and light. Sometimes microbes can accumulate the pollutant within their body. Pollutants that can mortify through microbial degradation are hydrocarbons, PCBs, PAHs, radionuclides, pharmaceutical substances, and so on.

Biodegradation is the eco-friendly approach of contaminant removal. The advantages of this method bring more acceptability to these methods. The field of microbiology is progressing in the research of modification of efficient microorganisms for the biodegradation process. Genetically modified microbes are adopted in the global research.

There are two kind of microbial degradation will occur;

i) **Aerobic degradation of pollutants**: It is the process of breaking down of OPs in the presence of oxygen, with the help of microbial community. So it depends on oxidative conditions. Bacteria which degrade pollutants in the presence of oxygen are called as aerobes. Cellular respiration is the source of energy for aerobes. Oxygen is the catalyst in the knockdown of tiny molecules into CO_2 and water. Energy will also be released during this process. It will not produce any kind of unpleasant derivative gases.

ii) **Anaerobic degradation of pollutants**: In the absence of aerobic conditions or oxygen, anaerobic microbes will increase in an area. This condition drives to the anaerobic digestion of OPs. Anaerobic microbes can break down organic materials in the absence of oxygen. As a result of chemical reactions, it can produce biogas. The resultant gas can be used for the power generation, which leads to sustainability. The major threat in the society is the waste generation and lack of proper waste management system. Anaerobic digestion is an eco-friendly approach of waste management that can produce energy for human purposes. This can reduce the overutilization of fossil fuels. Slurry obtained at the final stage can be used as a fertilizer. Iron-reducing *Geobacter metallireducens* is an example of anaerobic organism which performs biodegradation [9].

6.5.2 Microbes and biodegradation potential

6.5.2.1 Bacterial degradation

There are a lot of studies on the biological mortification or OP by using bacteria. Bacteria can biodegrade many hydrocarbons and other OPs aerobically or anaerobically. *Bacillus*, *Corynebacterium*, *Staphylococcus*, *Streptococcus*, *Shigella*, *Alcaligenes*, *Acinetobacter*, *Escherichia*, *Klebsiella*, and *Enterobacter* are some of the most effective bacterial strains with hydrocarbon-degrading capability. Gram-negative *Pseudomonas* bacteria are isolated and studied for the degradation of aromatic hydrocarbons. Various studies regarding the potential of bacterial strains have been done by many researchers and they conclude that the mixed microbial communities have the capacity to degrade the OP than the single organisms. It is due to the necessity of high-potential genetic material to degrade the OPs in the environment.

Aquatic and terrestrial environment containing xenobiotics such as PCBs can degrade into less toxic or nontoxic substances with the help of microorganisms. In case of degradation of PCBs by bacteria, it can be done by both aerobic and anaerobic activities. Bacterial degradation of pollutants includes chemical reactions that provide substances with decreasing toxicity in each stage. Anaerobic degradation of PCB involves dechlorination process by microorganisms. Dechlorinated PCB can be derived to various by-products such as acyl-CoA or carbon dioxide. These intermediate products are the result of BH pathway done by microbes. Lack of BH pathway can convert dechlorinated PCB into minerals by means of fungus or green plants. Reductive dehalogenation is the process used by the anaerobic microorganisms for the degradation of highly chlorinated PCBs. The genera *Pseudomonas*, *Burkholderia*, *Comamonas*, *Sphingomonas*, and so on are some of the bacterial strains that can be used for the degradation of PCBs.

Degradation of PCBs and biphenyls is possible in both aerobic and anaerobic degradation. Excessive chemical stability and chlorination makes the PCB into not suitable for energy source and can be utilized as an electron acceptor in the process of anaerobic respiration, respectively. The first stage of degradation is the dechlorination of PCB that makes it less toxic with less than five chlorines. It is followed by the biphenyl degradation pathway (BP pathway). Through this pathway, biphenyl converts into pyruvate and acyl-CoA which are the intermediate products of TCA cycle.

Heavy metals are another threat to the environment and are substances that resist normal degradation. But bacterial strains can be used in the bioremediation of heavy metals. By the activity, bacteria will convert the metal from its oxidation state to another which can easily bioremediate. Bacteria utilize the heavy metals as the electron acceptors for their anaerobic respiration. Microbial methylation with species like *Alcaligenes* and *Bacillus pumilus* are capable of degrading methyl mercury.

6.5.2.2 Fungi degradation

It is the process of cleaning up of environmental contaminants by the use of fungus (see Figure 6.2). As many of the green plants, fungi also received recognition because of its bioremediation potential, and the efficiency of bioremediation because of enzymes. Morphological characteristics such as hyphae can be used by the fungus to penetrate into the surface of pollutants, where the other microbes lack this property. Microfungi are the eukaryotic, aerobic microbes which work like bacteria, and also metabolize the dissolved organic matter from the environment. It can grow in low moisture content area having low pH which gave the better growth condition for the organic matter degradation. Because of its multienzyme complexes, fungi can degrade the polymeric complexes. The symbiotic association in the middle of fungi and roots of higher plants are called mycorrhiza, and the bioremediation with this is known as mycorrhiza remediation. Yeast also belongs to fungi, and several yeasts can grow with the use of aromatic compounds as growth substrates. *Candida lipolytica*, *C. tropicalis*, *Rhodotorula rubra*, and so on are some of the studied microfungi because of its potential to degrade diesel soil.

Growth of fungi in a contaminated area with toxic metals depends on various factors like structural properties of chemicals, physiological, morphological, and genomic modifications of fungi, persistence of metals, and bioavailability. For example, PAHs are the highly toxic pollutants that can persist in the environment for a long time. This can be generated as a result of industrialization and other human activities. It can degrade biologically by the use of fungal strains such as *Daedalea*, *Aspergillus*, *Penicillium*, *Paecilomyces*, *Coriolus*, *Pycnoporus*, *Pleurotus*, and *Fomitopsis*, and various studies show that fungal strains have the ability to degrade threatening

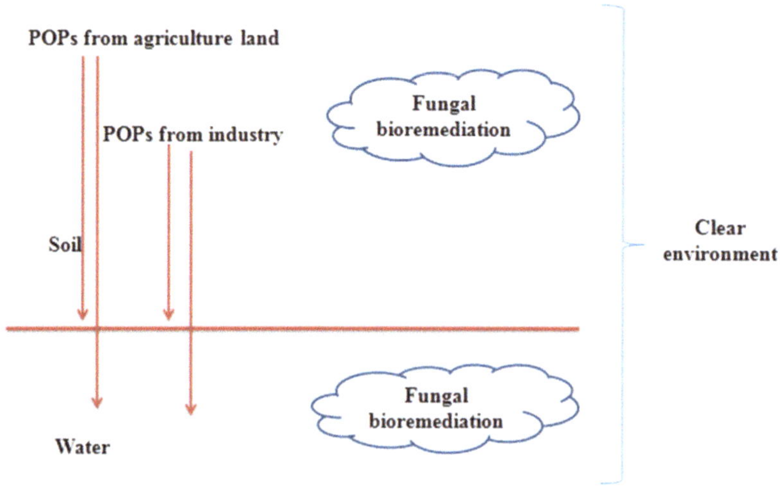

Figure 6.2: Fungal degradation of POPs.

xenobiotics in the soil and water medium. Previous studies in this topic explain the use of genera of white rot fungi (*Phanerochaete*, *Irpex*, *Polyporus*, *Stereum*, *Lentinus*, *Bjerkandera*, *Irpex*, *Pleurotus*, and *Phlebia*) in the field of bioremediation. *Coriolus versicolor*, *Trichoderma* sp., *Aspergillus niger*, and *Fusarium* sp. are very effective in the bioremediation of pyrene.

6.5.2.3 Algae degradation

Algae and protozoa are not used for degradation of pollutants as other microorganisms. Some studies tell that algal species like red algae, some green algae, and cyanobacteria are capable of degrading the crude oil and hydrocarbons and some uptake of PAHs.

Many species can survive the heavy metal concentration surrounding them. Marine algae such as *Chlorella*, *Anabaena*, and *Stigeoclonium* are used to remove the heavy metals. Algal species will absorb the heavy metal, and metal chelation in unicellular microalgae can be done. Cell wall components will absorb the heavy metal.

6.6 Advantages of microbial degradation

Pollution of natural resources is the complication faced by the modern world. Industrialization and exploitation of fossil fuels are some of the major reasons. The accumulation of pollutants in the terrestrial and aquatic ecosystem disturbs the well-being of humans, animals, and plants. Minamata disease, the Love Canal chemical waste dump, the Baia Mare cyanide spill, and so on are some of the examples happened in the past. The only way to reduce the negative impact of toxic pollutants is the bioremediation. Microorganisms are isolated by many researchers for the bioremediation studies. Strains of *Flavobacterium* sp. and *Pseudomonas* sp. from petroleum hydrocarbon contaminated soil sample were found to degrade hydrocarbon. Multiple strains are more effective than the single strain for the process.

6.6.1 Enzymatic degradation of organic pollutant

Enzymes are biological catalysts rendering favorable conditions to enhance the transformation of substrates to final products at agreeable conditions. Enzymes tend to decrease the activation energy of the process. These activities of enzymes are used in bioremediation. Bioremediation is the implementation of biological processes that unclutter the hazardous chemicals present in the environment. Each microorganism contains different kinds of enzymes and all of them have each property to degrade the pollutant.

6.6.2 Microbial enzymes which help in biodegradation

6.6.2.1 Microbial oxidoreductases

Oxidoreductases are the enzymes that help in the process of oxidative coupling used by the microorganisms in the detoxification. These enzymes can mediate the process of imparting energy for the cleavage of chemical bonds. The transformation of reduced organic substances into various chemical compounds through electron transfer can also be aided by these enzymes. As a result, the chemical reaction can deliver less toxic substances. It can detoxify various harmful substances like phenolic compounds by the process of humification. Bacteria can assist the electron transfer between organic compounds (electron donor) and radioactive metals (electron acceptor). In some cases, bacteria can run the process with the help of an intermediate donor of electron [10].

Oxygenase participates in the reaction that oxidizes or reduces the substrate through the movement of oxygen from the molecular oxygen species (O_2). Monooxygenases and dioxygenases are two types in this group, which are classified on the basis of oxygen utilizing for oxygenation. The degradation of halogenated organic compounds is achieved by specific oxygenases.

6.6.2.2 Microbial laccases

These are the enzymes that catalyze the oxidation reaction. It is one of the biocatalysts with a family of multicopper oxidases having a wide range of applications. Plants, fungi, insects, and bacteria can produce laccases. It is resistant to organic solvents. It is a well-known substance that can be used for the cleanup of lignin, reduced phenolics, and aromatic substrates biologically. The final product is the conversion of molecular oxygen into water. Laccases can be produced by various microorganisms. This can make use of catalyst of oxidation reaction of many OPs such as *ortho*- and *para*-diphenols, polyphenols, polyamines and aryl diamines, and many other inorganic compounds. Decarboxylation and demethylation are also the reactions mediated by laccase.

The major factor that affects the potential of laccase is pH. Halides, azide, cyanide, and hydroxide compounds can hinder the activity of laccase. Increased level of nitrogen can improve the production of laccase in fungi.

6.6.2.3 Microbial peroxidases

This is the redox reaction catalyzed by enzymes. Peroxidases are the donors of hydrogen peroxide oxidoreductases during the oxidation of lignin and other phenolic compounds. The reaction can be accelerated by the use of a catalyst or a mediator such as hydrogen peroxide. Immune system or hormone regulation can be done by these enzymes in vertebrates. In plants, peroxidase can guide the actions such as production of lignin and suberin. The processes of cross-linking of cell wall components in plants are also mediated by this. Peroxidases can also support the auxin metabolism. Peroxidases may be of two types: heme and nonheme peroxidases.

6.6.2.4 Microbial hydrolytic enzymes

Bacteria play a major character in the decomposition of organic contaminants. Extracellular enzymes can actively participate in the process of utilization of organic polymers. Hydrolytic enzymes are capable of disrupting the tough chemical bonds in xenobiotics. As a result of this process, toxicity of the substances will reduce. Pollution by oil spill, organophosphate, carbamate, and so on can mitigate by these enzymes in various environmental conditions. Condensations and alcoholysis are processes that can be mediated by hydrolases enzymes. This enzyme class is readily available, and biomass degradation can be done with enzymes such as hemicellulase, cellulase, and glycosidase. Amylases, proteases, lipases, DNases, pullulanases, and xylanases have high industrial demand like biomedical sciences and chemical industries [11].

6.6.2.5 Microbial lipases

Microorganisms are capable of production of lipase other than plants and animals. Lipids can be degraded by using this enzyme. Scientific research proved the ability of lipase to actively participate in the process of hydrocarbon degradation from the contaminated soil. Lipase of microbial origin, especially from bacteria, apart from plants and animals is more active and efficient for industrial applications. Alcoholysis, aminolysis, hydrolysis, and so on are the catalytic reactions by lipases.

It can also hydrolyze the triacylglycerols which is the major components of oil. This form can be hydrolyzed into glycerol, diacylglycerol, monoacylglycerol, and fatty acids. It can be used for many industries. Glycerol and fatty acids are widely used as raw materials. Monoacylglycerol is the substance with applications in various industries like food and medicine.

Prasad and Manjunath [12] concluded that lipase efficiency can be used as the measure to find the hydrocarbon toxicity in soil. Not only the bioremediation potential but also the pharmaceutical potential of lipase is very high.

6.6.2.6 Microbial proteases

Microbial proteases are used for the degradation of proteinaceous substance. This can hydrolyze the organic substances introduced into the nature through the death of animals, and also as by-products of some industries like poultry, fishery, and leather. These are the substances that hydrolyze peptide bonds. Other than biodegradation, it can be used in the field of detergent, food, and leather industries. Based on the catalytic activity of peptide chains in compounds, microbial proteases can be divided into endopeptidases and exopeptidases [13–15].

Genetically engineered microorganisms are the result of researches in biotechnology and these are the organisms that have the capacity to degrade various toxicants. Application of genetic engineering for the process of heavy metal removal has a considerable interest. *Alcaligenes eutrophus* is a modified organism that is used for the chromium removal.

6.6.2.7 Advantages of microbial degradation

- Natural process
- Microbes can degrade the pollutants and increase their number while the pollutant level increases
- It can be carried out on-site without any disturbance to the ecosystem and reduces the cost of remediation and transportation [16, 17]
- Biotransformation or complete degradation of pollutants is possible

- Used to clean up hazardous wastes and important method for the restoration of oil-polluted sites
- Fertilizers can be used as a nutrient source for the enhancement of fast growth in microbes, and no need for hazardous chemicals for its growth
- It is a simple process with less manpower and is cost-effective
- Sustainable approach
- Xenobiotics are degraded to less toxic forms
- Implementation is not difficult
- Efficient in remediating natural ecosystem from a contaminated condition and act as an environment-friendly option

There are some disadvantages for microbial bioremediation:
- It is limited to the substances that are biodegradable [18]
- Site factors are important for the success
- Time-consuming process

6.7 Factors affecting microbial degradation

Microbial degradation of OPs present in the environment can be affected by various factors such as microbial population, toxicity of pollutant and its end product, nutrients, oxygen, salinity, and temperature. All these factors will affect the performance of microbes in the degradation process [19–24] (Figure 6.3).

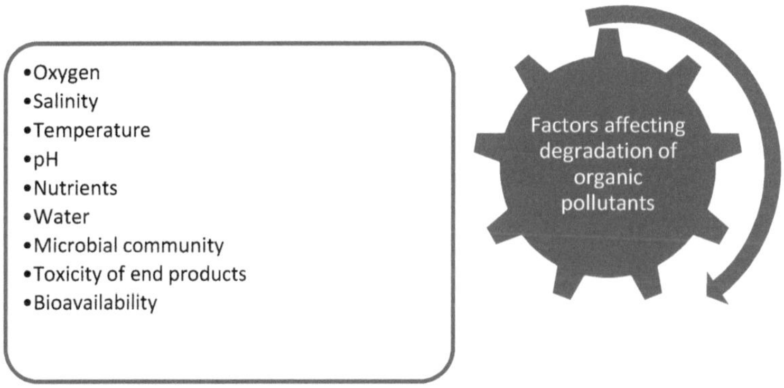

Figure 6.3: Factors affecting the degradation of organic pollutants.

- **Temperature**

For microbial degradation, temperature plays an important role. For the in situ bioremediation, most of the sites are not under the ideal temperature. The solubility and bioavailability of OPs like PAHs depend on the temperature of its surrounding medium. Oxygen solubility in soil is also proportional to temperature. Physical and chemical nature of pollutants and metabolic activity of organisms are also important factors. Many studies show that the microbial population and enzyme activity of microbes increase when reach the mesophilic and thermophilic temperature conditions. It can improve the hydrocarbon degradation in the soil. Studies suggest that the optimum temperature region between 0 and 40 °C supports the best growth of microorganisms.

- **Soil characteristics**

Soil is the base for biodegradation process of contaminated soil. Soil characteristics such as texture, pH, water-holding capacity, oxygen content, and soil permeability are the factors that determine the growth of plants and microbes in the soil surface. These characteristics are also important for the complete degradation of hydrocarbons in the contaminated soil. Each soil character is interconnected. Permeability determines the water content, oxygen, and nutrient availability. The soil having all the factors in optimum level can support the microbial growth.

- **pH**

Like soil conditions, pH is another factor that affects the microbial degradation. Leaching of OP is responsible for the increase in pH of the soil. This condition makes the platform of microbial growth inimical. But the leaching of coal creates an acidic medium in the soil. The optimum level of pH for the microbial biodegradation is around 6.5–7.5.

- **Oxygen**

Microbial degradation may be aerobic or anaerobic. The presence or absence of oxygen can lead to microbial degradation. Aerobic biodegradation depends completely on the availability of oxygen. About 2% oxygen is required in gaseous form for the easy degradation process. In case of PAH degradation, oxygen is necessary for the degradation of its aromatic rings in the initial stages. So it is important to keep the soil or water medium to be aerated for the aerobic metabolism of organisms. For this purpose, we can use various methods such as soil tilling or mixing in compost areas or contaminated sites. In the degradation pathway, a microbe will use the dissolved molecular oxygen present in the soil. In situ bioremediation mainly depends on the oxygen and aerobic metabolism.

- **Nutrient availability**

Every process mediated by living things demand energy. Microbes need energy or nutrients for the biodegradation. Nitrogen, phosphate, and potassium (N, P, and K) are the essential nutrient components that help the cellular metabolism. The level of organic carbons in the polluted site will vary with the nature of pollutant. Along with the carbon source, microbes need extra nutrient supplement for their growth.

For the enhancement of in situ bioremediation potential, it is recommendable to provide nutrients like nitrogen and phosphates. Enhanced metabolism in microbes can easily degrade the organic compounds like PAHs and petroleum compounds. Hydrocarbon degradation requires carbon along with basic nutrients. The capacity of microorganism to absorb the nutrients from the available form determines its growth and bioremediation potential.

– **Microbial population**

Selection or availability of appropriate microbes for the biodegradation is another factor. Some of the organisms can degrade any kind of toxicants. The microbial community, number of suitable organisms, and its catabolic activity will determine the rate of degradation. Microorganisms are found everywhere. Bacteria, fungi, algae, and protozoa can isolate from soil itself for the decontamination of soil. Biotransformation of contaminant depends on the chemical structure of pollutant, bioavailability to organisms, and organism's population. Favorable conditions make a rapid growth of microbes.

– **Bioavailability**

It is the amount of materials that can access the microbes from its surroundings. This factor also determines the potential of biodegradation.

– **Contaminant characteristics**

Biodegradation depends on the chemical structure of pollutants. Every OP has complex chemical structures and the microorganism should have the modification to degrade the substance. Complex contaminants like hydrocarbons and the degradation potential of each constituent is not same. Concentration and chemical nature are the controlling factors for the degradation of hydrocarbons. Solvent solubility of hydrocarbons is a major factor in the rate of microbial uptake.

– **Water**

Water is also an important factor that determines the degradation potential. A range of 50–70% soil moisture is required for the decontamination of soil. Moisture is the basic factor that controls the movement of nutrients. The effect of moisture can be different for aquatic and terrestrial ecosystems. In soil, an adequate level is needed for the transportation of nutrient, waste particles, and so on, but in the case of water-based ecosystem, the problems are less compared with soil [25].

– **Toxicity of end-products**

Toxicity and persistence are closely related, and the major objective of microbial biodegradation is to reduce the toxic level of contaminants and to transform them to less toxic by-products. A decrease in the degree of toxicity level determines the efficiency of bioremediation.

6.8 Future perspectives

Earth's ecosystem is being contaminated gradually by the activity of humans. Because of the harmful effects of contaminants present in the environment, environmental restoration has come out as a challenging issue to the scientific society. Advancement in environmental biotechnology can explain more about the potential of various microorganisms in pollutant degradation. Microorganisms are the principal agents, which can clean and modify the complex organic contaminants present in the environment. Microorganisms attack the pollutants by the enzymatic apparatus. The presence of harmful contaminants in the environment either induces or decreases the enzymatic function of microorganisms [26–27]. This capability largely depends upon the microbial community and structure of organic compounds. Because of the need of more strong genetically modified microbes for the purpose of biodegradation, scientists are researching on the invention of more potential microorganisms. Fast growth and evolution and vast functional and genetic diversity are the elements that make the microbes as excellent bioremediation agents.

Growing human population will produce various kinds of hazardous pollutants into the environment. If an environmental system can restore the polluted sites, it is better than any other remediation methods. Microorganisms have the potential to remediate the contaminated environment by various POPs. As a result of genetic engineering techniques, the genetic traits of microorganism can be modified or exchanged. These engineered organisms are called genetically engineered microorganisms. Introduction of this technology can improve the efficiency of bioremediation. It is the fruitful technique for the remediation of OPs.

6.9 Conclusion

POPs are a commination to the environment which is introduced into the environment by human activities such as industrialization, pesticide usage, fertilizers, and dumping of waste. Removal or recovery of POP-contaminated sites is a cumbersome task because of the resistance of POPs to various degradation processes. Uses of natural agents for the decontamination of polluted sites are a cost-effective, environment-friendly activity. Microorganisms are more effective agents for the removal of toxic contaminants from soil or water environment. The immense biodiversity of microbes in the environment provides appreciable potential for transformation of toxic compounds to less toxic by-products. It is a cost-effective, natural process of environmental restoration, and microbial biodegradation of pollutants will produce less toxic or nontoxic by-products. Microbial enzymes play a crucial role in biodegradation of soil contaminated with OPs such as diesel, petrol, or PAH compound. In future, these enzymes open further exploration by researchers, which is likely to open a new era of microbiology aiding various environmental cleanup technologies.

References

[1] Markiewicz., A., Borklund, K., Eriksson, E., Kalmykova., Y., Stromvall, A.-M., Siopi, A. Emissions of organic pollutants from traffic and roads:Priority pollution selection and substance flow analysis. Sci Total Environ. 2017, 1162–1174.

[2] Malmqvist, B., Rundle, S. Threats to the running water ecosystems of the world. Environ Conserv. 2002, 29, 134–153. doi: https://doi.org/10.1017/S0376892902000097.

[3] Ratnakar., A., Shankar, S., Shikha An overview of biodegradation of organic pollutants. Int J Sci Innov Res. 2016, 2347–4971.

[4] Melinda Haydee, K., Emoke Dalma, K. 2017. 10.5772/67755

[5] Pariatamby, A., Kee, Y.L. Persistent organic pollutants management and remediation. Proc Environ Sci. 2016, 842–848.

[6] United Nations Treaty collection: CHAPTER XXVII- ENVIRONMENT- 15. Stockholm convention on Persistent organic pollutants.

[7] Umar, M., Aziz, H.A. Phytocatalytic degradation of organic pollutants in water. 2012. doi: 10.5772/53699.

[8] Carberry, J.B., Wik, J. Comparison of ex situ and in situ bioremediation of unsaturated soils contaminated petroleum. J Environ Sci Health A. 2001, 36, 1491–1503.

[9] Kane, S.R., Beller, H.R., Legler, T.C., Anderson, R.T. Biochemical and genetic evidence of benzylsuccinate synthase in toluenedegrading, ferric iron-reducing Geobacter metallireducens. Biodegradation. 2002, 13, 149–154.

[10] Karigar, C.S., Rao, S.S. Role of microbial enzymes in the bioremediation of pollutants: A review. Enzyme. 2011. doi: 10.4061/2011/805187.

[11] Vasileva-Tonkova, E., Galabova, D. Hydrolytic enzymes and surfactants of bacterial isolates from lubricant-contaminated wastewater. Zeitschrift Fur Naturforschung. 2003, 58(12), 87–92.

[12] Prasad, M.P., Manjunath, K. Comparative study on biodegradation of lipid-rich wastewater using lipase producing bacterial species. Ind J Biotechnol. 2011, 121–124.

[13] Pieper, D.H., Dos Santos, V.A.M., Golyshin, P.N. Genomic and mechanistic insights into the biodegradation of organic pollutants. Curr Opin Biotechnol. 2004, 15:, 215–224.

[14] Diez, M.C. Biological aspects involved in the degradation of organic pollutants. J Soil Sci Plant Nutr. 2010, 244–267.

[15] Pathiraja., G., Egodawatta, P., Goonetilleke, A., Te'o, V.S.J. Solubilization and degradation of polychlorinated biphenyls (PCBs) by naturally occurring facultative anaerobic bacteria. 2018. https://www.x-mol.com/paperRedirect/849545.

[16] Tekere, M. Microbial biodegradation and different bioreactors designs applied. 2019. doi: 10/5772/83661.

[17] Zacharia, J.T. Degradation pathways of persistent organic pollutants (POPs) in the environment. 2018. doi: 10.5772/intechopen.79645.

[18] Van Hamme, J.D. (2004). Bioavailability and biodegradation of organic pollutants – a microbial perspective. Soil biology, Volume 2 Biodegradation and Bioremediation, eds. Singh, A., Ward, O.P.

[19] Kafilzadeh, F., Sahragard, P., Jamali, H., Tahery, Y. Isolation and identification of hydrocarbons degrading bacteria in soil around Shiraz Refinery. Afr J Microbiol Res. 2011, 3084–3089.
[20] Farhadian., M., Vachelard., C., Duchez, D., Larroche, C. In situ bioremediation of monoaromatic pollutants in ground water: A review. Bioresour Technol. 2008, 5296–5308.
[21] Muneer A, M., Dubey, A., Yadav, S., Kumar, A., Hashem, A., Abd-Allah, E.F. Understanding and designing the strategies for the Microbe-mediated remediation of environmental contaminants using omics approaches. Front Microbiol. 2018. doi: 10.3389/fmicb.2018.01132.
[22] Joutey., N.T., Bahafid, W., Sayel, H., El Ghachtouli, N. Biodegradation: Involved microorganisms and genetically engineered microorganisms. 2013. doi: 10.5772/56194.
[23] Parales, R.E., Bruce, N.C., Schmid, A., Wackett, L.P. Biodegradation, biotransformation, and biocatalysis (B3). Appl Environ Microbiol. 2002, 4699–4709.
[24] Sihag., S., Pathak, H., Jaroli, D.P. Factors affecting the rate of biodegradation of polyaromatic hydrocarbons. Int J Pure Appl Biosci. 2014, 185–202.
[25] Threats to the running water ecosystems of the world. Published online by Cambridge University Press:21 August 2002.
[26] Pinedo-Rivilla, C., Aleu, J., Collado, I.G. Pollutants biodegradation by fungi. Curr Org Chem. 2009, 1194–1214.
[27] Aksu, Z. Application of biosorption for the removal of organic pollutants: A review. Process Biochem. 2005, 997–1026.

Satarupa Dey and A. K. Paul

7 Bauxite residue (red mud): microbial resources and their possible exploitation in remediation and rehabilitation

Abstract: Bauxite residue (also called red mud) is the by-product of industrial extraction of aluminum by the Bayer process. Since the late nineteenth century, it has been produced across the world, with a current estimated deposit of about 3 giga ton that is increasing every year by 120 metric tons. In addition to its high pH and salinity, bauxite residues also contain many heavy metals and radioactive elements and are the main source of environmental pollution. Management and remediation by physicochemical methods are neither economically viable nor environment-friendly.

Culture-dependent and metagenomic approaches have explored the functionally and genetically diverse indigenous microbial communities of these extreme bauxite residues. This extremophilic microbiota has not only adapted to hostile environment but also plays a critical role in the biogeochemical processes. Attempts have recently been initiated to assess the potential of these diverse microbial resources for possible application in bio-amelioration, bioremediation and bioleaching. Implementation of these technologies, however, demands success in both pilot scale and field-level experimentations. This present review will provide updated information on microbial resources of bauxite residues and their potential role in the biomanagement of this unique environment.

Keywords: bauxite residue, microbiota, extremophiles bioremediation, bioleaching, rehabilitation

7.1 Introduction

Bauxite ore used for the extraction of alumina is highly heterogeneous in nature and contains aluminum hydroxide associated with silica, iron oxides, titania, and aluminum silicates in various amounts. Characteristically, the ores are of very low specific gravity, showing a range of 2.6 to 3.5, are pinkish in color, and contain low amount of iron. They are rich in hydrated aluminum oxides, which contain gibbsite [$Al(OH)_3$], boehmite [γ-AlO(OH)], and diaspora [α-AlO(OH)]. The other compounds

Satarupa Dey, Department of Botany, Shyampur Siddheswari Mahavidyalaya, Howrah, India
A. K. Paul, Microbiology Laboratory, Department of Botany, University of Calcutta, Kolkata, India, e-mail: amalk_paul@yahoo.co.in

https://doi.org/10.1515/9783110727227-007

present are hematite [Fe_2O_3], goethite [FeO(OH)], quartz [SiO2], kaolinite [$Al_2Si_2O_5(OH)_4$], and anatase [TiO_2] [1]. From this bauxite ore, alumina is extracted by Bayer process in which the ore is primarily crushed and digested with sodium hydroxide. The solid waste residue generated as a by-product of the extraction process is designated as the bauxite residue or red mud.

Bauxite residue or red mud so produced is extremely alkaline (pH 9–12), highly saline, and polymetallic in nature, containing residual minerals, trace elements, and rare earth elements [2]. It is produced at the rate of around 1–1.75 tons in production of per ton of alumina (Al_2O_3), depending on the type and quality of bauxite ore used [3]. This leads to an estimated global production of about 90 million tons of red mud per year, which is dumped in impoundments and landfills [4]. Lack of proper storage and accidental spillage often result in environmental contamination and social problems.

The high salinity, alkalinity, as well as high content of toxic heavy metals and radioactive elements of the red mud not only deteriorates the soil structure but also adversely affects plant growth. Moreover, continuous leaching of the red mud by rainwater often leads to ground water contamination. Treatment and appropriate utilization of the stored red mud is one of the major challenges for the alumina industry, worldwide. While revegetation and establishment of plant coverage on the amended red mud with sustainable plant communities have been demonstrated in a number of studies [5], efforts have also been made in utilization of red mud, including production of construction materials, preparation of new materials, pigments, paints, and coagulants for wastewater treatment. The bauxite residues, due to their high content of metal, have also been used not only for re-extraction of alumina but also for recovery of metal values including rare and rare earth metals [6].

In a global perspective, several efforts and technological developments have been made by the aluminum industry to ensure bauxite residue management by providing safe discharge and long term storage options, remediation solutions, and possibilities of reutilization with minimum social and environmental impacts. However, the microbiology of the red mud and its influence on red mud remediation have been not been studied in detail until now. In this review, we have tried to provide updated information on microbial resources of bauxite residues and highlight the different possible microbe-mediated strategies for red mud remediation and rehabilitation.

7.2 Genesis and characterization of bauxite residue

The alumina manufacturing industries use bauxite ores for the extraction of aluminum oxide, which in general is highly heterogeneous in nature, containing aluminum hydroxide associated with silica, iron oxides, titania, and aluminum silicates, in various amounts. The alumina refineries have two main options for the extraction of aluminum oxide from bauxite ores: a pyro-metallurgical sintering and the well-known Bayer process; although a combined process is also practiced in some refineries. While in China and Russia, alumina is produced primarily from the naturally occurring boehmite and diaspore bauxites following the lime sintering process, the other alumina-producing countries use the Bayer process for production of nearly 95% of the world production. The bauxite residue produced as a result of the sintering process is known as white mud (dicalcium silicate, Ca_2SiO_4) which, unlike the red mud, is used for manufacturing of cement. It is important to note that, faced with problems related to capital and operational costs, the sintered process is gradually being replaced by the Bayer process. Alternatively, new technologies are being developed to produce alumina from non-bauxite ore like kaoline.

In the Bayer process, the crushed raw bauxite and hot caustic liquor are mixed together in the presence of lime to complete the pre-desilication stage. Following this, the bauxite slurry is processed with hot sodium hydroxide (NaOH) solution at an elevated temperature and pressure, leading to the conversion of alumina to aluminum hydroxide Al(OH)3. The sodium aluminate liquor is separated by filtration, and the alumina is recovered through the process of precipitation and calcinations. The other insoluble elements of bauxite slurry result in the formation of bauxite residue also known as red mud [7]. On average, a typical Bayer plant generates about 1.75 tons of red mud per ton of alumina produced, which, on separation, is disposed of in impoundments.

7.2.1 Characterization of bauxite residue

Data pertaining to the analysis of the properties of bauxite residues of different origin and production process has been published by several authors [2] and is not in the preview of the present discussion. However, for ready reference of the readers, mention may be made of the most recent comprehensive review of Reddy et al. [8], where the authors have included critical discussions on the physical, chemical, mineralogical, and geotechnical properties of bauxite residues, in the context of Indian and global perspectives.

The physicochemical characteristics of bauxite residue vary considerably, depending primarily on the nature of the parent bauxite ores used and the methodology followed for extraction. The bauxite residue in general is highly alkaline (pH 10–12) in nature with high acid neutralizing capacity and has high specific gravity, ranging from 2.7–3.7 due to its iron-rich phases. The particle size or grain size distribution, one of the important features of red mud, also varies considerably, ranging from coarse sandy grains to submicron particle, while the median size particles range from 5–10 µm. The red coloration of the residue is due to high content (20–50% by wt.) of iron oxides. In addition to iron, it also contains oxides and salts of Al, Ti, Si, Ca, and Na, and a variety of trace elements, namely As, Cr, Cd, Ni, Pb, Zn, etc., depending on the composition of the parent ore used in alumina production [9]. Many of these trace elements are potentially toxic and are of concern when they exceed the environmental acceptability limit. The high salinity, alkalinity, electrical conductivity, and high sodium content, along with low levels of organic carbon and plant available N, P, K of bauxite residues not only deteriorates the soil structure but also inhibits plant growth. Other characteristics include high surface area (13–16 m^2/g) and a variety of major and minor residual minerals such as hematite, boehmite, quartz and goethite, tricalciumaluminate ($Ca_3Al_2O_6$), and Na, Al-silicate ($NaAlSiO_4$) as indicated in Table 7.1. It may also be emphasized that the bauxite residue also contains radioactive elements and rare earth elements that are of high commercial value. These include lanthanides and scandium, yttrium, thorium, uranium, ytterbium, dysprosium, samarium, and neodymium. Though the radioactivity in most of the cases is negligible, the amount and activity of the individual element differ in the bauxite ore used for alumina extracting.

Table 7.1: Some major and minor minerals occurring in bauxite residue.

THEAD	Mineral
Fe	Hematite [α-Fe_2O_3], Goethite [α-FeOOH], Magnetite [Fe_3O_4], Ilmenite [$FeO \cdot TiO_2$], Ferryhydrite [$Fe_2O_3 \cdot 0.5H_2O$], Maghemite [γ-Fe_2O_3]
Ti	Anatase [TiO_2], Rutile [TiO_2], Perovskite [$CaTiO_3$], Ilmenite [$FeO \cdot TiO_2$]
Si, Na, Ca	Quartz [SiO_2], Kaolinite [$Al_2Si_2O_5(OH)_4$], Sillimanite [Al_2SiO_5], Halloysite [$Al_2Si_2O_5(OH)_4$], Sodalite [$Na_8(Al_6Si_6O_{24})Cl_2$], Cancrinite [$Na_6Ca_2[(CO_3)_2|Al_6Si_6O_{24}] \cdot 2H_2O$]
Al	Gibbsite [α-$Al_2O_3 \cdot 3H_2O$], Boehmite [α- $Al_2O_3 \cdot H_2O$], Diaspore [β- $Al_2O_3 \cdot H_2O$]
REE	Monazite [(Ce, La, Pr, Nd, Th, Y)PO_4], Xenotime [YPO_4], Synchysite-[(Ce) CaCe(CO_3)2 F]

Compiled from Hammond et al. [10].

7.2.2 Biological characteristics of bauxite residue

Information on biological characteristics of bauxite residue is very rare; however, Caravaca et al. [11] reported maximum dehydrogenase activity in naturally vegetated bauxite residue, which can be correlated with the fast-growing bacterial count and subsequent biological oxidation. Dehydrogenase activity, which is largely used for evaluation of the effect of heavy metals on microbial communities also acts as an indicator of functional state of the bare bauxite residue [12]. In older bauxite residue samples, decrease in pH was associated with subsequent decrease in alkaline phosphatase, which helped in vegetation growth. Bauxite residue is characterized by low phosphorus content, which plays a vital role in promoting extracellular activity for the supply of inorganic phosphorus [13].

Mishra et al. [14] reported the biological properties of bare bauxite residue samples from Hindalco Industries Pvt. Ltd., Renukoot, where the activities of different enzymes such as dehydrogenase (0.69 µg TPF/g/h), alkaline phosphatase (99.16 µg PNP/g/h), and protease (123.12 µg Tyrosine/g/h) were evident. Usually, both low dehydrogenase and protease activities indicate poor functional state of the red mud that contains several heavy metals and is not vegetated. When the microbial count was taken into consideration, it was found to be very low (around 24.67 cfu × 10^{-4}), whereas, microbial biomass carbon was detected to be 157.78 µg/g of are bauxite residue. In the same study, the total organic carbon and total nitrogen were reported to be 0.53 and 0.56 mg/g respectively. Overall microbial load and activity along with the activity of different enzymes was reported to be much less in non-vegetated red mud samples, which improved due to development of vegetation on the red mud [14]. Recent studies on microbial activity as measured by fluorescein diacetate (FDA) hydrolytic activity was found to range from 0.32–0.96 µg fluorescein/g/h in red mud deposits of different deposition history collected from Alumina Plant at Muri, Jharkhand, India. In the same study, the microbial count was reported to be 1.20–8.96 (cfu/g × 10^2) and was dominated by the bacterial population only [15].

7.3 Residue deposits and associated environmental impact

The huge volume of red mud generated as a by-product of the alumina industries is less utilized and often poses unique challenges related to waste and environmental management. Disposal and storage of the bauxite residue waste are of serious concern, mainly due to its high pH, seepage of soluble compounds into underground water, river and land, spillage in the surrounding ecosystem, and the impact of the dust formed by drying of red mud on human and plant life.

Alumina industries across the globe follow several methods for storage as well as disposal of red mud, depending on the local availability of space and the environmental regulations. Jones and Haynes [16] have summarized the different techniques of disposal of bauxite residue along with their merits and demerits of methods. These include sea/lake disposal in a planned manner in countries with very little availability of free space/land. The red mud slurry that is properly processed is usually discharged directly in the deep sea. Though it does not have any impact on land, it exerts toxic effect on marine life and environment. Solar drying with intense irradiation by the sun and stacking in layers appear to be apparently simple and convenient methods and also reduce the area of disposal. In conventional slurry disposal, the bauxite residue slurry is simply pumped in impoundments known as red mud lakes and allowed to dry naturally. This process is economically viable but occupies substantial land area for storage and is of concern for creating environmental pollution including ground water contamination. Its improved version, the drained slurry disposal with increased consolidation of the residue and reduced potential for seepage, has also attracted attention. Dry mud disposal, the most popular method, involves dry stacking of red mud in layers in drying beds and allows drying by evaporation. In this process, the red mud is first dried and then stacked on a special liner. Once there is sufficient red mud stacked, it is covered by the topsoil to reduce contamination and dust formation. This process reduces the area of disposal but renders the land unsuitable for farming due to the presence of large amounts of sodium. In most cases, there is the risk of leaching of caustic soda in the groundwater, making rehabilitation of the land difficult.

Since different kinds of red mud storage and disposal practices are followed all over the world, the impact of red mud, particularly its toxic components, on environment and human beings are also diverse and have been reported from many alumina-producing countries such as Australia, Brazil, Canada, China, Hungary, India, and the United States. Red mud spillage from storage ponds is a potential threat to the surrounding ecosystem as the red mud spills are extremely alkaline in nature. Several such accidents have happened in different parts of the world, such as in Brazil in the Alunorte refinery, in the year 2009. In October 2010, Hungary declared a state of emergency after red mud flooded the town of Kolontar, killing 10 people. In India, such an accident happened in Lanjigarh village in India's eastern state of Odisha – a pond storing waste from Vedanta Aluminum Ltd.'s alumina refinery started overflowing into a nearby river on May 16th 2011. Similar incidents also occurred in China on May 7th 2019 at Xinfa Aluminum factory.

The aftereffect of red mud spill has been well studied in Hungary, where red mud spill resulted in increasing the salinity of the soil as well as altering its cation exchange capacity. The soil structure also changes and becomes sodic in nature [17], which may inhibit the growth of vegetation. Red mud spills also lead to generation of dust of PM_{10}, which developed mild respiratory symptoms in healthy individuals [18, 19]. However, no long term cytogenetic or genotoxic effect was visible in the population [20].

7.4 Microbial community analysis: Biochemical and molecular approaches

Microbial communities of bauxite residues, as in other extreme environments, represent a heterogeneous assemblage of functionally and genetically diverse microbial groups that not only interact among themselves but also play a critical role in all biogeochemical processes, help in nutrient enrichment, and subsequently promote the growth of vegetation. Further, the microbial communities with inherent properties of responding to ecological changes and adapting themselves with the environment may act as a vital resource for recovery from ecological catastrophe and habitat restoration [21].

The extreme alkalinity and salinity, poor water retention ability, and almost negligible organic matter along with nutrient deficiency are considered to be the main limiting factors for microbial and vegetation growth in red mud. However, there have been several reports, which have established the ubiquity of alkaliphilic bacteria and archaea of different physiological groups in naturally occurring hyperalkaline environments like soda lakes and soda deserts, as well as anthropogenic alkaline sources [22, 23]. In the recent past, attention is being focused on exploring their industrial and biotechnological applications.

The bauxite residues produced artificially by the Bayer process under high temperature and pressure differ significantly from natural alkaline environments in terms of their physicochemical characteristics, biological diversity, and productivity. The chances of occurrence of microbes in red mud immediately after its disposal and dumping are extremely rare and can only be traced after prolonged storage and human interference. This extreme environment favors the selective growth and development of bacteria that are tolerant to alkaline pH, high levels of sodium, heavy metals, and electrical conductivity and can sustain low levels of organic carbon, essential nutrients, and growth factors. Moreover, as survival strategies, these microbes are likely to develop multiple adaptive features to overcome the multi-stressed conditions. They are reported to produce acid, which may help in bioremediation of the red mud by lowering the pH in the range between 5.5 and 9.0 and can also reduce electrical conductivity to < 4 milli siemens/cm (mS cm-1), which can promote plant growth.

While the soil microbiota activity is traditionally studied by substrate respiration and/or enzyme activity, bacterial communities in bauxite residues are assessed by the recently developed biochemical and molecular technology tools. In their comparative studies on the development of bacterial communities in bauxite residues of different restoration histories, Schmalenberger et al. [24] adopted the 16S rRNA gene fragment analysis (454 FLX pyrosequencing) and revealed significantly lower α-diversities in the unamended residues. Members of the family Chitinophagaceae, Beijerinckiaceae, Xanthomonadaceae, and Acetobacteraceae, were found predominantly in unamended residues, showing its close relatedness with the communities prevailing in alkaline lakes, hypersaline mats, and similar other environments with high salinity, pH, and poor organic content.

In a novel attempt, microbial community structure and development in bauxite residues as engineered terrestrial haloalkaline environments were compared with those of naturally occurring alkaline systems. Microbial community structure of bauxite residues collected from geographically distant sites (Germany, Ireland, and Australia) along a remediation gradient were directly assessed, following the cultivation-independent approaches. The bacterial communities in bauxite residues were similar to those reported from natural alkaline environments and were dominated by Proteobacteria, Firmicutes, Actinobacteria, and Bacteroidetes. Likewise, the fungal community was dominated by Ascomycota, as observed in other hypersaline environments [25].

It is apparent that most of the microbial community development studies were conducted in bauxite residue under artificial restoration processes and very little is known about the gradual changes in the development of microbial community structure under natural conditions of restoration. In a recent publication, Wu et al. [26] reported the effect of natural restoration and weathering processes on the diversity and composition of the bacterial community in a bauxite residue disposal area with weathering history of 1–25 years. Application of Illumina high-throughput sequencing technology revealed significant development of bacterial community during long term weathering processes. Diversity analysis has clearly indicated the dominance of Firmicutes and Actinobacteria in unweathered residue, whereas the natural processes of weathering encouraged the development of populations of Proteobacteria, Chloroflexi, Acidobacteria and Planctomycetes, and Proteobacteria, consistent with previous studies [24, 25, 27]. Redundancy analysis (RDA) indicated that total organic carbon (TOC) was the dominant factor affecting microbial structure. These studies have further extended the understanding of the correlation between microbial variation and residue properties during natural weathering processes in bauxite residue disposal areas.

7.5 Inventory of cultivable microbial diversity of bauxite residue

Research related to assessment and documentation of microbial diversity of red mud broadly involves microscopic methods, cultivation methods following specific substrate utilization, molecular, and genomic methods. Technological advancement in microscopic methods have no doubt revolutionized microbial diversity assessment but are dependent on other forms of analysis for a comprehensive idea of the diversity. Use of culture-dependent approaches with respect to direct cultures, enrichment cultures, or selective cultures are time-consuming and provide information only on 1% of the population. The molecular and genomic methodologies include fatty acid analysis, protein profiles, and nucleic acid sequence-based analysis such as the denaturing gradient gel electrophoresis (DGGE), automated ribosomal intergenic spacer region analysis [ARISA], including the high throughput sequencing tools, for example, the 454 FLX pyrosequencing. In spite of the availability of a wide variety of techniques, only a few of them have been applied for diversity analysis of the bauxite residue. As a result, published data on the cultivable and non-cultivable diversity of bacteria of bauxite residue are scanty [28]. The microbiota of red mud that has been analyzed so far is mostly dominated by oligotrophic alkalitolerant and metal resistant bacteria.

It was Grant et al. [29], who, in their extensive review on alkaliphiles, have highlighted that among the many man-made alkaline environments, the bauxite-processing red mud tailings also support the growth of alkaliphiles, although very little or no information was available about the microbial diversity, at that time. Search for novel alkaliphiles from bauxite-processing waste by Agnew et al. [30] appeared to be the first systematic survey of microbial diversity in red mud samples, following the aerobic enrichment culture technique. Using the mud samples from an undefined bauxite-processing red mud tailing pond, the authors isolated and characterized three novel *Bacillus* isolates designated as strains JaA, JaD, and JaH. Among them, strains JaD and JaH were obligate alkaliphiles, and JaH, in particular, was identified as a novel species, *Bacillus veddari*. Moreover, the red mud samples were claimed to be the source of bacteriophages that were active against alkaliphiles, as they also reported the isolation of a temperate bacteriophage that was active against strain JaD, which could be effective in horizontal gene transfer among the indigenous microbiota.

Targeted towards the bacterial amelioration of bauxite residue wastes, Hamdy and Williams [28] analyzed both viable and injured bacterial cells from Alcoa bauxite residue lake deposit at Mobile, Alabama, USA using minimal and enriched media. Viable cells were tested by methylene blue reduction test, while the metabolically inert cells were repaired by several enrichment media and other added nutrients. The injured cells were grown on enriched media such as tryptic soy broth, and/or tryptic soy agar. The treated bauxite residue samples were reported to yield more than 150 bacterial cultures, mainly predominated by members of the genera *Bacillus, Lactobacillus*, and *Leuconostoc*. However, *Micrococcus, Staphylococcus, Pseudomonas, Flavobacterium, Enterobacter*, and *Proteus* were not uncommon.

The red mud deposits as they are commercially produced and characterized, with low levels of organic carbon and nitrogen content along with essential micronutrients, limit the growth of heterotrophic microbial population. Krishna et al. [31], in their attempt to isolate xylanase-producing alkaliphilic bacteria from artificial environment obtained 57 isolates from red mud samples collected from the National Aluminum Company (NALCO), Damanjodi, Orissa using different media and growth conditions. The most prominent extracellular xylanase- producing strain, designated as RM1, represented the first report of isolation of actinomycetes, *Kocuria* sp., from bauxite residue and appeared to be an effective one involved in xylan degradation. Subsequent to this, Arora et al. [32], from the same laboratory, using red mud samples of the same impoundment, also reported another alkalitolerant and xylanolytic bacterial strain identified as *Paenibacillus montaniterrae* RMV1. Taxonomic identity of the strain was confirmed, based on both biochemical and 16S rDNA sequence analysis, and comparison with validly published and phylogenetically related species of *Paenibacillus* such as *P. septentrionalisa* and *P. agaridevoransa.*

Systematic culture-dependent and culture-independent approaches were also adopted by Krishna et al. [27] for diversity analysis of samples from the same industry

(NALCO). Oligotrophic bacteria were isolated and enumerated by the standard dilution and plating method on alkaline nutrient agar, alkaline tryptic soy agar, and Horikoshi agar media and were found to belong to phylum Firmicutes, Actinobacteria, and Proteobacteria. Based on 16 S rDNA sequence analysis, the isolates were assigned to *Agromyces*, *Bacillus*, *Chungangia*, *Kokuria*, *Microbacterium*, *Planococcus*, *Pseudomonas*, and *Salinococcus*. The culture-independent approach, on the other hand, has led to the isolation of total DNA from the red mud, amplification of 16S rRNA genes, and construction of 16S rDNA clone library. Selected 16S rDNA clones were segregated into 16 operational taxonomic units (OTUs), a majority of which were assigned to betaproteobacteria and were followed by gammaproteobacteria and bacteroidetes.

In an attempt to isolate and characterize bacterial strains in bauxite residue from the region of Poços de Caldas, Minas Gerais, Brazil, Nogueira et al. [33] followed the enrichment culture technique using oxalate as the sole source of carbon. They have reported the presence of five different isolates that showed > 99% similarity to different alkaliphilic species of *Bacillus*-based 16S rRNA gene sequence analysis but differ significantly in their biochemical attributes. The isolates were identified as *Bacillus cohnii, Bacillus pseudofirmus,* and *Bacillus clarkii*. Later, Wu et al. [34] also reported the presence of *Bacillus* strains from the samples of bauxite residue of Guangxi Province, China.

Twenty-two different chemoheterotrophic bacterial strains were isolated by Qu et al. [35] from the red mud samples collected from the Chinalco in southwest of China, with a deposition history of over 20 years. Isolation of bacteria was achieved by serial dilution and direct plating method on nutrient agar, tryptic soya agar, and Horikoshi agar media, supplemented with 2% (v/w) red mud. One of the selected strains, RM-B07, was found to have the ability of maximum acid production during growth in the presence of red mud. Deploying the morphological, physio-biochemical, and 16S rRNA gene sequence homology, the strain RM-B07 was identified as *Acetobacter* sp., which supported the predominance of Acetobacteriaceae in red mud impoundments [24].

A total of 12 isolates were obtained from the red mud samples of different deposition history collected from Hindalco plant, Muri, Jharkhand, India, by the authors [15], using Horikoshi medium and Peptone Yeast extract glucose medium (pH 10). The isolates were reported to be alkaliphilic in nature and tolerant to different levels of NaCl, along with resistance to toxic metals such Pb, Cr, Fe, Ni, Mn, and Al. These isolates mostly belonged to the genera *Bacillus*, along with members of other genera such as *Paenibacillus, Halomonas, Ochrobactrum, Salinococcus, Pseudomonas,* and *Kurthia*. Most of these isolates exhibited the ability to produce acid during growth, and at least some of them were capable of producing extracellular polymeric substances, which indicated their potential for bioremediation. The detailed information on culturable diversity of microbiota derived from red mud samples of different origin are represented in Table 7.2.

Table 7.2: Culturable microbial diversity, as reported from red mud samples of different origin, indicating their method of isolation and characteristic features.

Source// location of red mud	Methodology for isolation	Microbial diversity	Characteristic features	Reference
Bacteria				
Red mud tailing pond, location not specified	Enrichment in alkaliphilic oxalate medium	*Bacillus* strain JaA, JaD and JaH *Bacillus vedderi* JaH	Obligate alkaliphile, pH for growth: range 8.9–10.5; optimum, 10.0	[30]
BR lake deposits, Mobile, Alabama, USA.	Minimal medium, tryptic soy broth and agar	*Lactobacillus, Leuconostoc, Bacillus*	NR	[28]
Red mud pond, NALCO, Damanjodi, Orissa, India	Nutrient media, tryptic soya media and Horikoshi media	*Kocuria* sp	Extracellular xylanase production	[31]
Red mud pond, NALCO, Damanjodi, Orissa, India	Nutrient agar medium supplemented with 1% xylan	*Paenibacillus montaniterrae*	Growth associated xylanase production at pH 7.0–10.5 and temperature 25–45 °C.	[32]
Red mud pond, NALCO, Damanjodi, Orissa, India	Alkaline nutrient agar, alkaline tryptic soy agar, and alkalophilic Horikoshi agar	*Agromyces indicus, Bacillus litoralis, B. anthracis, Chungangia koreensis Kokuria flava, K. Polaris, Microbacterium hominis, Planococcus plakortidis, Pseudomonas alcaliphila, Salinococcus roseus*	Tolerance to 3 or 6% of NaCl	[27]
Bauxite residue, Minas Gerais, Brazil	Horikoshi medium (pH 10.5)	*Bacillus cohnii, B. pseudofirmus, B. clarkia*	Alkaliphilic, pH 10.5 at 30 °C	[33]

Table 7.2 (continued)

Source// location of red mud	Methodology for isolation	Microbial diversity	Characteristic features	Reference
Bauxite residue deposit, Guangxi Province, China	Enrichment in medium containing 2% glucose, 0.5% peptone and 0.5% NaCl	*Bacillus* sp EEEL02	Survived at 0–25% NaCl and a temperature range of 25–45 °C at pH 12.0	[34]
Chinalco in southwest of China	Nutrient agar, tryptic soya medium and Horikoshi medium	*Acetobacter* sp. RM-B07	NR	[35]
Hindalco, Muri, Jharkhand, India	Horikoshi medium, peptone yeast extract glucose medium	*Bacillus* sp., *Paenibacillus* sp., *Halomonas* sp., *Ochrobactrum* sp, *Salinococcus* sp, *Pseudomonas* sp, *Kurthia* sp.	Tolerate 0–20% NaCl, temperature range 32–37 °C at pH 4–12	[15]
Fungi				
Stored tailings from Germany and Australia	NR	Ascomycota Basidiomycota	Alkaliphilic, salt tolerant	[25]
Bauxite residue deposit, Guangxi Province, China	PDA at initial pH of 9.0	*Penicillium oxalicum* EEEL01, *Aspergillus* sp., *Mycosphaerella* sp.	Alkaliphilic and acid producing	[36]

NALCO, National Aluminum Company Limited, India
NR, not reported

7.6 Physiology, biochemistry, and adaptive strategies of microbiota

Microorganisms retrieved from the red mud samples with extremely alkaline pH and high concentration of soluble ions so far analyzed belong to the class of "alkaliphiles" and, depending on their pH and salt requirement for growth, could be segregated as true alkaliphiles and haloalkaliphiles. The true alkaliphiles as a physiological group require a pH ≥ 9 for optimal growth, whereas haloalkaliphiles thrive at pH higher than 9 and salinity of 5 M NaCl. In view of very poor microbial load and the requirements of extreme growth and nutritional conditions, early attempts at isolation of indigenous alkaliphilic microbes from red muds involved the application of enrichment culture techniques, using specific culture media such as alkaliphilic oxalate medium [30] and Horikoshi medium. Later on, minimal medium and tryptic soy broth at alkaline pH [28], nutrient broth (pH 10.0), and tryptic soy broth (pH 10.0) [27, 31, 35] have been tried by several others. The enrichment medium used by Wu et al. [34], however, contained glucose, peptone, and 0.5% NaCl. Most of the isolates obtained from the red mud samples were obligate alkaliphilic in nature and were able to tolerate a pH range of 7.0–10.5 [27, 32, 33], and NaCl concentration ranging from 5–10% (w/v). The temperature for growth of the isolate ranged from 25–45 ºC [31, 32]. These features bear a striking similarity to alkaliphilic microbes derived from soda lakes and desert.

The red mud isolates appeared to be versatile for carbon source utilization for growth and metabolism. They were able to utilize a wide range of carbon substrates such as dextrin, tween 40, tween 80, D-fructose, D-gluconic acid, D-glucose, maltotriose, D-mannitol, D-mannose, palatinose, D-psicose, D-sorbitol, sucrose, D-trehalose, pyruvic acid, glycerol, adenosine, inosine, thymidine, a-cyclodextrin, acetic acid, a-hydroxybutyric acid, p-hydroxyphenylacetic acid, a-ketovaleric acid, L-lactic acid, D-malic acid, L-malic acid, pyruvic acid methyl ester, succinic acid monomethyl ester, propionic acid, succinamic acid, succinic acid, N-acetyl-L-glutamic acid, L-alanine, L-asparagine, L-glutamic acid, L-serine, putrescine, and 20-hydroxyadenosine. In the case of *Bacillus* and *Kocuria* sp. the preferred carbon sources were reported to be glucose and glycerol [30, 31].

Agnew et al. [30] also reported the use of complex organic sources such as yeast extract, peptone, and casamino acid by *Bacillus veddari* JaH as nitrogenous source. However, different species of *Bacillus* from bauxite residue samples of Minas Gerais, Brazil were able to grow on glucose, fructose, sucrose, mannitol, or yeast extract, but none was found to grow on xylose or lactose [33]. Recently, Wu et al. [34] have reported an alkaliphilic acid producing bacterium, *Bacillus thuringiensis* EEEL02, which preferred glucose and peptone as carbon and nitrogen sources, respectively.

The isolates obtained from red mud samples are known to be a repository of different types of extracellular enzymes. Isolates like *Kocuria* sp. [31] and *Paenibacillus montaniterrae* [32] were reported to produce extracellular xylanases. It was reported that these red mud isolates also produced a wide range of enzymes such as urease, extracellular acid phosphatase, and cellulose [27, 35].

Most of these isolates were found to have an acid producing ability, which is very vital from the point of view of bioleaching of metals from the bauxite residue. Such a phenomenon of acidification was first reported by Hamdy and Williams [28] in species of *Bacillus*, *Lactobacillus*, *Leuconostoc*, *Micrococcus*, *Staphylococcus*, *Pseudomonas*, *Flavobacterium*, and *Enterobacter* isolated from bauxite residues and was recommended for their use in bioremediation purpose. Krishna et al. [27] did an extensive survey of both culturable as well as non-culturable microbes from red mud sample and demonstrated a wide range of isolates that were able to produce organic acids, mainly acetic acid and succinic acid. Nogueira et al., [33] isolated *Bacillus cohnii*, *B. pseudofirmus*, and *B. clarkia*, which were reported to produce a wide range of organic acids including considerable amounts of acetic, isobutyric, and isovaleric acids. Similarly, acetic acid and propionic acid were the major acid metabolites produced by fermentation, which result in pH reduction by *Bacillus thuringiensis* EEEL02 and *Acetobacter* sp. RM-B07 [34, 35].

It is apparent that there is ample scope in exploring the metal tolerance profile, mechanisms of tolerance, and adaptive strategies of these isolates in order to exploit their metabolic potentials in red mud utilization and management.

7.7 Microbial recovery of metal values from bauxite residue

In as much as the composition of red mud is concerned, it consists of high amounts of complex oxides of aluminum, iron, titanium, and silicon along with lower amounts of other valuable components, such as rare earth elements (scandium, yttrium, lanthanum, neodymium, and gadolinium) and radionuclides, radium and thorium [37]. Due to its diverse metal contents, red mud has been referred to as "polymetallic raw material" by metallurgists and could serve as a potential source of these metal values.

Bio hydrometallurgy, the process of recovering metal values from metal bearing low grade ores and wastes by the use of microorganisms, has attracted a lot of interest in recent years and could play an effective and beneficial role in the waste treatment processes, in future. These microbe-mediated technologies are environmentally benign, low energy requiring, cost-effective green process, and have been established as an attractive alternative to the presently used conventional physicochemical methods.

The success of this microbe-mediated leaching technology, however, depends broadly on (i) the type of microorganisms involved and (ii) the cultural techniques or bioleaching process followed for the extraction of metal values. The use of autotrophic or heterotrophic organisms for metal leaching depends on the nature of the low-grade ore and the waste material. Autotrophic microbes, the acidophilic chemolithoautotrophs, in particular, are the major group of organisms used in bioleaching of metals from industrial wastes under acidic conditions [38] and are not suitable for leaching of highly alkaline red mud. On the contrary, heterotrophic bacteria and fungi are effective for bioleaching in alkaline environment [39] including the red mud because of their rapid growth and adaptability in such extreme conditions [25]. So far as the leaching process is concerned, scientists have three major options to judge the metal extraction efficiency. These include one-step, two-step, and spent medium bioleaching process. While in one-step and two-step processes, a direct physical contact is established between the microbial cells / mycelia mass and the leaching materials, in spent medium bioleaching, the cell-free culture filtrate containing the metabolic products is used to interact with waste materials in the leaching process.

The mechanisms of metal leaching from industrial wastes by heterotrophic microbes have been widely studied and documented to be an indirect process based on solid phase dissolution or transformation, mediated by the production of organic acids [38]. Studies of Vachon et al. [40] appeared to be the first report on bioleaching of aluminum from the red mud by heterotrophic microorganisms. They used sewage sludge bacteria (indigenous thiobacilli) and few strains of fungi including *Aspergillus niger, Penicillium simplicissimum, Penicillium notaturn,* and *Trichoderma viride.* For bioleaching, *Penicillium simplicissimum* was the most suitable fungal strain capable of leaching nearly 75% of Al, when red mud was used at 10 % v/v concentration. Further, this fungal metabolite driven aluminum recovery from red was considered to be a promising one, even at industrial scale.

Studies on bioleaching of Al were also conducted by Ghorbani et al. [41], using indigenous fungi, *Aspergillus niger* and *Penicillium simplicissimum,* which were isolated from soil and water samples. Under shake flask culture, about 2082 mg of Al_2O_3/L was solubilized at 15% pulp density of red mud. Similarly, Urik et al. [42] have also investigated bioleaching of Al from red mud using 17 fungal species belonging to genera *Penicillium, Aspergillus, Eurotium.* and *Emericella.* In batch culture, the most efficient strain, *Aspergillus niger* G-10, was reported to solubilize approximately 141 mg/L of aluminum from 0.2 g of red mud, following acidification of the culture medium through extracellular production of a mixture of organic acids such as citric, oxalic, and gluconic acid.

A number of different conditions such as one-step, two-step, and spent medium bioleaching were used in bioleaching of various heavy metals from red mud samples (collected from Chinalco in Guiyang, China) using *A. niger* [37]. The heavy metal leaching efficiency at different bioleaching conditions were different with identical pulp density (1%), and spent medium bioleaching was identified as the best method that used citric acid as the main lixiviant excreted by *A. niger.* Among the different metals, Pb and Zn were released most efficiently (80%) and was followed by Cu (67%), Ni (50%), As (44%), Ba (31%), Cr (26%), and Fe and Zr (11%). Leaching of metals leads to significant decrease in toxicity as well as changes in the micromorphology of the red mud.

Red mud, as a polymetallic raw material, has also been recognized as a strategic resource of rare earth elements (REE) mainly because of their multifarious applications in industry. Over the years, attention has been focused on the chemical leaching of REE using sulfur dioxide [43] and different inorganic acids, including HNO_3 [44], H_2SO_4, and HCl [45]. Qu and Lian [46] were probably the first to initiate bioleaching of REE from bauxite residues using an indigenous fungal strain, *Penicillium tricolor* RM-10. The fungi isolated from the red mud showed moderate to good survival capability in red mud and was used for one-step and two-step bioleaching processes. Oxalic and citric acids excreted by the fungi were identified as the main lixiviants for the leaching process. Though the one-step bioleaching process was effective, highest extraction yields of REE and radioactive elements were achieved under two-step process.

Faced with the several disadvantages of using fungi in bioleaching, Qu et al. [35] opted for chemoheterotrophic bacteria, in leaching of major, rare earth, and radioactive elements from red mud (RM). The efficacy of *Acetobacter* sp. isolated from red mud impoundment was explored under one-step, two-step, and spent medium bioleaching process. Under one-step and two-step processes, *Acetobacter* sp. exhibited excretion of high amount of acetic, oxalic, and malic acid, but high metal leaching performance of red mud was recorded under one-step process, and the leaching ratios of Al, Lu, Y, Sc, and Th were 55%, 53%, 61%, 52%, and 53% respectively, at 2% pulp density. Since a direct contact between RM and microbe is established in this efficient one-step leaching process, the authors find it to be of great potential in in situ heap and dump leaching. Moreover, the decline in radioactivity of the red mud following bioleaching has led the authors to recommend the use of bioleached red mud directly as construction material, based on the radioactive regulations of China.

Recently, Zhang et al. [47] have also reported the extraction of Al and REs from bauxite residues by both aerobic and anaerobic bi-stage bioleaching, using *Acidianus manzaensis*. The ability of bioleaching was enhanced following addition of pyrite to red mud, which resulted in 85.1, 82.4, 86.8, 85.3, and 78.6% leaching of Al, Ce, Gd, Y, and Sc, respectively, after 22 days of incubation. The results showed that during the aerobic bioleaching stage, with increase in mass ratios of pyrite/RM, the extraction rates of Al and REs increased; the highest rates were achieved at pyrite/RM mass ratio 2:1 after 22 days of incubation. Table 7.3 represents a comprehensive list of different microorganisms and their efficiency of leaching metal values from bauxite residues.

7.8 Bioremediation and rehabilitation of bauxite residue

Microbe-mediated remediation of red mud has been reported to be achieved by exploiting the wide range of metabolic activities of the indigenous or introduced bacteria, archaea, and fungi. The different strategies that are followed for bioremediation of red mud deposits or the contaminated lands include: i) reduction of highly alkaline pH of red mud by production of organic and inorganic acid or by production of carbon dioxide; ii) aggregation of particles facilitated by production of extracellular polymeric substances by different groups of microbes, which helps in development of well-aerated soils, increases the stability, and promotes plant growth; and iii)

Table 7.3: Bioleaching of heavy metal values and rare earth elements from red mud.

Microbial strains	Bioleaching process used	Leaching efficiency of metal(s)	Pulp density (%)	Organic acid(s) Produced	References
Indigenous thiobacilli	One-step	6265 mg of Al/L	5	Not reported	[40]
Aspergillus niger	Spent medium	4810 mg of Al/L	3	Not reported	
Penicillium simplicissimum	One-step Spent medium	1880 mg of Al/L 10,585 mg Al/L	3	Mainly citric acid	
Penicillium notatum	Spent medium	4730 mg of Al/L	3	Not reported	
Trichoderma viride	Spent medium	5245 mg of Al/L	3	Not reported	
Aspergillus niger and *Penicillium notatum*	Spent medium	2082 mg of Al_2O_3/L 4 mg/L of Titanium	15	Citric and oxalic acid	[41]
Aspergillus niger	One-step Two-step, and Spent medium	80% of Pb and Zn, 67% of Cu, 44% of As, 50 % of Ni, 26 % of Cr. about 11% of Fe and Zr	1	Citric, gluconic, and oxalic acid	[37]
Penicillium tricolor RM 10	One-step Two-step, and Spent medium	220.3 (one step), 164.5 (two step), 245.1 (spent medium) Bq/kg of Ra 169.2 (one step), 151.6 (two step), 188.9 (spent medium) Bq/kg of Th 250.1 (one step), 207.8 (two step), 288.3 (spent medium) Bq/kg of K	10	Citric and oxalic acid	[46]
Penicillium crustosum G-140	One step	127.7 mg/L of Al	0.2 g	Citric, gluconic, and oxalic acid	[42]
Aspergillus niger G-10	One step	141 mg/L of Al			

Table 7.3 (continued)

Microbial strains	Bioleaching process used	Leaching efficiency of metal(s)	Pulp density (%)	Organic acid(s) Produced	References
Acetobacter	One-step	Al, Lu, Y, Sc, and Th were 55%, 53%, 61%, 52%, and 53%	2%	Succinic, lactic, malic, tartaric, citric, oxalic and acetic acid	[35]
Acidianus manzaensis	Aerobic and anaerobic bi-stage bioleaching	85.1% of Al, 82.4% of Ce, 86.8% of Gd, 85.3% of Y and 78.6% of Sc	0.6 g	Not reported	[47]

reduction in exchangeable sodium percentage, which helps in aggregate formation, making the bauxite residue more suitable for growth of vegetation.

7.8.1 Reduction in pH

Lowering the pH of the bauxite residue can be done following the production of both organic and inorganic acids, as well as carbon dioxide. Organic acids are usually produced by fermentative metabolism of associated microbial community as well as decomposition of organic matter and plant residues. The different organic acids that have been identified to play a pivotal role in bioremediation include acetic acid, propionic acid, lactic, and citric acids. The amount of the acids produced varies quantitatively depending on the microbe(s) involved, their enzymatic constitution, and the available nutrient resources. The organic acids so produced decrease the pH by reacting with various sources of alkalinity present in the red mud.

Lactic acid and propionic acid are the main organic acids produced by bacteria belonging to the genera *Streptococcus*, *Pediococcus*, *Lactobacillus*, *Leuconostoc*, and *Propionibacter* following anaerobic fermentative metabolism. On the other hand, citric acid is predominantly produced aerobically by fungi such as *Penicillium* spp, *Aspergillus* spp., *Candida* spp. and bacteria such as *Arthrobacter* spp. However, the main drawback in utilizing these microbial species is their requirement of a strict physicochemical environment and culture conditions, which are difficult to be fulfilled in the bauxite residue deposits.

Several fungal strains have been reported to produce acids, which, in turn, helped in reducing the pH of the red mud samples. Krishna et al. [48] reported a strain of *Aspergillus tubingensis,* which was able to tolerate high pH and was effective in reducing alkalinity of the culture medium. Later, Babu and Reddy [49] documented that application of *Aspergillus tubingensis* along with gypsum and sludge could help in neutralizing bauxite residues and, consequently, help in growth of vegetation. Similarly, Courtney et al. [50] have reported the occurrence of fungi and arbuscular mycorrhizal fungi in amended bauxite residues and suggested that they play an important functional role in rehabilitation of the red mud residues.

It was also evident that when *Penicillium oxalicum* was inoculated in bauxite residue, the pH was reduced from 10.26 to 6.48 in 11 days [51]. Similar report of reducing the residue pH from 10.26 to 5.62 within 5 days of incubation was evident from the findings of Wu et al. [34], when the acid producing bacterium *Bacillus thuringiensis* EEEL02 was used. Such reduction of pH by the isolate was assisted by the production of metabolites such as acetic acid, propionic acid, and CO_2. Hamdy and Williams [28] also demonstrated that bacteria grown in essential nutrients and hay formed organic acids, which resulted in lowering of pH from 13.0 to 7.0.

Several microorganisms are known to undergo chemoautotrophic or chemoheterotrophic metabolism and produce inorganic acids such as H_2SO_4, in situ. The microbes, which are mainly iron and sulfur-oxidizing, produce sulfuric acid, and they assist in microbe-driven remediation of red mud by lowering the pH [52]. Several reports state that sulfur-oxidizing bacteria obtained from alkaline and saline soda lakes are capable of tolerating high pH and high salinity [52]. A mixed culture of various *Thiobacillus* spp. was applied to decrease the pH of bauxite residues similar to the practice followed successfully in heap leaching of copper [53].

Carbon dioxide is generated by microbes during respiration and fermentation, which gradually dissolves in water to form carbonic acid. This carbonic acid, a weak acid, often dissociates to generate HCO_3 that reacts with metal to form metal carbonate, which can help in remediation of bauxite residues by reducing the pH. Addition of CO_2 to red mud slurry in storage facilities has been implemented for remedial purposes, which reduces pH, increases shear strength, and neutralizes tricalcium aluminate ($Ca_3Al_2(OH)_{12}$), which is one of the major sources of alkalinity in bauxite residues. In most of the cases, microbial communities of bauxite residues are carbon limited and require carbon source to drive CO_2 production, and as the microbial communities in the bauxite residues diversifies from autotrophs to heterotrophs, bacterial CO_2 production increases.

7.8.2 Formation of aggregates of red mud particles

Formation of aggregates usually occurs by production of extracellular polymeric substances (EPS) by different microbes. The EPS mainly consists of polysaccharides, proteins, nucleic acid, and lipids. These are released by bacterial and fungal cells, which provide stability by improving the cohesion within the cells and act as protective agents against extreme environmental conditions [54]. A number of different factors such as temperature, pH, and nutrient concentration have a profound effect on EPS as well as organic acid production in bacteria and fungi. Production of EPS can facilitate bauxite residue remediation by improving particle aggregation, which may subsequently help in development of well-aerated soil. Other advantages of EPS production are conferring insulation and resistance to extreme physicochemical factors such as temperature, pH, and salinity to the organisms [54] by secretion of different thermoprotectors and osmoprotectors. It also limits proton permeability, which favors growth at an unfavorable pH. The carboxyl group present in the polysaccharide of EPS facilitates complex formation with metal ions including Na, Pb, Ni, and Cd and could be of great help in bioremediation of bauxite residues [54].

7.8.3 Reduction in exchangeable sodium percentage (ESP)

Red mud samples are characterized by presence of high concentration of Na; thus, in order to make red mud samples suitable for plant growth, the exchangeable sodium percentage must be reduced to ≤ 9.5 %. Increasing the Ca/Na ratio by addition of gypsum helps reduce the exchangeable Na percentage, improves ionic balance, and significantly decreases pH. Solubilization of Ca from Ca-containing minerals such as calcite ($CaCO_3$), tricalcium aluminate ($Ca_3Al_2(OH)_{12}$), cancrinite ($Na_6Ca_2Al_6Si_6O_{24}(CO_3)_2.2H_2O$), or perovskite ($CaTiO_3$) helps in increasing the amount of Ca. Bacteria, fungi, and other microorganisms can increase the concentration of ions in the soil solution through solubilization of the mineral or organic components of the soil. Fungal species including *Phoma glomerata*, *Epicoccum nigrum*, *Penicillum chrysogenum*, *Cladosporium herbarum*, *Altemria tenuis*, and *Aspergillus ustus* can solubilize calcite and dolomite crystals and reduce the pH.

7.9 Conclusion

The present comprehensive review based primarily on the survey of literature currently available is a reflection of some specific aspects of bauxite residue or red mud, which includes their genesis from different bauxite ores and production methods, disposal and management strategies, physicochemical and biological properties with special emphasis on the development of diversity and structure of microbial communities, and their exploitation in bioprospecting of the residue. The global production and accumulation of the bauxite residue representing an extreme environment for growth and survivability of all living forms is also of serious concern, not only from the point of view of environmental contamination but also of generated social problems.

Microbiological studies of bauxite residues, a comparatively underexplored domain have revealed the development of selected group of microorganisms having close similarity and relatedness of the microbial populations of naturally occurring hypersaline and alkaline environments. While cultivation-dependent and metagenomic studies have highlighted and focused attention on the development of microbial communities during the long-term storage and weathering of bauxite residue, understanding of the metabolic potentials and physio-biochemical functions of these heterotrophic microorganisms will help apprehend the nutrient cycling, bio hydrometallurgy in recovering of metal values and rare earth elements and rehabilitation of the anthropogenic environment. However, a coordinated approach by the industries and their management, environmentalists, engineers, and scientific researchers seems to be necessary for successful and effective implementation of commercially viable and environment-friendly process in bauxite residue biomanagement and rehabilitation, in future.

List of abbreviation

ARISA	automated ribosomal intergenic spacer analysis
Cfu	colony forming unit
DGGE	denaturing gradient gel electrophoresis
FDA	fluorescein diacetate activity
Gt	giga ton
M	molar
Mt	metric ton
mS cm-1	milli siemens/cm
μg TPF/g/h	microgram triphenyl formazan/gram/hour
μg PNP/g/h	microgram / *p*-nitrophenol /gram/hour
OTU	operational taxonomic unit
RDA	Redundancy analysis
TOC	total organic carbon
Wt	weight

References

[1] Balasubramaniam K.S., Paropkari A.L. Mineralogy and genesis of bauxites from Nagardaswadi plateau". Kolhapur district. Can Mineral. 1975, 13(3), 222–226.
[2] Samal S., Ray A.K., Bandhopadhyay A. Proposal for resources, utilization and processes of red mud in India – A review. Int J Miner Process. 2013, 118, 43–55.
[3] Zhang R., Zheng S., Ma S., Zhang Y. Recovery of alumina and alkali in Bayer red mud by the formation of andradite-grossular hydrogarnet in hydrothermal process. J Hazard Mater. 2011, 189, 827–835.
[4] Amritphale S.S., Bhasin S., Chandra N. Energy efficient process for making pyrophyllite based ceramic tiles using phosphoric acid and mineralizes. Ceram Int. 2006, 32, 181–187.
[5] Courtney R., Mullen G., Harrington T. An evaluation of revegetation success on bauxite residue. Restor Ecol. 2009, 17(3), 350–358.

[6] Klauber C., Gräfe M. Power G. Bauxite residue issues: II. Options for residue utilization. Hydrometallurgy. 2011, 108(1–2), 11–32.
[7] Kelessidis V.C., Tsamantaki C., Dalamarinis P. Effect of pH and electrolyte on the rheology of aqueous Wyoming bentonite dispersions. Appl Clay Sci. 2007, 38(1), 86–96.
[8] Reddy P.S., Reddy N.G., Serjun V. et al. Properties and Assessment of Applications of Red Mud (Bauxite Residue): Current Status and Research Needs. Waste Biomass Valor. 2020, https://doi.org/10.1007/s12649-020-01089-z
[9] Liu Y., Naidu R., Ming H. Red mud as an amendment for pollutants in solid and liquid phases. Geoderma. 2011, 163, 1–12.
[10] Hammond K. Recovery of Value-added Products From Red Mud and Foundry Baghouse Dust. Colorado School of Mines. 2014.
[11] Caravaca F., Masciandaro G., Ceccanti B. Land use in relation to soil chemical and biochemical properties in a semiarid Mediterranean environment. Soil Tillage Res. 2002, 68, 23–30.
[12] Oliveira A., Pampulha M.E. Effects of long-term heavy metal contamination on soil microbial characteristics. J Bio Sci Bioeng. 2006, 102, 157–161.
[13] Gianfreda L., Antonietta R.M., Piotrowska A., Palumbo G., Colombo C. Soil enzyme activities as affected by anthropogenic alterations: Intensive agricultural practices and organic pollution. Sci Total Environ. 2005, 341, 265–279.
[14] Mishra T., Singh N.B., Singh N. Restoration of red mud deposits by naturally growing vegetation. Int J Phytorem. 2016, 19(5), 439–445.
[15] Dey S., Paul A.K. Evaluation of physio-biochemical potentials of alkaliphilic bacterial diversity in bauxite processing residues of diverse restoration history. Environ Sustain. 2020, https://doi.org/10.1007/s42398-020-00152-8. In press.
[16] Jones B.E.H., Haynes R.J. Bauxite processing residue: A critical review of its formation, properties, storage, and revegetation. Crit Rev Environ Sci Tecnol. 2011, 41, 271–315.
[17] Bolt G.H., Bruggenwert M.G.M. Soil Chemistry: A. Basic Elements, Amsterdam, Elsevier Scientific Publishing Company, 1976, 179–180.
[18] Gelencsér A., Kováts N., Turóczi B., Rostási Á., Hoffer A., Imre K., Nyirő-Kósa I., Csákberényi-Malasics D., Tóth Á., Czitrovszky A., Nagy A., Nagy S., Ács A., Kovács A., Ferincz Á., Hartyáni Z., Pósfai M. The red mud accident in Ajka (Hungary): Characterization and potential health effects of fugitive dust. Environ Sci Technol. 2011, 45(4), 1608–1615.
[19] Czövek D., Novák Z., Somlai C., Asztalos T., Tiszlavicz L., Bozóki Z., Ajtai T., Utry N., Filep A, Bari F., Peták F. Respiratory consequences of red sludge dust inhalation in rats. Toxicol Lett. 2012, 209, 113–120.
[20] Mayes W.M., Burke I.T., Gomes H.I. et al. Advances in understanding environmental risks of red mud after the Ajka spill, Hungary. J Sustain Metall. 2016, 2, 332–343. https://doi.org/10.1007/s40831-016-0050-z.
[21] Harris J. Soil microbial communities and restoration ecology: Facilitators or followers? Science. 2009, 325, 573–574.
[22] Joshi A.A., Kanekar P.P., Kelkar A.S., Shouche Y.S., Vani A.A, Borgave S.B., Sarnaik S.S. Cultivable bacterial diversity of alkaline Lonar Lake, India. Microb Ecol. 2008, 55, 163–172.
[23] Yang C., Niu Y., Su H., Wang Z., Tao F., Wang X., Tang H., Ma C., Xu P. A novel microbial habitat of alkaline black liquor with very high pollution load: Microbial diversity and the key members in application potentials. Bioresource Technol. 2010, 101, 1737–1744.
[24] Schmalenberger A., O'Sullivan O., Gahan J., Cotter P.D., Courtney R. Bacterial communities established in bauxite residues with different restoration histories. Environ Sci Technol. 2013, 47, 7110–7119.

[25] Santini T.C., Warren I.A., Kendra K.E. Microbial diversity in engineered haloalkaline environments shaped by shared geochemical drivers observed in natural analogues. Appl Environ Microbiol. 2015, 81(15), 5026–5036.

[26] Wu H., Tang T., Zhu F., Wei X., Hartley W., Xue S. Long term natural restoration creates soil-like microbial communities in bauxite residue: A case for 50-year field study. Land Degrad Dev. 2020, https://doi.org/10.1002/ldr.3728

[27] Krishna P., Babu A.G., Reddy M.S. Bacterial diversity of extremely alkaline bauxite residue site of alumina industrial plant using culturable bacteria and residue 16S rRNA gene clones. Extremophiles. 2014, 18, 665–676.

[28] Hamdy M.K., Williams F.S. Bacterial amelioration of bauxite residue waste of industrial alumina plants. J Ind Microbiol Biotechnol. 2001, 27, 228–233.

[29] Grant W.D., Mwatha W.E., Jones B.E. Alkaliphiles: Ecology, diversity and applications. FEMS Microbiol Rev. 1990, 6(2-3), 255–269. https://doi.org/10.1111/j.1574-6968.1990.tb04099.x.

[30] Agnew M.D., Kova S.F., Jarrell K.F. Isolation and characterization of novel alkaliphiles from bauxite-processing waste and description of *Bacillus vedderi* sp. nov., a new obligate Alkaliphile. Syst Appl Microbiol. 1995, 18, 221–230.

[31] Krishna P., Arora A., Reddy M.S. An alkaliphilic and xylanolytic strain of actinomycetes *Kocuria* sp. RM1 isolated from extremely alkaline bauxite residue sites. World J Microbiol Biotechnol. 2008, 24, 3079–3085.

[32] Arora A., Krishna P., Malik V., Reddy M.S. Alkalistable xylanase production by alkalitolerant *Paenibacillus montaniterrae* RMV1 isolated from red mud. J Basic Microbiol. 2014, 54, 1023–1029.

[33] Nogueira E.W., Hayash E.A., Alves E., Lima C.A.D., Adorno M.T., Brucha G. Characterization of alkaliphilic bacteria isolated from bauxite residue in the southern region of minas gerais, Brazil [J]. Braz Arch Biol Technol. 2017, 60. doi: 10.1590/1678-4324-2017160215.

[34] Wu H., Liao J., Zhu F., Millar G., Courtney R., Xue S. Isolation of an acid producing *Bacillus sp.* EEEL02: Potential for bauxite residue neutralization [J]. J Cent South Univ. 2019, 26(2), 343–352. doi: https://doi.org/10.1007/s11771-019-4006-x.

[35] Qu Y., Li H., Wang X., Tian W., Shi B., Yao M., Zhang Y. Bioleaching of major, rare earth, and radioactive elements from red mud by using indigenous chemoheterotrophic bacterium *Acetobacter* sp. Minerals. 2019, 9, 67.

[36] Liao J., Jiang J., Xue S., Qingyu C., Wu H., Manikandan R., Hartley W., Huang L. A novel acid-producing fungus isolated from bauxite residue: The potential to reduce the alkalinity. Geomicrobiol J. 2018. doi: 10.1080/01490451.2018.1479807.

[37] Qu Y., Lian B., Mo B., Liu C. Bioleaching of heavy metals from red mud using *Aspergillus niger*. Hydrometallurgy. 2013, 136, 71–77.

[38] Burgstaller W, Schinner F. Leaching of metals with fungi. J Biotechnol. 1993, 27, 91–116.

[39] Debaraj M., Jin Kim D., Ahn J.G., Rhee Y.H. Bioleaching A microbial process of metal recovery A review. Met Mater Int. 2005, 11, 249–256. doi: 10.1007/BF03027450.

[40] Vachon P., Tyagl R.D., Auclair J.C., Wilkinson K.J. Chemical and biological leaching of Al from red mud. Environ Sci Technol. 1994, 28, 26–30.

[41] Ghorbani Y., Oliazadeh M., Shahvedi A. Aluminum solubilization from red mud by some indigenous fungi in Iran. J Appl Biosci. 2008, 7, 207–213.

[42] Urik M., Bujdos M., Milova-Ziakova B., Mikusova P., Slovak M, Matus P. Aluminium leaching from red mud by filamentous fungi. J Inorg Biochem. 2015, 152, 154–159.

[43] Fulford G.D., Lever G., Sato T. Recovery of rare earth elements from Bayer process red mud. US Patent US5030424, 1991.

[44] Ochsenkuhn-Petropoulou M.T., Hatzilyberis K.S., Mendrinos L.N., Salmas C.E. (2002) Pilot-plant investigation of the leaching process for the recovery of scandium from red mud. Ind Eng Chem Res. 2002, 41, 5794–5801. doi: 10.1021/ie011047b.
[45] Borra C.R., Blanpain B., Pontikes Y., Binnemans K., Van Gerven T. Smelting of bauxite residue (red mud) in view of iron and selective rare earths recovery. J Sustain Metall. 2015, 2, 28–37. doi: 10.1007/s40831-015-0026-4.
[46] Qu Y., Lian B. Bioleaching of rare earth and radioactive elements from red mud using *Penicillium tricolor* RM-10. Bioresour Technol. 2013, 136, 16–23.
[47] Zhang D., Chen H., Rui Nie Z., Yuan Xia J., Lan Li E., Ping Fan X., Zheng L. Extraction of Al and rare earths (Ce, Gd, Sc, Y) from red mud by aerobic and anaerobic bi-stage bioleaching. Chem Eng J. 2020, 401. https://doi.org/10.1016/j.cej.2020.125914.
[48] Krishna P., Reddy M.S., Patnaik S.K. *Aspergillus tubingensis* reduces the pH of the Bauxite Residue (Red Mud) amended soils. Water Air Soil Pollut. 2005, 167, 201–209. https://doi.org/10.1007/s11270-005-0242-9.
[49] Babu A.G., Reddy M.S. *Aspergillus tubingensis* improves the growth and native mycorrhizal colonization of Bermuda grass in bauxite residue. Bioremediat J. 2011, 15(3), 157–164.
[50] Courtney R., Harris J.A., Pawlett M. Microbial community composition in a rehabilitated bauxite residue disposal area: A case study for improving microbial community composition. Restor Ecol. 2014, 22(6), 798–805.
[51] Liao J., Jiang J., Xue S., Qingyu C., Wu H., Manikandan R., Hartley W., Huang L. A novel acid-producing fungus isolated from bauxite residue: The potential to reduce the alkalinity. Geomicrobiol J. 2018. doi: 10.1080/01490451.2018.1479807.
[52] Sorokin D.Y., Kuenen J.G., Muyzer G. The microbial sulfur cycle at extremely haloalkaline conditions of soda lakes. Front Microbiol. 2011, 2, 1–16.
[53] Groudev S.N., Groudeva V.I. Microbial communities in four industrial copper dump leaching operations in Bulgaria. FEMS Microbiol Rev. 1993, 11, 261–267.
[54] Strelkova E.A., Pozdnyakova N.V., Zhurina M.V., Plakunov V.K., Belyaev S.S. Role of the extracellular polymer matrix in resistance of bacterial biofilms to extreme environmental factors. Microbiology. 2013, 82, 119–125.

Oluka N. Ngofa and George Z. Kyzas

8 Advancements in application of dispersants to oil spills

Abstract: Dispersants are gradually being recognized as an effective mechanism in oil spills management and includes surfactants or ("surface-active agents") and "solvents" that minimize interfacial tension between oil and water. They have been established as very effective in oil spill scenarios due to the dispersants synthesis materials applied, the likes of comb polycarboxylic acid (CPCA) dispersants, sodium dioctyl sulfosuccinate (SDOSS), ethoxylated sodium mono-alkyl sulfosuccinate, alkenyl succinimides synthesis (ASIs), alkyl poly-glycoside glycerol ethers, Schiff base, etc. The structures and properties of modern day dispersants are very important to be categorized into a vast assortment of "oil" types based on potency, procedure, and the right application ratios, due to the advancement of this type of applications. This process is interesting because it concentrates on some of the low toxic non-ionic surface-active agents and solvents as oil spill dispersants, which can lead to the minimization of environmental damage to marine life.

Keywords: Dispersants, Oil spills, Synthesis, Application methods

8.1 Introduction

Global emphasis has been laid to the consequences of marine oil spills contamination arising from land drainage, offshore oil exploration and production activities, vessels and pipelines incidents, maritime activities, and unauthorized effluent water discharges, both by university researchers and government authorities [1].

In their severity and degree of damage, oil spills are distinct. This is due to the change in varieties of crude, location, and the weather factors involved. Physical, chemical, and biological mechanisms control the propagation and actions of spilled oil in the water or aquatic ecosystem. These processes rely upon the "properties" of the oil, hydrodynamics, weather conditions, and climate. Advection, diffusion, and spreading are the physical processes; evaporation, dissolution, and emulsification are the chemical processes, while biodegradation is the primary biological mechanism [2–4].

Oluka N. Ngofa, Department of Chemistry, International Hellenic University, GR-654 04 Kavala, Greece, e-mail: oluka4freedom@yahoo.com
George Z. Kyzas, Department of Chemistry, International Hellenic University, GR-654 04 Kavala, Greece, e-mail: kyzas@chem.ihu.gr

https://doi.org/10.1515/9783110727227-008

Oil spillage is harmful to the economy and environment of many countries of the world, so a variety of techniques of cleaning have been developed. These methods can be divided into four main categories: (a) to "extract" the oil from the surface of the ocean by booms and skimmers methods; (b) to "enable" the oil to go through biodegradation; (c) to regulate the consuming (in situ burning) of the oil, permitting the oil to varnish into the environment (this component contaminates the air); and (d) to use chemical dispersants [2, 5, 6]. Methods mentioned above are aimed at extracting oil from the ocean surface.

Dispersants application in oil spills management essentially pushes the oil from the water surface to the "water column," seabed, or the environment. In a particular situation, decisions about applications of dispersants to combat an "oil slick" must be made. The time after the spills, the air conditions, the climate, and different oil types that have been spilt should be taken into consideration in the "decision-making" process. Inside an hour or two (2) after the underlying spill, using dispersants can be more successful [7–9]. "Surfactant and solvent" materials that can be synthesized are most successful or available chemical types of dispersants commercially used for advancement in the applications and effectiveness in the utilization of "dispersants" in oil slick reaction.

8.2 Techniques for oil-spills management

When crude oil, which basically contains hydrocarbons, in the range of 150–1800 mg/L, for example, BTEX ("benzene-toluene-ethylbenzene-xylene") that have some H_2O "dissolvability" and are toxic, is released into the marine ecosystem (environments) or land, it will definitely be subjected to a variety of biological, physical, and chemical changes in the ecosystems and is basically referred to as oil spills [5, 10, 11].

Oil spills are very dangerous and harmful occurrences for the marine environment (ecosystem) and affect marine life forms, which will get unnecessarily threatened due to the toxicity of unrefined hydrocarbon. It should be noted that ever since the exploration of "raw petroleum" from the oceanic resources, oil spills happen due to accidents on the shorelines, and this can be attributed to a variation in the types of oil, oil spills locations, and also the presence of weather conditions [12–14].

Crude oil (raw petroleum) is essentially one of the most important contributor of toxins in the marine climate (biological system) or maritime zones, and it contributes nearly three to four million metric loads of oil pollutants to the ocean annually; oil spills vary in the extent of effects or damages they can cause and in their severity. Irrespective of this, oil spills scenarios are very concerning as they can inflict a lot of damage to the environmental well-being. And therefore, it has become very important to employ various techniques for oil spills cleanup management [15–18].

Techniques for the management or response to oil-spills occurrence comprises almost ten methods or techniques, but in this chapter, we will focus on applications of "booms", "skimmers", adsorbents (sorbents), burnings (in situ), and very importantly, the chemical technique "dispersant" [19–23].

8.2.1 Use of booms for oil spills cleanup

The utilization of booms in "oil slick" situations is alluded to as one of the easiest or more famous ways of managing oil spills from spreading. To keep the crude oil (oil) from spilling across the sea surface, boom devices or strategies operate as a barrier.. Basically, boom machinery floats on the water surface; its enclosure is not designed to clean up oil leaks, but helps trap them. The best choice to carry out the cleanup exercise will be determined by how one of the spills was contained by the cleanup crew [24]. Booms are of two main categories namely, curtain and fence booms.

Usually, a boom is a mechanical intervention mechanism carried to a spill site and it is used to [25]:

1. Contain slicks for removal by burning or skimmers.
2. Exclude slicks and safeguard vulnerable shorelines and facilities from selected areas.
3. Deflect or redirect slicks to a collection area or away from sensitive resources.
4. Ensure that the addition of sorbent materials produces oil.

8.2.1.1 Curtain booms

Impervious, nonabsorbent, floating devices are "curtain booms." They are made of "polystyrene," "polyurethane," bubble wrap or stopper and have an enormous oval, froth-filled chamber that stays over the water and has adaptability or flexibility. The "breadth" of the chamber goes from 100–500 mm and length of the skirt varies from 150–800 mm. These devices are very reliable in clam water in the offshore scenario, with high towing versatility and work better than "fence boom"; however, they are used more to clean and store skirts that stay under the water [26–28].

8.2.1.2 Fence booms

A fence boom is a floating fence-like structure in the vertical plane consisting of rigid or semi-rigid materials that are free to bend in the horizontal plane. As 60% of the "booms" remains under the water and 40% is over the water surface, the aircraft conforms to the water flow and creates a vertical shield against floating oil [24, 27, 29].

8.2.2 Use of skimmers for oil spills cleanup

A collection of mechanical devices used to physically eliminate oil spills from the water surface is categorized as skimmers. A "transport line," set to bring the spilled oil into a tank where it is put away for extraction and recuperation, is used by many designs. For the recovery of spilled oil, other skimmer systems use "suction" and "weir" skimmers. In general, they are only considered successful in quiet water.

Obviously, it relies upon the type of oil" extracted, marine environment in the process of cleanup exercise, and the ice or debris present in the water; different types have different advantages and disadvantages [30, 31].

8.2.2.1 Weir skimmers

A barrier or framework installed at the oil/water interface is used by "weir skimmers." Oil floating on the water surface will pour over the dams and be stuck inside a well, taking as little water as possible with it. It is then possible to drain the trapped oil and water mixture out to a collection tank for processing or disposal through a pipe or hose. These skimmers are vulnerable to floating debris being jammed and clogged [31].

No weir skimmer is reliable in the steep wave, although swelling alone does not normally interfere with the operation of the skimmer. Few weir skimmers have an onboard pump to overcome friction losses along transfer hoses such that the recovered oil forces along the hose rather than relying on suction [30].

8.2.2.2 Oleophilic (oil-attracting) skimmers

"Oleophilic" skimmers use oleophilic substance belts, disks, or continuous mop chains to scrub the "oil" from the surface of the "water." Then the oil is pumped out of a recovery-tank or scraped off. Oleophilic skimmers have the advantage of flexibility, making them useful on spills' of any thickness. Other styles operate best on water that is clogged with debris or rough ice, such as the chain or rope-mop skimmer [31].

8.2.2.3 Suction skimmers

The simplest concept, in terms of operating theory, is a suction mechanism, whereby oil is recovered directly from the water surface by pumps or air suction systems. Vacuum tracks or trailers, which incorporate the elements of recovery, preservation, transport, and isolation of oil and water, are also accessible locally to the spill-site either commercially or by municipal or agricultural associations and are ideally suited to the recovery of oil on or near the coastline. While suction skimmers are normally successful, they are vulnerable to debris clogging and require continuous professional observation. On clear water, where oil has accumulated against a boom or fence, suction skimmers perform better [30, 31].

8.2.3 Use of adsorbent (sorbent) for oil spills cleanup

One of the most popular techniques or methodology in oil spills cleanup is "sorption," which has been described or noted to be successful. Adsorption is a simple, relatively low-cost technique. Absorbing properties of a wide range of porous sorbent materials, especially in the areas of oil spill cleanup have been studied for use in organic removal. It is further processed in some of these products and material to be utilized as sorbents; the oil can preferentially be in the form of liquids, i.e., hydrophobic and oleophilic [32, 33].

Several parameters governed by the adsorbent structure play a crucial role in the process, such as hydrophobicity, porosity, ideal pore sizes, and surface area. Surface area, a significant property of the adsorbent, helps in efficient separation of oil from water, as well as the high carbon or oxygen content, as this factor contributes to a strong recovery of oil from water [34, 35].

Sorbent substances can be used in adsorption or, less often, in absorption. During adsorption process, the oil is preferentially attracted to the surface, while absorbents absorb oil or other substance to be retrieved into the body of the substance. Adsorbents are the most common oil spill response products available; a few are individual absorbents. Oil removal can effectively happen in two ways, namely, absorption and adsorption. In pore spaces, absorbents permit oil to be absorbed into the substance they are made of, while adsorbents draw oil on their surfaces but do not cause the materials to be absorbed [36–38].

8.2.3.1 Adsorbents

Hydrophobic sorbents are highly useful in managing oil spills" when they operate during the skimming process as a final cleanup step for the residual oil. They facilitate the conversion of liquid into a semi-solid form for the complete disposal of crude. The adsorbent materials for oil spill cleanup are natural organics, synthetic materials, natural inorganics; adsorbent nanotubes, and nanoparticles [39, 40].

Activated carbons are the most frequently used adsorbents for both drinking and wastewater treatment [41–43]. This includes a large range of porous solids resulting from the carbonization of a wide variety of organic precursors and controlled activation such as coal, peat, coke, wood, fruit stones, nutshells, and a wide variety of organic residues [44, 45].

In literature, two categories of activation processes are identified, namely physical and chemical activation. Direct gasification using steam or carbon dioxide of a pre-carbonized material at temperatures between 850" and 1,100 °C involves the physical process, until being heated at temperatures about 700–900 C, while impregnation of the carbonized substance with a strong dehydrating agent, usually phosphoric acid involves chemical activation [46].

8.2.3.1.1 Synthetic adsorbents

Synthetic adsorbents are composed of polymer particles with spherical crosslinks that have a porous structure. The hydrophobic interactions between the compounds and the synthetic adsorbents primarily result in the adsorption of compounds to synthetic adsorbents. Synthetic adsorbents have an extremely porous structure due to special processing processes, allowing the adsorption of a wide variety of target compounds in significant amounts [47, 48].

The most widely used manufacturing adsorbents are synthetic adsorbents (Figure 8.1). It is commonly used for; polyester foam, polypropylene, and polystyrene and are available as rolls or booms, or sheets and can also be used on the water surface as powder. A maximum adsorption of oil ("4.5 g g^{-1} light cycle oil" or "light gasoline oil") has been recorded for polypropylene [49]. Open-cell polyurethane foams were able to absorb ultralight oil from oil–water mixtures 100 times their weight, as stated [50].

Because of their hydrophobic and oleophilic nature, synthetic adsorbents can adsorb 70–100 % of their oil weights. These formulations of synthetic adsorbents can be reused several times. The main disadvantages are their conservation since their use and non-biodegradability [51, 52].

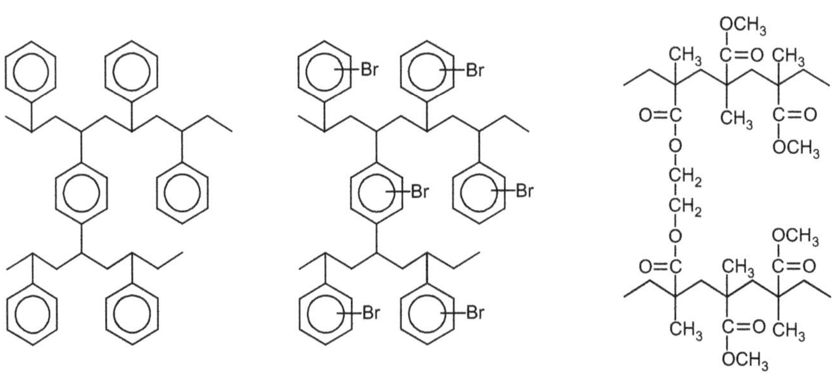

Polystyrenic adsorbent — Modified polystyrenic adsorbent — Polymethacrylic adsorbent

Figure 8.1: Chemical structure of synthetic adsorbents.

8.2.3.1.2 Natural organic adsorbents

Natural sorbents exhibit restricted oil capacity, strong buoyancy, and low-cost. The method of sorbent-disposal is a crucial factor when a sorbent-material is selected. Natural sorbents are known to be favored on the basis of water content criterion. The use of powdered or low-ash coal advanced technology allows cleaning of water from oil products. The carbon-containing materials such as brown coal (lignite), the remnants of dead plants, peat, and wood residue form the coal [53–55].

They include kapok, peat moss, milkweed, vegetable fibers, saw dust, and straw [56]. It has been shown that milkweed and cotton fibers absorb 74–85% of crude oil from the surface of an artificial sea water bath containing crude oil [57].

Sawdust reached the maximum adsorption limit of3.6 g/g sawdust, while "oleic corrosive" joined sawdust reached 6 g/g within 5 min, with the maximum adsorption of6.7 g^{-1} peat moss within 5 min [20].

Natural organic adsorbents are basically less expensive, readily available, and their weight is 3–15times their adsorbing efficiency. The main disadvantages are that they are energy consuming and adsorb water along with oil that leads to their sinking; It is also very difficult to remove adsorbents after spreading on the oil spill water and they must be disposed of [51, 58].

Organic adsorbents such as activated carbons composed mainly of imperfect concentrated aromatic sheets with exposed surface areas have a pronounced hydrophobic quality and interactions between these surfaces and nonpolar organic species occur predominantly through the forces of Van der Waal [46].

8.2.3.1.3 Natural inorganic adsorbents

In contaminated waters, the development of adsorbent showing a strong adsorption capacity and a low cost for the removal of various pollutants were particularly essential. Natural adsorbents such as charcoal, clay crystals, clay, and zeolite adsorbents have also been thoroughly studied to provide an alternative to activated charcoal in the treatment of surface and ground water and effluents] [59].

Natural inorganic sorbents are also less expensive and their absorption-capacity is 4–20 times their weight. The main risks when using them are that they are not certified for the water surface, some natural inorganic adsorbents such as clay and vermiculite are loose and very difficult to apply in windy conditions, and that, if inhaled, they are associated with potential health risks [51].

The presence of functional groups carbonyl, carboxyl, hydroxyl, or inorganic inclusions gives the adsorbent surface a polar nature that enhances its ability to interact, such as metallic cations, with polar and charged species [46]. On surfaces of inorganic adsorbents such as; zeolites and clay, this hydrophilic character is much more pronounced due to their inorganic existence [42].

8.2.3.1.4 Nanotubes and nanoparticles adsorbent

As a modern concept in adsorbents, carbon nanotubes (CNTs) have attracted a lot of interest due to their extremely high adsorption capacity for separation between oil and water. The high hydrophobicity of CNTs makes them strong candidates for enhancing the wastewater de-oiling process [60]. For extracting oil particles from water, the updated CNTs have proven to be very promising.

As a new adsorbent type, the interest in CNTs in the field of oil adsorbent materials lies in their light-weight, excellent hydrophobic properties, flexibility of surface modifications, and functionalization due to their extremely high adsorption capacity to distinguish oil-water. CNTs and multiwall carbon nanotubes [MWCNTs] have thus evoked a great deal of interest [61]. CNTs are excellent candidates for enhancing the water de-oiling process because of their high hydrophobicity.

The purification and treatment of oil–water by using CNTs as adsorbents has been established by researchers, and it is a new interesting area of study. The reported oil–water separation experiments were mainly concerned with cleaning up oil spills from the ocean floor [62].

8.2.3.2 Absorbent

The absorbents available for environmental response or methodology are manufactured from engineered polymers with a high surface area to allow efficient absorption. For volatile materials, absorbents can be used as they may reduce the material's surface area [49, 63]. Absorbent materials are technologically capable of recovering light fuel oils and some crude oils, the time needed for absorption may be longer than feasible or desirable, and, as a result, they are more suitable for recovering especially hazardous and noxious substances from low-viscosity liquids and spilt chemicals discussed in the separate ITOPF paper on the response to marine chemicals. Therefore, in oil spill reaction, absorbents are less frequently encountered than adsorbents [22, 64], in coping with the prevalent environmental catastrophe of oil spills and hydrophobic oil absorbents of interconnected porous surface, which have been widely used [64].

8.2.4 Use of in situ burning for oil spills cleanup

In situ burning is the oldest technique in the management of oil spill recovery, which implies the controlled burning of oil slicks to remove the spill of offshore oil before it approaches the coastline. Burning technique has been successfully used in many large spill responses, including the2010 Deepwater Horizon spill, where an estimated 200,000 to 300,000 barrels of oil were burned on shore [19].

Additionally, when effective, no further downstream oil separation or removal processes are required after the application of in situ burning. Once it is burned, the oil is differently eliminated. Generally, it is considered successful in most cases; several considerations are also involved in determining to use of in situ burning examples including: environmental concerns (toxic or hazardous emissions are generated by burning oil); technological constraints (accessibility of fire-resistant containment booms and appropriate spill site igniters); and operational requirements (effectiveness of the specific spill oil technique); and benefits of in situ burning include: extremely fast and efficient oil removal (efficiencies often exceed 90% with removal rates of2000 m^3/h [65, 66].

In-place burning oil spills frequently establish a comprehensive plume of smoke containing black soot and other combustion materials produced in the burning process. Lack of awareness of the magnitude of the smoke plume area affected by combustion of crude oil spills and risk of toxic combustion materials being carried in the plume have generated widespread questions about the consequences of deliberate combustion of significant crude oil spills. The application of deliberate burning to oil spills has also been hampered by unresolved questions about personnel and equipment safety from heat and thermal radiation produced by large fires [67].

Local authorities need to provide mechanisms in the decision-making process for authorizing deliberate burning of oil spills to evaluate the possible "benefits" of burning in terms of oil extraction and the possible consequences in terms of the plumes of smoke emitted by the burning of oil.

8.3 Dispersants

In certain scenarios, such as unfriendly atmosphere conditions or deep water, dispersants are gradually being recognized as an effective response mechanism in oil spills management [68]. A well-known chemical agent, dispersants are more soluble in water and slightly soluble in oil. By encouraging the breakdown of oil spills into fine droplets, these chemicals alter the fate of oil emissions. The fine oil droplets produced by the dispersant function are suspended in the water of the mass and dispersed. Therefore, oil penetration into the column of seawater will increase. In the water column, the scattered oil fragments decompose quicker than the levels of oil and create environments that are less hazardous to the environment [26, 69].

Dispersants are often a liquid combination of surface-active agents (surfactants) and solvents that are meant to speeded and break up fine droplets of oil spills that spontaneously spread in the ocean. The complex parts of the surfactants are "cleanser-like" atomic structures that have all oils searching for lipophilic areas and hydrophilic water-chasing. Mixtures of surfactants greatly reduce "interfacial-tension" between water and oil; consequently, oil will be separated into fine-droplets, which will basically disperse the oil droplets into the water-column. It should be acknowledged that due to natural harm it could cause, the use of dispersants is a delicate subject, for example, a country like Greece always considers the use of dispersants as a final option in an intervention for oil spill outbreak. Especially in the event of large oil spill scenarios, the aspersions are still applicable [26, 70, 71].

8.3.1 Brief history of dispersants usage in oil spills management

Chemical dispersant agents that were effectively used as oil removing substances in operating rooms of bilges, engine, and tanker chambers were produced. In late 1967, this dispersant, often referred to as a "detergent" was used to disperse crude oil that was spilled off the England Coast from the "Torrey Canyon" tanker; the spill was estimated to be around one million barrels or 42,000,000 gallons of unrefined petroleum products from Kuwait. In a phenomenal attempt to disperse the oil, about 10,000 barrels or 420,000 gallons of various chemical moisturizers or cleaning solvents were added to the water and shoreline, and it was discovered to be extremely toxic to the aquatic environment. This was largely due to the aromatic hydrocarbons in the solvent and alkylphenol surfactants [72, 73].

Occurrences of "Torrey Canyon spill" saw the utilization of chemical dispersants as one of the important response methodologies that had a significant and unwanted effect and created several concerns regarding the environmental effects that have continued until the present moment. The Torrey Canyon tanker spill was a major opportunity for Canada, Europe, and the United States to endorse strong testing and support services to assess the adequacy of the chemical dispersant, establish provincial implementation laws, and build up the second and third generations of chemical dispersants [74].

The "Deepwater Horizon" (DH) oil platform, located in the Gulf of Mexico, approximately 40 miles off the coast of Louisiana, erupted on April 20, 2010, resulting as one of largest aquatic oil spills" in history [75–77]. Macondo well spill was estimated to be around 4.9 to 5.8 million barrels of crude oil prior to stabilization for approximately three months. This raised highly interconnected environmental, economic, and human health threats [78]. Over 2.1 million gallons of dispersant was used in between May 15 and July 12 for the Deepwater Horizon oil spill on the surface of about 1.4 m gallons and wellhead of about 0.77 m gallons of the shoreline [79]. Dispersant was applied in both phases to reduce the interfacial tension between oil and water, thus decreasing the size of oil droplets generated by ejection of oil and gas (deep-water) or wave action (surface) from the wellhead [80]. Applying dispersants on the "surface" of the seawater will prevent the development of large emulsions or slicks capable of covering and harming fragile marine environment. Oil mixes below the surface of seawater followed addition of dispersant that will be degraded or dissolved into the water-column [81, 82].

The U.S. "Environmental Protection Agency" (EPA) recognizes 23 affirmed industry dispersants authorized for use in strategies of response to oil spills, including Corexit EC9500B, Corexit EC9500A and Corexit EC9527A, which are applicable and used around the globe, with combinations of surface-active agents and different proportions of solvents [23].

8.3.2 How dispersants work

Oil appears to often show a coherence or reluctance to break up due to interfacial-tension between oil and water that shows coherent strength as crude oil is often thought of as unrefined petroleum spilled on sea or water. In essence, oil and water do not combine well. Since oil spreading in a calm environment is drifting [83], the activity of wave mixture will consolidate water and oil in two separate stages.

Natural dispersal: The waves can allow the oil spill to turn into a tiny droplet of oil that is conveniently suspended in water in this case. Most of these oil droplets are sufficiently large to buoy and easily return to the surface, because a little bit of the oil is a kind of small droplet, and the push is almost impartial. These extraordinarily tiny droplets of oil will stay dispersed in water, will stay virtually infinite, and will be repeatedly extracted by wave action under pressure [69, 84].

Water-in-oil emulsion: Conversely, the process of blending waves may generate water- in-oil droplets in the form of water–oil emulsions, also referred to as "chocolate-mousse." The emulsions have a viscosity that is higher than that of the oil. As the emulsion volume normally includes up to 75% of water, it will eventually rise too many times the measure of the oil spilled. Because of the rise in viscosity, the volatile constituents in the oil appear to evaporate and resist the effects of waves in turning oil into tiny droplets, as well as increasing formation of emulsion, because water droplets are exhausted more slowly than oil, with a higher consistency [69, 85, 86].

8.3.3 Physical characteristics of dispersants

Most "physical" properties of dispersants can have functional implications on the use of these materials. For this reason, most countries around the global have some viscosity and/or pour point, flash point, and stability or shelf life criteria in their approval process [87, 88].

8.3.3.1 Specific gravity

It is the ratio between the weight of solid or a liquid and the weight of an equivalent amount of H_2O at a given temperature. In general, conventional dispersants have smaller specific gravities (ranging from 0.80–0.90) than concentrates that range from 0.90–1.05.

8.3.3.2 Pour point

Pour point is the temperature that this liquid is not going to flow below. The pour point of most dispersants is well below 0 °C (–40––10 °C) and can never solidify under most conditions.

8.3.3.3 Viscosity

A liquid's viscosity is known as its flow resistance. Temperature is basically dependenton viscosity of the dispersants (Table 8.1).

Table 8.1: Viscosity, a standard dispersant viscosity spectrum, affects the size of the "dispersant-droplets." In this respect, some countries may need certain restrictions on the viscosity of the dispersant.

Typical viscosity profiles of dispersants cP/temperature °C	0 °C	20 °C
Conventional	10 to 50	5 to 25
Concentrates	60 to 250	30 to 100

8.3.3.4 Flash point

The lowest temperature at which, when exposed to a flame, vapors above the toxic and carcinogenic substance ignite in the air. The flash point of most dispersants is over 60 °C and can be considered nonflammable. Any nation can restrict the flash point for practical safety purposes.

8.3.3.5 Others

The components of certain dispersants can allow the packages (drums or containers) in which the commodity is stored to become corrosive over extended periods.

Regulations relating to dispersants in some countries also mandate that the substance does not contain certain components.

8.3.4 Compositions of dispersants

The modern chemical treatment of oil spills known as "dispersants" consists of a combination of surfactants and solvents that minimize interfacial tension between oil and water. It is defined as the change in free energy that occurs at the interface between two immiscible phases of matter arising from the difference in 25 contact zones, namely liquid-to-liquid, solid-to-liquid, and liquid-to-gas [82].

8.3.4.1 Surfactant

The well-known main factors or components of chemical dispersants are: surfactants (surface-active agents). Surfactants used other than in oil spill management cause deterioration in the aquatic ecosystems. That has been a significant problem since the fifties. The compounds contained both an oil-compatible (lipophilic) and water-compatible (hydrophilic)] group of hydrophilic-lipophilic balance (HLB) [71, 89].

The real syntheses of the particular dispersant items can, obviously, differ significantly, However, in view of a review of literature, the use of surfactants in current dispersants are large mixes of anionic and non-anionic types: anionic surfactants incorporate salts of dialkyl-sulfosuccinates and of alkyl benzene sulfonic-acid; and the non-anionic surfactants incorporate sorbitan esters of greasy acids, polyalkoxylated greasy alcohols, polyalkoxylated sorbitan esters of unsaturated fats, polyethylene glycol esters of oleic corrosive and tall oil esters. Sorbitan monolaurate, ethoxylated sorbitan trioleate, ethylene/propylene oxide condensates, ethoxylated tridecylphosphate, sodium dioctyl sulfosuccinate, sodium lauryl sulfate, and isopropylamine dodecyl benzene sulfonate are basic examples of surfactants used [71, 90].

8.3.4.2 Solvent

Solvent also involves the formulation of chemical dispersants, and certain surface agents that are used in the formulation of oil spill dispersants are viscous as well as other types of solvents that are required to minimize viscosity, by which conventional spray equipment can properly adhere the mixture. Solvents can also work to lower the freezing point to reduce the use of temperatures and to improve the penetration/mixing of surfactants into different viscous oils. The current surfactant has demonstrated very low toxicity, allowing aqueous solvents or dearomatized hydrocarbons replaced by the current formulations, resulting in very low toxicity dispersants relative to early formulations [91].

The disintegration of additives and surfactants into a homogenous mixture of dispersants is mostly encouraged by water-based (aqueous) solvents. The use of dissolvable hydroxyl (solvent) or hydroxy mixes (compounds) can be critical because basic alcohols form hydrogen-bonds when reacting with H_2O, and the resulting properties of the mixes appear to be relatively "water-miscible," in general [73, 82].

8.3.5 Structure and properties

Surfactant and solvent are the two major chemical types of oil dispersants commercially used for effectiveness in oil spill scenarios for the decrease of the oil water (O=W) interfacial-tension. Surfactants help disperse oil into tiny droplets in water. Solvents are applied in the case of the dissolution of additives and surfactants to result in a homogeneous system. The additives serve many purposes such as to increase the biodegradability of scattering, and improving the dissolution of the dispersants into the oil spill and to increase the long-term stability of the dispersion. The surfactant molecule has two components: anionic and the non-anionic [92]. This structure gives the ability to the molecule to make a dispersant without any formation (Table 8.2).

Table 8.2: Structure and properties.

Names	Properties	
Polyisobutylene [93]	[Molecular-Formula] Mf	(C_4H_8) n
	[Melting point] Mp	54–56 °C
	[Boiling point] Bp	300 °C
	[Density]	0.92 g/mL at 25 °C(lit.)
	[Refractive index]	n20/D 1.51
Maleic anhydride [94]	MF	$C_4H_2O_3$
	Mw	98.1
	Bp	202 °C
	Mp	53 °C
	[Density]	1.5 g/cm^3
	[Solubility in water]	reaction

Table 8.2 (continued)

Names	Properties	
	[Vapor-pressure]	Pa at 25 °C: 25
	[Flashpoint] FP	102 °C c.c.
	Explosive limits, vol % in air	1.4–7.1
Polyethylene glycols [95]	MF	$C_{2n}H_{4n+2}O_{n+1}$
	[Molar mass]	44.05 n + 18.02 g/mol
	[Density]	1.125
	Mw	104.15
	BP	250 °C
	MP	−65 °C
	FP	171 °C
	Signal-word	Danger
Penta-methylene-hexamine (PMHA) [96–98]	MF	$C_{10}H_{28}N_6$
	BP	662–734 °F at 760 mm Hg
	FP	> 200 °F
	Solubility	> 100 mg/mL at 70 °F
	Density	1 at 68 °F
Sorbitan oleate [96, 99]	MF	$C_{24}H_{44}O_6$
	Mw	428.6 g/mol
	Color/Form	Amber liquid
	[Solubility]	Dispersible in water
	[Viscosity]	1,000 Centipoises
Ethylene glycol [100]	MF	$C_2H_6O_2$
	[Molar mass]	62.07 g/mol
	[Appearance]	Clear, colorless liquid
	[Odor]	Odorless
	[Density]	1113.20 kg/m^3
	MP	−12.9 °C
	BP	197.3 °C
	[Viscosity]	6.1 c. P (@20 °C)

Table 8.2 (continued)

Names	Properties	
Alkyl polyglycoside [101–103]	Superior wetting properties, low surface tension. Good compatibility with all other types of surfactants. Low gel phenomena, tolerance to hard water. Good detergency property. Mild, low-toxic, low-irritating. Readily biodegradable.	
Polyoxyethylene (P) sorbitol tetra-oleates [93, 104, 105]	Formula	$C_{90}H_{174}O_{20}$
	Mw	1576.33 g/mol
	BP	> 100 °C(lit.)
	[Density]	1 g/mL at 25 °C(lit.)
	FP	113 °C
	[Solubility]	Propylene-glycol: insoluble
Dioctyl sodium sulfosuccinate (DOSS) [96, 106]	Formula	$C_{20}H_{37}NaO_7S$
	Molar mass	444.5583 g/mol
	Melting point	153 °C
	Classification	Sulfonic acid
	Soluble	In water
Linear alkylbenzene sulfonates acid [104]	M. Formula	$C_{18}H_{30}O_3S$
	[Molar Mass]	326.49 g/mol
	MP	10 °C
	[Density]	.06 g/ml
	BP	315 °C
	[Appearance]	Brown liquid
	[Solubility]	Soluble
Sodium lauryl sulfate [96, 98, 107, 108]	M. Formula	$C_{12}H_{25}NaO_4S$
	Mw	288.38 g/mol
	MP	399 to 405 °F
	Density	> 1.1 at 68 °F
	Surface Tension	39.5 dyn/cm at 25 °C
	Solubility	Less than 1 mg/mL at 66° F

Table 8.2 (continued)

Names	Properties	
	Odor	Faint odor of fatty substances
	Color/Form	White or cream-colored crystals, flakes, or powder
Sorbitan monolaurate [104]	Chemical formula	$C_{18}H_{34}O_6$
	Molecular Weight	346.464 g · mol – 1
	Specific Gravity	
	Boiling Point	516.12 °C [@760.00 mm Hg (est)]
	F. Point	> 230.00 °F. [TCC (> 110.00 °C)]
	Density	1.032 g/cm^3
	Solubility in water	insoluble
Isopropyl amine dodecylbenzene sulfonate [96]	Molecular Formula	$C_{21}H_{39}NO_3S$
	Molecular Weight	385.6 g/mol
	Soluble	In water
Phthalic anhydride	M. formula	$C_8H_4O_3$
	Molar mass	48.1 g/mol
	Appearance	White flakes, solid
	Density	1.53 g/cm^3
	MP	131 °C
	BP	295 °C

8.3.5.1 Synthesis

8.3.5.1.1 Synthesis of comb poly-carboxylic acid (CPCA) dispersants

The CPCA dispersant's shown in Figure 8.2 can be assembled by a radical polymerization under special conditions from macro-monomer methoxy poly (ethylene-glycol) methacrylate "m-PEGMA" and Methacrylic-acid "MAA". m-PEGMA was prepared by MAA condensation and m-PEG (Mw 1/4. 350, 550, or 750) in the presence of catalyst methane sulfonic acid, "MSA" and radical polymerization inhibitor hydroquinone in the presence of methyl-ethyl ketone, "MEK" at 80 °C for 7 h. The substances were extracted twice in "n-hexane" by precipitation and were dried in an oven. Therefore, polymeric surfactants that can be used as anionic dispersants are the CPCADs. In the presence of MSA as a catalyst, m-PEGMA was synthesized by esterification of MAA and m-PEG [109].

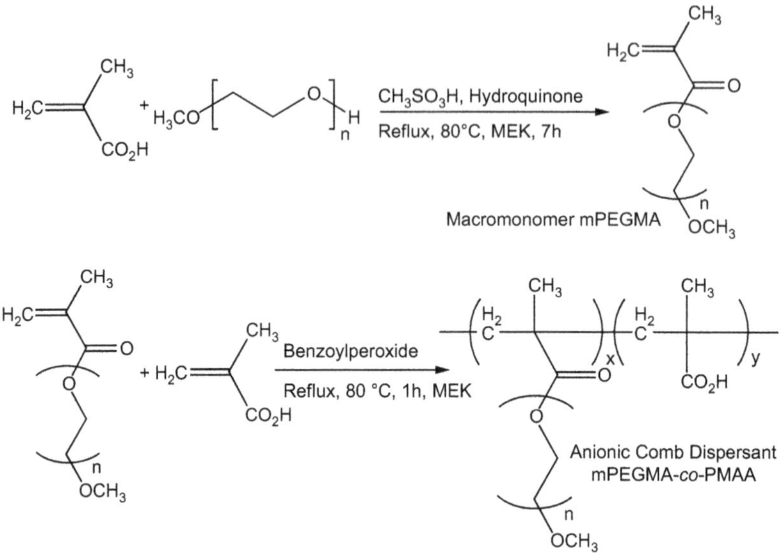

Figure 8.2: Synthesis of CPCA dispersants.

8.3.5.1.2 Synthesis of alkyl polyglycoside butyl ethers

Enhanced foaming is a property of alkyl polyglycolides that are frequently in demand. However, this potential is viewed as a disadvantage in many applications. The development of alkyl polyglycoside derivatives is of concern, too. Alkyl polyglycoside butyl ethers have been synthesized with the purpose of combining good cleaning efficiency with just a minor propensity to vacuum about foam [110].

The aqueous solution reaction is a drawback on an industrial scale, as without additional work-up stages, distilled water-free products cannot be obtained. Figure 8.3 indicates that water-free process was therefore created. Initially, alkyl polyglycoside was added to the reactor with an abundance of butyl chloride at 80 °C heated temperature. There was initiated reaction by potassium-hydroxide, when added as the catalyst. The mix reaction was then "neutralized" at the conclusion of the reaction, the precipitate of potassium chloride is filtered out, while butyl chloride is distilled out in excess. Both alkyl polyglycoside and alkyl polyglycoside butyl ethers make different products. The ratio of alkyl monoglycoside, alkyl mono glycoside mono-butyl ether and alkyl monoglycoside poly-butyl ether is 1:3:1.5 according to GC research. Figure 8.6 shows the alkyl-polyglycoside etherification reaction in the direction for °C. The quality of mono glycoside dropped from about 70% to less than 200/0.0. At the same moment, the mono ether value increases to 500/0.0. The more monobutyl ether available, the more it is possible to form polybutyl ethers from it. There is substantial formation of polybutyl ethers only after 24 h. The content of polyethers increases with time, as predicted. A value of 20% is, however, not met [111, 112].

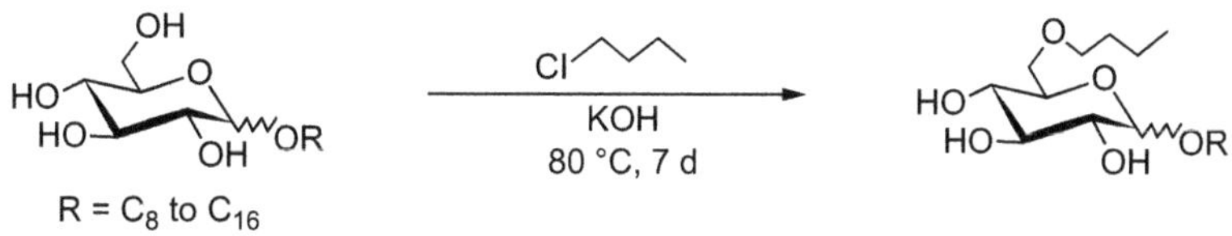

Figure 8.3: Synthesis of alkyl-polyglycoside-butyl-ethers.

One to three butyl groups per unit of alkyl glycoside are the typical degree of etherification. For a C alkyl polyglycoside, the best effects are achieved. The findings escalate in the case of alkyl polyglycoside butyl ethers, where n = 8 or 16. From these three examples, it is obvious that alkyl polyglycoside derivatives are readily available. It is, therefore, decided that the basic application envisaged requires the surface-active features of these derivatives.

8.3.5.1.3 Synthesis of alkyl polyglycoside glycerol ethers [111, 113]

Alkyl polyglycoside glycerol synthesis of ethers was performed by three different approaches [Figure 8.4]; only the alkyl mono glycoside is seen as the educt instead of the alkyl polyglycoside" mixture. The etherification of alkyl polyglycoside with glycerol by process "A" continues under reaction conditions, in the presence of simple catalysts, and "B" process take place in an opening ring of an epoxide. An alternative system, "C" reacted with glycerol carbonate, which is followed by the removal of CO and is likely to continue as an intermediate stage through epoxy. The reaction involved in nucleophilic substitution by method A is easy to regulate. Initially, in the ratio of 1:2, alkyl polyglycoside and glycerol were added into the reactor, melted in an atmosphere of nitrogen. 2% by weight potassium hydroxide was added when the reaction mixture was mixable.

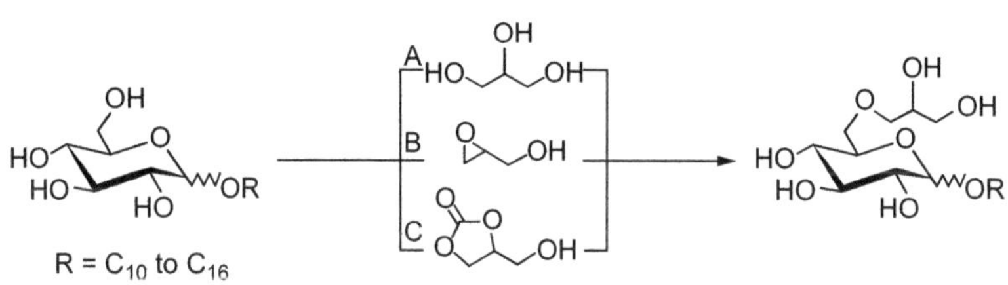

Figure 8.4: Synthesis of alkyl polyglycoside glycerol ethers.

The reaction mixture is then heated at 200 °C for a span of 7 h after which the formulated water is distilled continuously to transfer the equilibrium to the product side, as far as possible. As predicted, in addition to the mono,glycerol ether, alkyl polyglycoside bi- and triglycerol ethers are formed. Another secondary reaction is glycerol self-condensation to form oligoglycerols that can react in the same manner as glycerol to the alkyl polyglycoside. These higher levels of "oligomers" can be important because they further increase the [hydrophilicity] and the solubility of the products in H_2O. The products may be dissolved in H_2O after etherification and bleached in a known way, such as with hydrogen peroxide. The degree of etherification of the products under these reaction conditions

The alkyl polyglycoside that was used is independent of the alkyl chain length. The percentage content of mono-, di- and triglycerol ethers in the crude product mixture for four different lengths of the alkyl chain is shown in [Figure 8.7]. The Clz alkyl polyglycoside reaction produces a standard outcome. Mono-, di- and triglycerol ethers are formed in the ratio of approximately 3:2:1, according to a GC. The average glycerol-ether content is about 35%.

8.3.5.1.4 Synthesis of Sodium dioctyl Sulfosuccinate (SDOSS)

A blended aqueous solution of different surfactants was prepared from commercial SDOSS and an anionic–nonionic surfactant moiety E (n) SMOSS, E (n) SMLSS, or E (n) SMCSS. 0.2, 0.4, 0.6, and 0.8 can represent the molar fraction, alternately, and their respective mixture was diluted stepwise with H_2O and stirred for 30 min in a thermostat at 30 °C for the establishment of their equilibrium, and the surface tension [ST] was measured at 25 °C for these surfactants individually, where n can be 9, 14, 23, or 90.

8.3.5.1.5 Synthesis of monoalkyl sulfosuccinate

0.1 mol of Sulfosuccinic acid was reacted with three different fatty-alcohols at 0.1 mol, namely, octyl, lauryl, or acetyl alcohol, in the presence of 100 mL xylene as solvent, and 0.1% of p-toluene sulphonic acid as a catalyst around four hours at 140 °C of heating and reflux. After complete evacuation of the theoretical measure of H_2O of about 1.8 mL, the solvent was stripped out using a rotary evaporator. The reaction mixture was then transferred to a three-necked flask equipped with a stirrer, condenser, and a dropping funnel. NaOH at 30% conc. was added dropwise with vigorous stirring and heated on a H_2O bath until slightly alkaline [phenolphthalein indicator]. In the wake of cooling, the mixture was moved to an isolating funnel and left until it separated into two layers. The fluid-layer, which contains [SMASS], was washed for few times with oil ether at 60–80 °C to remove the unreacted fatty alcohols. After H_2O evaporation, the product was refluxed in excess absolute [ethyl-alcohol] for 30 min. Recrystallization, cooling, and then filtration were conducted to isolate the products [114].

8.3.5.1.6 Synthesis of ethoxylated sodium monoalkyl sulfosuccinate

SMASS of 0.1 mol was responded with 600 14 ethylene oxide units polyethylene glycol of 100 ml xylene presence as solvent and with 0.1% of p-toluene sulphonic acid catalyst. At 140 °C heated reaction ingredients with 2 h continuous stirring and refluxing and response mixture was dissipated, and the residue organic was disintegrated in 30 ml of isopropanol, and then poured into supersaturated NaCl solution at 90 °C, the mixture is heated for 1 h with isolated organic layer, and isopropanol refined off to give the relating surfactants [E (14) SMOSS, E (14) SMLSS, and E (14) SMCSS] [114], where n can be 9, 14, 23 or 90.

8.3.5.1.7 Alkenyl succinimides synthesis (ASIs)

4.00 g, 15 mol [2-dodecenyl]-succinic anhydride of a flack was charged at heated temperature of 60 °C, under a nitrogen source. The "n-alkyl amine" was added with 1.1 equip and the resulting mixture was then heated for 24 h at130 °C, stirring under the nitrogen flow. With flash column GC on silica gel, the crude oil product was purified using 20% of ethyl acetate in petroleum spirit [40–60 °C].

ASI-Asi, [N-propyl, n-butyl, n-pentyl, n-hexyl, and n-heptyl amines] were prepared with dispersant analogues submitting the corresponding [N-alkyl ASI]. Transparent and colorless-liquids were given with 98% of good yields purity for all alkyl ASIs as calculated by flame ionization detection gas chromatography (GCFID). Proton nuclear magnetic resonance spectroscopy [1H NMR], and GC with electron ionization mass-spectrometry [GC-EI-MS] and electrospray ionization mass-spectrometry [ESI-MS] have characterized the distilled sample [115, 116].

8.3.5.1.8 Synthesis of Schiff base

In room novel crude oil dispersants (861 temperature) for 1 h, 20.634 g, 0.2 mol of diethylene-triamine mixture, and 48.848 g, 0.4 mol of salicylaldehyde in 100 mL methanol was stirred to give a yellow liquid. 67.4 g, 97.00% of pure Schiff base A was obtained after methanol evaporation.

b, b0-dichlorodiethyl ether and polyethylene glycol usage of Schiff base ethoxylation. With NaOH as a catalyst, four different Mw of "PEG" [200, 600, 1,000 and 2,000 g = mol] were reacted with compounds A and b, b0-dichlorodiethyl ether to create the dispersants by the following process: 0.1 mole A, 0.1 mole b, b0-dichlorodiethyl ether, 0.1 mole b, 0.1-dichlorodiethyl ether, 0.0-dichlorodiethyl ether were applied to three-neck round-bottom flask of 250 ml, mechanical stirrer, thermometer was used fitted with a condenser. When both base ends are ethoxylated, the ratios will be doubled. It was heated up slowly to a temperature of 343 K with a steady stirred up reaction. This temperature was maintained for 5 h. The progression was accompanied by the determination of the NaCl material, which rises steadily at the end of the reaction to achieve a constant value. An equivalent vol of saturated NaCl solution with a dilute HCl, which was neutralized, was then treated with the product. With an increased mixture temperature at 363 K and held for 1 h at 323 K with a constant weight, the upper "waxy" layer was then removed and dried in a vacuum-oven. By using different Mw of PEG, and hydrophilic = lipophilic balances

[HLBs] were obtained. Based on the Mw of the polyethylene glycol added, the formulated surfactants were designated as A200, A600, A1000 and A2000 [117].

8.3.5.1.9 Synthesis of alkyl polyglycoside carbonates

APGs carbonates have been prepared by transesterification of diethyl carbonate monoglycoside. In the interests of research and development, it is proven to use diethyl carbonate by combining the reactants, so it can act both as transesterification component and also a solvent. Mole of 2% at 50% solution of NaOH is applied dropwise with stirring at 120 °C to this mixture; and it is allowed to cool to 80 °C after 3 h (hours) under reflux and neutralized with 85 9′0 phosphoric acid. In a vacuum, the residual diethyl carbonate is distilled. One hydroxyl group is ideally esterified in these reaction conditions. 1:2.5: 1 (Monoglycoside: mono-carbonate: polycarbonate) is the ratio of remaining educts to products [111, 118].

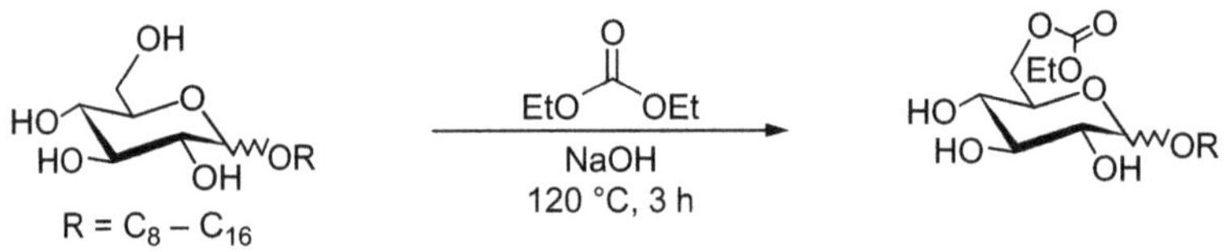

Figure 8.5: Synthesis of alkyl polyglycoside carbonates.

In this reaction, solvents with a "relatively" high degree of substitution are often formed in addition to mono carbonate. The response of professional management added degree of carbonate can be managed. A distribution of mono-, di- and tricarbonate of 7, 3, and 1 is obtained for a C, monoglycoside under the reaction conditions just stated (Figure 8.9). The key product is C, monoglycoside dicarbonate, if the time is extended to 7 h during the reactions, 2 moles of ethanol are distilled out at same time factor. If it is raised to 10 h and 3 moles of ethanol are distilled out, tricarbonate is the main component eventually obtained. Thus, carbonate degree addition, and hence the hydrophilic/lipophilic balance of the alkyl polyglycoside compound, can simply be changed by adjusting the time and the volume of the distillate reaction [111, 118].

8.3.6 Characterization

The characterization of the Schiff base materials undergoing purified synthesized compounds was studied using the FTIR spectrophotometer of the ATI Mattson Genesis Sequence. The analyzed samples were done as thin film between two disks of KBr. In CDCl3, surfactants, which have been prepared were basically analyzed and dissolved by means of JNM-EX model NMR spectrometer "Joël" 270-MHz as "spectroscopic-spectrometer," techniques that are used in determination of HLB and chemicals composition by comparing the integrals trace of various peaks [117].

1 H-NMR and GPC can be used to characterize CPCA dispersants. With a calculated value based on their molecular-weights, m-PEG volume, and values of acid, the dispersion of CPCA dispersants varies. Titanium dioxide (TiO) dispersion has been studied in traditional solvent-based paint formulations. The physiochemical and mechanical properties of CPCAD-containing surface coatings, for example, hardness, contrast ratio, and gloss were studied. In graft copolymeric surfactants, the CH_2 protons of m-PEG segments were the 1 H-NMR spectrum at 3.5–3.6 ppm. There were CH_2 protons at 1.8 to 2.0 ppm on backbone. The substitute CH_3 protons on backbone were also between 1.2 to 1.3 ppm. The CH_3 methoxy m-PEG protons were at 3.6 ppm [109].

The characterization of industrial SDOSS and [E (14) SMOSS, E (14) SMLSS, or E (14) SMCSS], different surfactants of aqueous mixed solution were prepared. The separate molar fractions 0.2, 0.4, 0.6, and 0.8 were diluted stepwise with water alternately and with their respective mixture. To determine their equilibrium, the individual surfactant materials undergo stirred activities in thermostats at 30 °C for 30 min, and surface tension (ST) were calculated at 25 °C [119], using the "dognon-abribat" tensiometer (serial number-03295.00, "Prolabo") for ST measurements. Double-distilled water (conductivity (1.1×10^6) mho cm^{-1} at 25 °C) was measured against the instrument. The solutions calculated were freshly prepared before each assessment [120].

$$\begin{array}{l} NaSO_3 - \underset{|}{CH} - COOR \\ \qquad\quad\; CH_2 - CO\,(OCH_2\,CH_2)_{14} - OH \end{array}$$

where:

$R = C_8H_{17}$ E(14)SMOSS

$R = C_{12}H_{25}$ E(14)SMLSS

$R = C_{16}H_{33}$ E(14)SMCSS

FTIR and 1-HNMR spectroscopic analysis validate the structure of the prepared surfactants. The presence of the "ester-band", SO_3Na, OH, ethereal bands, both from FTIR and 1 H-NMR showed that surfactants prepared were basically fully "ethoxylated-sulfosuccinic acid mono-alkyl ester." This confirmation was given for the evanescence's of the 1 H-NMR COOH and SO_3H protons confirmation [114].

Method of preparation of sorbitol esters was explicitly defined, and the preparation reaction was characterized. FTIR was used to validate the chemical composition of sorbitol esters of oleic acid. In this regard, the sorbitol spectrum displaying at 3448: 1098: 2932: and 1256 cm^{21} was assigned to vibrations of -OH, -CH_2- "stretching and bending." New sets showing the sorbitol chemical modification by esterification with three moles of oleic acid to form STO sorbitol trioleate ester is seen in FTIR spectrum. The development of sorbitol oleate ester is shown by the presence of a new set at 948 cm^{21}, $(CH_2)_7$ bending, and 1600 cm^{21}, for -C55C- stretching. And it emerged as a product of the oleic acid alkyl group. The tight stretching band of ester group-COO in 1736 cm^{21} showed that the ester was successfully prepared between sorbitol and oleic acid [121, 122].

GC-FID: ASIs, squalene concentrations, and their concentrations were characterized utilizing "Shimadzu GC" (gas chromatograph)- 17-A with "Phenomenex" ZB-5HT column (30 m × 0.25 mm × 0.25 μm), and with calculated "autoxidation" products, before "quantitation" samples. The correctly measured sample (0.1 g to 4 digits) was dissolved in ethyl-acetate (1,000 cm^3) for all GC-FID samples to obtain a satisfactory sample of viscosity and concentration, diluted sample was inserted around (1 μL) @ 370 °C into the inlet. Product 12 was quantified an initial oven temperature at 50 °C for 1 h, using split ratio of 2:1 min with 5 °C min^{-1} ramp to 340 °C and was kept for 41 min. A 10:1 split ratio was used to measure all other organisms, allowing the quantification products containing (> 0.05%) of the reacted substance quantity. With a 5 °C min^{-1} ramp at 340 °C, the initial oven temperature of 50 °C for (1 min) and held also for around (10 min). External calibration by the form of the effective carbon number was used to measure properties of tidy N-butyl ASI autoxidation [123, 124].

8.4 Application

Onshore and offshore marine environment can be seriously affected by oil spills. For many years, oil slicks removal from the surfaces of the sea have been researched and a variety of countermeasures for "operational" spills has been established [14, 57, 125–127]. The application of chemical dispersants has been successful in major or large oil spills in recent years due to the establishment of the methodology of the chemical dispersants effectiveness in responding to oil spills, after few days of the occurrence.

Oil and water interfacial tension can be minimized by dispersants' comprising solvents and surfactants to drastically minimize micron-scale oil droplet sizes under mixing wave propagation encountered in the field [82, 128, 129].

Modern-day dispersants have been confirmed to be less hazardous (toxic) compared to the oils used to disperse them. Basically, it can improve separation rate of crude oil's toxicities, low molecular weight aromatic components into columns of water. In essence, however, within the few hours prior to dispersant application, much or all these components may have already evaporated. In the case of crude oil, the rise in the disintegration of these parts is likely to be small due to the use of dispersant [71].

The aim of any spray technique used for dispersant application, notwithstanding the vessel, boat, or airplane on which devices are connected, is to apply dispersants in as efficient quantities as required on the environment to be treated in the optimal ratio. Almost all the devices used in dispersant spray consist of storage tanks, and transfer pump (usually for spray-arms) and nozzles, which are usually deposited on the spilled area. When spills are affected by oil and are hit by the spray containing dispersants, the size of the dispersant droplets formed by the nozzles is arranged so that there is a slight overlap –these components vary in size and efficiency due to different spraying units used when applying dispersants in oil spill scenarios [71, 131, 132].

8.4.1 Vessel dispersants spraying in oil spills management

The application of dispersants by vessels spraying is usually slower, with small rates of transit, which are basically subjected to restrictions and swathe (space) widths due to the passage of vessels caused by sea conditions. However, in the "critical" hours or few days after oil spillage incident that take place at certain times, vessel application is an avenue to apply dispersant effectively and efficiently. In places where vessels are abundantly present, vessels provide many advantages over aircraft systems, in ease of deployment, lower cost, reliability, and high degree of spray control [133].

Technologically, vessels can devote considerable time in effectively spreading dispersant in one approach to thicker segments of an oil spill without wasting time re-positioning or turning-around, which may require aircraft applications. There is the possibility of channeling vessels to areas of a slick. As oil spills in a marine ecosystem do not spread uniformly, much of these spill volumes ranging from 90–95 percent can be concentrated in thicker zones (> 1 mm thickness) (5 to 10 percent of the spill area) that are surrounded by sheens, and this is an essential trait for successful dispersant application [134].

There are three main types of implementation schemes for vessel dispersants application. To spread diluted or undiluted (neat) dispersant, spray boom (arms) devices are used in spray arms overboard. Fire monitoring devices dilute dispersants that is applied using firefighting foam systems and some basic chemical delivering alternative systems (chemical injection positive "pressure-pump" units). And the newest idea is similar to fire detectors – single nozzle undiluted dispersant application devices that are exclusively prepared to apply undiluted dispersant in "low pressure–high flow" or "high pressure–low flow" operating state.

8.4.2 Aerial dispersants spraying in oil spills management

Aircraft dispersant applications are another preference for considerably larger spills and in environments that are unsuitable for vessel dispersant applications. The significantly higher aircraft movement means that the site treatment rate is about

15 times greater than that of vessels, so there is a strong positive use of aerial spraying for oil spills involving massive areas or locations. Aircraft dispersion mechanisms are equipped and approved for specific types and models of aircraft. And for aerial spraying, Type III dispersants are better suited as the smaller concentrations (commonly 1:20 dispersant to oil) permits best utilization of the restricted payload.

8.4.3 Environmental hazard of dispersants on oil spills management

It has been shown that dispersants and dispersed oil have negative effects on marine life, going from fish to corals to winged creatures (birds). Dispersants contain toxic substances from dispersed oil that are more hazardous to aquatic life than untreated oil, either separately or in blend with oil droplets and synthetic dispersant substances. Not properly tested short-term and long-term dispersants impact aquatic environments. As the EPA acknowledges, the "long-term impact on marine activity of dispersants remain unclear" [104, 135].

Therefore, the environmental consequences of the use of dispersants primarily pertain to: (i) the toxicity (poisonousness) of dispersants; (ii) effect on the microbial crumbling of oil spilled; and (iii) effect on the environments of seabirds and well-evolved marine creatures (marine mammals) [87].

8.4.3.1 Corals

Dispersants and scattered oil have an amazingly negative impact on corals, provoking researchers to advocate for a prohibition on dispersant utilization of coral reefs. By expanding death rates, diminishing choosing reefs, and adjusting conduct, dispersants and scattered oil harm the beginning stages of corals. They evade preparation of developing eggs and block the development of youthful life phases of reef-building corals, as evidenced by one of the dispersants used in the BP spill reaction, Corexit-9527.

8.4.3.2 Fish

Through releasing harmful crude oil products into the sea or the marine environment, the use of dispersants creates a dangerous atmosphere for fish, if not checked. According to various laboratory studies that have examined several organisms, scattered oil has been demonstrated to be poisonous to fish at all life stages, from eggs to larval fish to grown-ups.

8.4.3.3 Birds

Research findings also shown that dispersed oil affects the thermal insulation of seabird feathers than just untreated oil dispersed by Corexit 9527 (9527A) making birds rather vulnerable to hypothermia and death. It has also found that dispersants and oil spread have toxic effects on bird eggs that are worse than those of untreated oil. Dispersed oil exposed to birds that return to their nests risks the death of their shells.

8.4.3.4 Human health at risks

Dispersants often pose critical threats to human health. According to the manufacturer's safety data sheet, (MSDS), one of the dispersants used in the BP spill, Corexit 9527A (9527A), contains the toxin 2-Butoxyethanol that can damage the red blood cells (hemolysis), liver or kidney on excessive exposure. The 11-million-gallon oil slick created by the Exxon Valdez spill was sprayed with Corexit 9527, and cleanup workers reportedly experienced health problems afterwards, with blood in their urine as well as kidney and liver disorders, linked to 2-Butoxyethanolol.

8.4.4 Effectiveness of dispersants on oil spills management

Modern dispersants can be useful or successful in an extensive variety of oil types, based on the potency of the procedure and the right application ratios. Generally, one ingredient dispersant can disperse or scatter approximately 20–30 pieces of oil. A still greater dispersion is possible – more than hundred portions of oil per part of the dispersants [130, 136].

If the oil is light and the energy is high in the ocean, heavy (for example, low API gravity) and weathered oils are more dispersion resistant and will need a higher "dispersant-to-oil ratio." Chemical dispersion can be completely resisted by extremely thick, non-spreading oils and waxy oils whose pour point is over the climatic water-temperature.

Dispersant effectiveness is measured by oil percentage spread following the application of dispersants into the water-column. In recent years, utilization of chemical dispersants for remover of oil spills has gained increasing attention for many purposes. There have been reports that modern-day dispersants are more effective and less toxic. Concentration levels can be very low under scattered spills and may have a very short duration, so that many marine species do not reach levels of acute toxic effects, especially at depths greater than five meters [130].

The potency of the dispersant is typically physically measured, and there are problems identified with real spread oil" quantities in the water column. Laboratory analysis of water samples taking many tests and actual spills have measured the level of oil in the water, and the available data indicates that oil concentrations after dispersion are not high. They rarely reach 100–200 ppm in the top few meters of the water column and typically range from 20–60 ppm for a limited time (approximately 1 to 2 h) [137].

Using boundaries, for example, the amount of oil scattered in the water, droplet size, and stability, effectiveness is measured.

8.4.5 Toxicity of oil affected using dispersants

A small amount of synthetic substances that is harmful to numerous amphibian life forms are contained in oils of various types. Any of the lower molecular weight substances that are more acutely toxic – "BTEX" synthesis – are often, to some degree, reactive and water soluble. Freshly spilled crude oils are substantially more intensely harmful than current oil slick dispersants [87].

PAHs (Polycyclic Aromatic Hydrocarbons) have higher molecular weight intensities that are found in low fixations in numerous oils and trigger uncertainty about toxicity. PAHs are carcinogenic in nature, and prolonged exposure may have other effects.

As the surface slick is dispersed into tiny oil droplets found in the water column, filter-feeding creatures such as oysters, clams, scallops, and copepods may absorb these oil droplets. There is no chance of dispersed oil quantities exceeding thresholds for extended periods, which could have serious consequences for most aquatic species, if dispersants scatter (disperse) oil in water where there is adequate profundity and water trade to instigate fitting dilution. It has been established from laboratory evaluation and implementation that the environmental effect is lower than predicted and has nullified the overall benefits arising from the use of dispersants [87].

8.5 Discussion

When there is an occurrence of oil spill, the first response is to design a methodology to use in order to minimize the environmental damages cause by the spillage to the degree of sensibly reasonable safety regulations. The alternatives include mechanical recuperation that considers common expulsion of oil from the location, cleanup of the area or seashore that may be affected due to oil spill. This process is brilliant because it concentrates some of the low toxic nonionic "surface-active" agents of dispersants and limits ecological harm. Some materials and techniques used to confront the oil slick effect have been recorded previously, while in various pieces of the spill this can be applied. This work concentrates on the advancements of chemicals treatment (dispersants) applications in various oil spills.

8.6 Conclusions

This chapter focuses on individual materials mentioned in the above sections, be it surfactants, solvent used as a dispersant, as well as use combined definitions and used to get superior of dispersion ability. Dispersion oil ratios (DORs) are very important to acquire a higher scattering (dispersed) ability of dispersants. International America Environmental Protection Agency (AEPA) recommend the dispersion oil ratios of around (1:10, 1:25), but other figures can also be applied. Also, interfacial tension plays a vital part in dispersion ability. Therefore, the "maximum" dispersion coefficient can be entered against the "minimum" interfacial tension. There are several cleanup procedures available for both offshore and onshore management of oil-contaminated locations; however, the recovery effectiveness of these techniques relies upon the kind of crude-oil, surface, environmental penetration profundity (depth), area susceptibility, and chemical noxious (toxicity). Modern-day dispersants can be applicable in larger or deep oil spills areas because of the less toxic nature, but it is required to be approved in most counties to be able to regulate and monitor the advancement in the application of dispersants. Note, if oil spills or spillages should arise, nevertheless, responses ought to be taken quickly to diminish conceivable ecological consequences.

References

[1] Lucas, Z., MacGregor, C. Characterization and source of oil contamination on the beaches and seabird corpses, Sable Island, Nova Scotia, 1996–2005. Mar Pollut Bull, 2006, 52(7), 778–789.

[2] Tkalich, P., Huda, M.K., Gin, K.Y.H. A multipurpose oil spill model. J Hydraul Res, 2003, 41(2), 115–125. doi:10.1080/00221680309499955.

[3] Lewis, A., Aurand, D. *Putting dispersants to work: overcoming obstacles*. in *1997 International Oil Spill Conference Technical Report IOSC*. 1997.

[4] Clark, R.B. Marine Pollution., 3rd edition, Oxford, Oxford University Press, 1992, 50–60.

[5] Nomack, M., Cleveland, C. Oil spill control technologies. Encyclopedia of Earth. http://www.eoearth.org/articles/view/158385, 2010.

[6] Raisbeck, J.M., Mohtadi, M.F. The environmental impacts of oil spills on land in the arctic regions. Water Air Soil Pollut, 1974, 3(2), 195–208.

[7] Mackay, D., Wells, P.G. Effectiveness, behavior, and toxicity of dispersants. In: International oil spill conference. American Petroleum Institute, 1983.

[8] McAuliffe, C.D., et al. The 1979 Southern California dispersant treated research oil spills. In: International Oil Spill Conference. American Petroleum Institute, 1981.

[9] Kinner, N., Ballestero, H. Chemical dispersants for oil spill application. J Ocean Technol, 2012, 7(4), 1–11.

[10] Whiteside, S.. *SPE Paper 26399, presented at the 68th Ann. Conf. and Exhibition of Society of Petroleum Engineers*, 1993, Texas.

[11] Aminian, K., Ameri, S. Evaluation of the petroleum technology-based dry soil barrier. J Pet Sci Eng, 2000, 26(1–4), 83–89.
[12] Jackson, J.B., et al. Ecological effects of a major oil spill on Panamanian coastal marine communities. Science 1989, 243(4887), 37–44.
[13] Onwurah, I.N.E., et al. Crude oil spills in the environment, effects and some innovative clean-up biotechnologies. Int J Enviro Res (IJER), 2007, 1(4), 307–320.
[14] Fingas, M. The Basics of Oil Spill Cleanup, Boca Raton, CRC Press, 2012.
[15] Sandrin, T.R.; Dowd, S.E.; Herman, D.C.; Maier, R.M. Chapter 6 - Aquatic Environments. In: Maier, R.M., Pepper, I.L., Gerba, C.P., eds. *Environmental Microbiology (Second Edition)*. San Diego, Academic Press, 2009, 103–122.
[16] Hokstad, J.N., et al. Chemical and toxicological characterization of water accommodated fractions relevant for oil spill situations. In: SPE international conference on health, safety and environment in oil and gas exploration and production. 2000, Society of Petroleum Engineers.
[17] Amro, M.M. Treatment techniques of oil-contaminated soil and water aquifers. In: International conf. on water resources & arid environment. 2004.
[18] Chapman, P.M., Riddle, M.J. Toxic Effects of Contaminants in Polar Marine Environments, ACS Publications, 2005.
[19] Allen, A.A., et al. The use of controlled burning during the Gulf of Mexico Deepwater Horizon MC-252 oil spill response. In International Oil Spill Conference Proceedings (IOSC). American Petroleum Institute, 2011.
[20] Banerjee, S.S., Joshi, M.V., Jayaram, R.V. Treatment of oil spill by sorption technique using fatty acid grafted sawdust. Chemosphere, 2006, 64(6), 1026–1031.
[21] Geraci, J.R., Aubin, D.J.S. Synthesis of Effects of Oil on Marine Mammals, Department of the Interior, Minerals Management Service, Atlantic OCS Region, 1988.
[22] ITOPF. *Use of sorbent materials in oil spill response*, In *ITOPF, Technical Information Paper No. 8*2012.
[23] Kleindienst, S., Paul, J.H., Joye, S.B. Using dispersants after oil spills: Impacts on the composition and activity of microbial communities. Nat Rev Microbiol, 2015, 13(6), 388–396.
[24] Schulze, R. Oil Spill Response Performance Review of Booms, US Minerals Management Service, 1998.
[25] Exxonmobil. *Oil spill response field manual revised*. 2014.
[26] Ventikos, N.P., et al. A high-level synthesis of oil spill response equipment and countermeasures. J Hazard Mater 2004, 107(1–2), 51–58.
[27] OSS. *Oil spill solution. Retrieved on 26thFebruary*. 2010; Available from: http://www.oilspill solutions.org/booms.htm.
[28] GPC. *Curtain Booms*. Global Spill Control, Retrieved on 7th March form 2010; Available from: http://www.globalspill.com/uploads/downloads/pa.
[29] Potter, S., Morrison, J. World Catalogue of Oil Spill Response Products, 9th edition, Ottawa, Ontario, Canada, S.L. Ross Environmental Research Ltd., 2008, 1–42.
[30] ITOPF. *Use of skimmers in oil pollution response*, in *ITOPF, Technical Information Paper No 5*2012.
[31] USEPA. *Emergency response*. Available from: https://www.epa.gov/emergency-response /skimmers.
[32] Karan, C.P., Rengasamy, R., Das, D. Oil Spill Cleanup by Structured Fibre Assembly, India, NISCAIR-CSIR, 2011.
[33] Cojocaru, C., Macoveanu, M., Cretescu, I. Peat-based sorbents for the removal of oil spills from water surface: Application of artificial neural network modeling. Colloids Surf A Physicochem Eng Asp, 2011, 384(1–3), 675–684.

[34] Kyzas, G., Kostoglou, M. Green adsorbents for wastewaters: A critical review. Materials, 2014, 7, 333–364.
[35] Sayed, S., El Sayed, A., Zayed, A. Oil spill pollution treatment by sorption on natural Cynanchum acutum L. plant. J Appl Sci Environ Manag, 2003, 7(2), 63–73.
[36] Rotar, O.V., et al. Cleanup of water surface from oil spills using natural sorbent materials. Procedia Chem Vol. 10: Chemistry and Chemical Engineering in XXI century, 2014., 2014, 10, 145–150.
[37] Carmody, O., et al. Adsorption of hydrocarbons on organo-clays – Implications for oil spill remediation. J Colloid Interface Sci 2007, 305(1), 17–24.
[38] Annunciado, T., Sydenstricker, T., Amico, S. Experimental investigation of various vegetable fibers as sorbent materials for oil spills. Mar Pollut Bull, 2005, 50(11), 1340–1346.
[39] Nguyen, H.N., Pignatello, J.J. Laboratory tests of biochars as absorbents for use in recovery or containment of marine crude oil spills. Environ Eng Sci, 2013, 30(7), 374–380.
[40] Songsaeng, S., Thamyongkit, P., Poompradub, S. Natural rubber/reduced-graphene oxide composite materials: Morphological and oil adsorption properties for treatment of oil spills. J Adv Res, 2019, 20, 79–89.
[41] Montgomery, J.M. Water Treatment: Principles and Design, New York, John Wiley & Sons, 2005.
[42] Yang, R.T. Adsorbents: Fundamentals and Applications, New Jersey (NJ), John Wiley & Sons, 2003.
[43] Mckay, G. Use of Adsorbents for the Removal of Pollutants from Wastewater, Boca Raton, FL, CRC Press, 1995.
[44] San Miguel, G., Fowler, G.D., Sollars, C.J. A study of the characteristics of activated carbons produced by steam and carbon dioxide activation of waste tyre rubber. Carbon, 2003, 41(5), 1009–1016.
[45] Pollard, S. et al. Low-cost adsorbents for waste and wastewater treatment: A review. Sci Total Environ 1992, 116(1–2), 31–52.
[46] Bansal, R.C., Donnet, J.-B., Stoeckli, F. A Review Of: Active Carbon, New York, Marcel Dekker, Inc, 1988, 482.
[47] Corporation, S.M.D.o.M.C. *Diaion™ technical manual synthetic adsorbents Diaion™ & Sepabeads™*, 2018; Available from: https://www.diaion.com/en/products/synthetic_adsorbents/pdf/technical_manual.pdf.
[48] Adachi, T., Isobe, E. Fundamental characteristics of synthetic adsorbents intended for industrial chromatographic separations. J Chromatogr A, 2004, 1036(1), 33–44.
[49] Teas, C., et al. Investigation of the effectiveness of absorbent materials in oil spills clean up. Desalination 2001, 140(3), 259–264.
[50] Jarre, W., Marx, M., Wurmb, R. Polyurethanschäume mit hohem ölabsorptionsvermögen. Die Angewandte Makromolekulare Chemie: Appl Macromo Chem Phys, 1979, 78(1), 67–74.
[51] USEPA. *Sorbents*. Emergency management. Retrieved on 7th March 2011a; Available from: http://www.epa.gov/OEM/content/learning/sorbent.
[52] Deschamps, G., et al. Oil removal from water by sorption on hydrophobic cotton fibers. 2. Study of sorption properties in dynamic mode. Environ Sci Technol 2003, 37(21), 5034–5039.
[53] Alcañiz-Monge, J., Linares-Solano, A., Rand, B. Water adsorption on activated carbons: Study of water adsorption in micro-and mesopores. J Phys Chem B, 2001, 105(33), 7998–8006.
[54] Bandosz, T.J., et al. Molecular models of porous carbons. Chem Phys Carbon 2003, 28, 41–228.
[55] Maciá-Agulló, J., et al. Activation of coal tar pitch carbon fibers: Physical activation vs. chemical activation. Carbon 2004, 42(7), 1367–1370.

[56] Karakasi, O., Moutsatsou, A. Surface modification of high calcium fly ash for its application in oil spill clean up. Fuel, 2010, 89(12), 3966–3970.
[57] Choi, H.M., Cloud, R.M. Natural sorbents in oil spill cleanup. Environ Sci Technol, 1992, 26 (4), 772–776.
[58] Ghaly, R., et al. Remediation of diesel-oil-contaminated soil using peat. Energy Sources 1999, 21(9), 785–799.
[59] Babel, S., Kurniawan, T.A. Low-cost adsorbents for heavy metals uptake from contaminated water: A review. J Hazard Mater, 2003, 97(1–3), 219–243.
[60] Kayvani Fard, A., et al. Enhancing oil removal from water using ferric oxide nanoparticles doped carbon nanotubes adsorbents. Chem Eng J 2016, 293, 90–101.
[61] Gui, X., et al. Magnetic and highly recyclable macroporous carbon nanotubes for spilled oil sorption and separation. ACS Appl Mater Interfaces 2013, 5(12), 5845–5850.
[62] Gu, J., et al. Janus polymer/carbon nanotube hybrid membranes for oil/water separation. ACS Appl Mater Interfaces 2014, 6(18), 16204–16209.
[63] Chatterjee, P.K. Chapter XII – products and technology perspective. In: Chatterjee, P.K., Gupta, B.S., eds. Textile science and technology, Amsterdam, Oxford, Elsevier Science, 2002, 447–477.
[64] Tan, J., et al. Synthesis of honeycomb-like carbon foam from larch sawdust as efficient absorbents for oil spills cleanup and recovery. Materials 2018, 11(7), 1106.
[65] Walton, W.D., Jason, N.H. In Situ burning of oil spills workshop proceedings [microform]. In: In Situ Burning of Oil Spills Workshop (1998: New Orleans, La.), 1999. For sale by the Supt. of Docs., USGPO.
[66] Buist, I. Window of opportunity for in situ burning. In: Situ burning of oil spills workshop proceedings, 1999.
[67] Evans, D.D., et al. In situ burning of oil spills. J Res Natl Inst Stand Technol 2001, 106(1), 231–278.
[68] Sammarco, P.W., et al. Distribution and concentrations of petroleum hydrocarbons associated with the BP/deepwater horizon oil spill, Gulf of Mexico. Mar Pollut Bull 2013, 73 (1), 129–143.
[69] Peyvandi, A.G. Determination of environmental considerations for the use of dispersant species in different areas of the Caspian Sea. EurAsian J BioSci, 2019, 13(2), 651–662.
[70] Vergetis, E. Oil Pollution in Greek Seas and Spill Confrontation Means-methods, Greece, National Technical University of Athens, 2002.
[71] Fiocco, R.J., Lewis, A. Oil spill dispersants. Pure Appl Chem, 1999, 71(1), 27–42.
[72] Portmann, J. The effect of china clay on the sediments of St Austell and Mevagissey Bays. J Mar Biol Assoc U K, 1970, 50(3), 577–591.
[73] Cormack, D. Using Oil Spill Dispersants on the Sea: Committee on Effectiveness of Oil Spill Dispersants, National Research Council, Washington DC, National Academy Press, 1989, xvi+ 336. Price $39.95 paperback. ISBN: 0-309-03882-0 1989 Pergamon.
[74] ITOPF. Mechanism of Dispersion and Dispersant Composition, London, The International Tanker Owners Pollution Federation Limited, 2011.
[75] McCauley, L.A. Will the BP oil spill affect our health?: At an institute of medicine meeting in June, experts examined potential short- and long-term effects. Am J Nurs, 2010, 110(9), 54–56.
[76] Palinkas, L.A. A conceptual framework for understanding the mental health impacts of oil spills: Lessons from the Exxon Valdez oil spill. Psychiatry(New York), 2012, 75(3), 203–222.
[77] Ingersoll, C., Locke, R.M., Reavis, C. BP and the Deepwater Horizon Disaster of 2010, Case Study, MIT Sloan School of Management, 2012.

[78] Griffiths, S.K. Oil release from macondo well MC252 following the deepwater horizon accident. Environ Sci Technol, 2012, 46(10), 5616–5622.
[79] www.whitehouse.gov/blog/issues/Deepwater-BP-oil-spill., (accessed November 3, 2010).
[80] McNutt, M.K., et al. Review of flow rate estimates of the deepwater horizon oil spill. Proc Indian Natl Acad Sci 2012, 109(50), 20260–20267.
[81] Kujawinski, E.B., et al. Fate of dispersants associated with the deepwater horizon oil spill. Environ Sci Technol 2011, 45(4), 1298–1306.
[82] Council, N.R. Oil Spill Dispersants: Efficacy and Effects, Washington, DC, National Academies Press, 2005, 396. https://doi.org/10.17226/11283.
[83] Lee, K. How Oil Dispersants Work, Nova Scotia Dartmouth, Nova Scotia Canada B2Y 4A2 Canada/ Centre for Offshore Oil, Gas and Energy Research (COOGER), 2010.
[84] Fletcher, L.M., Forrest, B.M., Bell, J.J. Natural dispersal mechanisms and dispersal potential of the invasive ascidian Didemnum vexillum. Biol Invasions, 2013, 15(3), 627–643.
[85] Vanderhoff, J.W., Wiley, R.M. Water-in-oil Emulsion Polymerization Process for Polymerizing Water soluble Monomers, Google Patents, 1966.
[86] Opawale, F.O., Burgess, D.J. Influence of interfacial properties of lipophilic surfactants on water-in-oil emulsion stability. J Colloid Interface Sci, 1998, 197(1), 142–150.
[87] REMPEC. *Guidelines for the use of dispersants for combating oil pollution at sea in the Mediterranean region. Part II: Basic information on dispersants and their application.* 2011; Available from: https://www.rempec.org/en/knowledge-centre/online-catalogue/guidelines-for-the-use-of-dispersants-for-combating-oil-pollution-at-sea-in-the-mediterranean-region-part-ii.
[88] Merlin, F. Draft guidelines for the use of dispersants for combating oil pollution at sea, Part 1, basic information on dispersants and their application, as presented to the OPRC-HNS technical group of IMO, session 16. Agenda Item, 2014, 3.
[89] Mackay, D., Hossain, K. Interfacial tensions of oil, water, chemical dispersant systems. Can J Chem Eng, 1982, 60(4), 546–550.
[90] Mackay, D. Effectiveness of Chemical Dispersants under Breaking Wave Conditions, in the Use of Chemicals in Oil Spill Response, Philadelphia, ASTM International, 1995.
[91] Fingas, M. Oil spill dispersants: A technical summary. In: Mervin Fingas (ed.) Oil spill science and technology. Gulf Professional Publishing, 2011, 435–582. doi:10.1016/b978-1-85617-943-0.10015-2
[92] Al-Sabagh, A. *Synthesis of some surfactants for pollution control technology*, 1992, PhD dissertation, College of Science, Al-Azhar University, Cairo, Egypt.
[93] chemicalbook. Available from: https://www.chemicalbook.com/ChemicalProductProperty_EN_CB9702767.htm.
[94] PubChem [Internet]. Bethesda (MD): National Library of Medicine (US), N.C.f.B.I. *PubChem Compound Summary for CID 7923, Maleic anhydride.* 2004; Available from: https://pubchem.ncbi.nlm.nih.gov/compound/Maleic-anhydride.
[95] glycol)", P.e. *ChemSrc.* 7 January 2020.; Available from: https://www.chemsrc.com/en/cas/25322-68-3_766012.html.
[96] PubChem. Available from: https://pubchem.ncbi.nlm.nih.gov/compound/Sodium-dodecyl-sulfate.
[97] Guard, U.-U.C. Chemical Hazard Response Information System (CHRIS)-Hazardous Chemical Data. Commandant Instruction 16465.12 C, Washington, DC, US Government Printing Office, 1999.
[98] Program, N.T. *Institute of Environmental Health Sciences, National Institutes of Health (NTP). 1992.* National Toxicology Program Chemical Repository Database. Research Triangle Park, North Carolina: NTP via https://canieochemicals.noaa.gov/chemical/20568.

[99] Remington, J.P., et al. Remington's Pharmaceutical Sciences, Mack Pub., Beltsville, Maryland and Washington, D.C. 1985.
[100] Eshgarf, H., Afrand, M. An experimental study on rheological behavior of non-Newtonian hybrid nano-coolant for application in cooling and heating systems. Exp Therm Fluid Sci, 2016, 76, 221–227.
[101] Sulek, M., et al. Alkyl polyglucosides as components of water based lubricants. J Surfactants Deterg 2013, 16, 369–375.
[102] Carnero Ruiz, C. (ed.). Sugar-Based Surfactants: Fundamentals and Applications (1st ed.). Boca Raton, CRC Press, 2008. https://doi.org/10.1201/9781420051674
[103] Tesmann, H., et al. Alkyl polyglycosides in personal care products. In: Alkyl polyglycosides: Technology, properties and applications, VCH Verlagsgesellschaft mbH, 1996, 71–98.
[104] Available from: https://www.worldofchemicals.com/chemicals/chemical-properties/linear-alkyl-benzene-sulfonic.
[105] Tsutsumi, H., Nakayama, H., Shinoda, K. Some characteristics of polyoxyethylene sorbitol tetra-oleate: Oligomer type emulsifier. J Am Oil Chem Soc, 1978, 55(3), 363–366.
[106] Nyankson, E., et al. Potential application of dioctyl sodium sulfosuccinate salt (DOSS)–saponin binary dispersant in oil spill remediation: Synergistic interaction between DOSS and saponin. Water Air Soil Pollut 2020, 231(2), 76.
[107] Karaffa, L.S. The Merck Index: An Encyclopedia of Chemicals, Drugs, and Biologicals, N.J., U.S.A., RSC Publishing, 2013.
[108] Guard, U.C. Chemical Hazard Response Information System (CHRIS)-Hazardous Chemical Data, Commandant Instruction 16465.12 C., *Washington, DC:, US Government Printing Office*, 1999.
[109] Najafi, F., et al. Synthesis and characterization of comb polycarboxylic acid dispersants for coatings. J Appl Polym Sci 2012, 126(3), 877–881.
[110] El?-A1 0364852, B., 1988.
[111] Rhode, O., Weuthen, M., Nickel, D. New nonionic derivatives of alkyl polyglycosides – synthesis and properties. In: Alkyl polyglycosides: Technology, properties and applications, New Jersey, USA, 1996, 139–149. doi:10.1002/9783527614691.ch8.
[112] Henkel, *WO 93106115*. 1991
[113] Al, H. *DE 4335956*, 1993.
[114] Al-Sabagh, A., et al. Synthesis of ethoxylated alkyl sulfosuccinate surfactants and the investigation of mixed solutions. J Surfactants Deterg 2007, 10(1), 3–8.
[115] Information. Available from: https://pubs.acs.org/doi/suppl/10.1021/acs.iecr.9b02780/suppl_file/ie9b02780_si_001.pdf
[116] Ruffell, J.E., et al. The autoxidation of Alkenyl succinimides – Mimics for polyisobutenyl succinimide dispersants. Ind Eng Chem Res 2019, 58(42), 19649–19660.
[117] El-Saeed, S.M., et al. Synthesis and characterization of novel crude oil dispersants based on ethoxylated Schiff base. Int J Polym Mater 2008, 57(9), 860–877.
[118] Henkel. *WO 93120089*. 1992.
[119] Al-Sabagh, A., Azzam, E., El-Din, M.N. Synthesis and evaluation of ethoxylated alkyl sulfosuccinates as oil spill dispersants. J Dispers Sci Technol, 2008, 29(6), 866–872.
[120] Kano, Y., Akiyama, S. Estimation of surface tension and surface segregation of poly (ethyl acrylate)/poly (vinylidene fluoride-co-hexafluoro acetone) blends. Polymer, 1996, 37(20), 4497–4503.
[121] Al-Sabagh, A.M., et al. Synthesis of some oil spill dispersants based on sorbitol esters and their capability to disperse crude oil on seawater to alleviate its accumulation and environmental impact. J Dispers Sci Technol 2007, 28(5), 661–670.

[122] Hreczuch, I. Ethoxylated rapeseed oil acid methyl esters as new ingredients for detergent formulations. Tenside, Surfactants, Deterg, 2001, 38(2), 72–79.
[123] Jorgensen, A.D., Picel, K.C., Stamoudis, V.C. Prediction of gas chromatography flame ionization detector response factors from molecular structures. Anal Chem, 1990, 62(7), 683–689.
[124] Scanlon, J.T., Willis, D.E. Calculation of flame ionization detector relative response factors using the effective carbon number concept. J Chromatogr Sci, 1985, 23(8), 333–340.
[125] Jenssen, B.M. Effects of oil pollution, chemically treated oil, and cleaning on thermal balance of birds. Environ Pollut, 1994, 86(2), 207–215.
[126] Setti, L., Mazzieri, S., Pifferi, P.G. Enhanced degradation of heavy oil in an aqueous system by a Pseudomonas sp. in the presence of natural and synthetic sorbents. Bioresour Technol, 1999, 67(2), 191–199.
[127] Payne, J.R. Oil-ice Sediment Interactions during Freeze-up and Break-up, US Department of Commerce, National Oceanic and Atmospheric Administration, 1989.
[128] Lessard, R.R., DeMarco, G. The significance of oil spill dispersants. Spill Sci Technol Bull, 2000, 6(1), 59–68.
[129] Venosa, A.D., King, D.W., Sorial, G.A. The baffled flask test for dispersant effectiveness: A round robin evaluation of reproducibility and repeatability. Spill Sci Technol Bull, 2002, 7 (5–6), 299–308.
[130] Board, O.S., Council, N.R. Oil Spill Dispersants: Efficacy and Effects, Washington, DC, National Academies Press, 2005. https://doi.org/10.17226/11283.
[131] Board, O.S.E. National academies of sciences, and medicine. In: The use of dispersants in marine oil spill response, Washington, DC, National Academies Press, 2020. https://doi.org/10.17226/25161.
[132] Wang, W., et al. PIV investigation of oil–mineral interaction for an oil spill application. Chem Eng J 2011, 170(1), 241–249.
[133] Motolenich-Salas, K.M., Clark, J.R. Vessel dispersant application in oil spill response: Research review. In: International oil spill conference, American Petroleum Institute, 2005.
[134] Ross, S. The case for using vessel-based systems to apply oil-spill dispersants. In: Arctic and marine oil spill program technical seminar, Ministry of Supply and Services, Canada, 1998.
[135] Kilduff, C., Lopez, J. Dispersants: The lesser of two evils or a cure worse than the disease. Ocean & Coastal LJ, 2010, 16, 375.
[136] Chandrasekar, S., Sorial, G.A., Weaver, J.W. Dispersant effectiveness on oil spills–impact of salinity. ICES J Mar Sci, 2006, 63(8), 1418–1430.
[137] (IPIECA), I.P.I.E.C.A. *Dispersants and Their Role in Oil Spill Response, IPIECA Report Series*. 2001. Volume 5: pp. 36.

Arundhati Pal and A. K. Paul

9 Serpentinophytes-associated microbiota: a review of applications in environmental biotechnology

Abstract: Serpentine plants commonly known as "serpentinophytes" represent a specialized vegetation growing on naturally metal percolated serpentine habitats that are derived from metamorphosis and pedogenesis of serpentinite or ultramafic rocks. The serpentine-obligate species are mostly endemic and characteristically symbolize metal-resistant and metal hyperaccumulator species; on the contrary, the serpentine facultative species show complex morpho-functional adaptations.

Serpentinophytes, in general, and the metal hyperaccumulators, in particular, have received much attention not only for their unusual, extreme habitats but also for their associated microbiota. Microorganisms colonizing this unique group of plants revealed novel features concerning their contribution to the phenomenon of hyperaccumulation, enhancement of metal bioavailability and plant-mediated remediation technologies such as phytoextraction and phytoremediation. The structural and functional diversity of metal resistant microbial communities associated with the serpentinophytes and their interactions have been analyzed following the modern approaches. In this review, due attention has been paid to provide an insight of the plant-microbe interaction pertaining to the serpentine soil associated plants with special reference to their heavy metal resistance, uptake, and accumulation. Finally, the potential of the serpentine microbial communities for possible application in bioremediation and restoration of metal of contaminated environment has been highlighted on the basis of the recent research findings.

Keywords: serpentinophytes, metal hyperaccumulators, serpentine microbiota, metal resistance, bioremediation, phytoextraction, biomining

9.1 Introduction

Microbial assemblage on and inside plants are directly involved in the stimulation of plant health as well as protection against numerous biotic and abiotic stresses. The structure and functions of microbial community associated with plants

Arundhati Pal, Post Graduate Department of Botany, Serampore College, 9, William Carey Road, Serampore, Hooghly, West Bengal 712201, India
A. K. Paul, Microbiology Laboratory, Department of Botany, University of Calcutta, 35, Ballygunge Circular Road, Kolkata, West Bengal 700019, India, e-mail: amalk_paul@yahoo.co.in

https://doi.org/10.1515/9783110727227-009

from different ecological niches has been an important field of research, which not only enumerates the microbiome of the host but also focuses attention on functioning of their metabolites that act as phytostimulants for improved growth and uptake of nutrients [1, 2]. It has been observed that the metabolic profile of the host plant as well the ecosystem influence the phylogenetic composition of the microbiota and plays a pivotal role in selection of rhizospheric and endospheric microbial community structure. Plant-associated bacteria and fungi resemble a mutualistic assembly, which enables evolution of diverse biochemical mechanisms that promote plant growth and survival under edaphic stress. Microorganisms, by virtue of their adaptability to adverse environmental conditions, have developed strategies for reclamation and ecological restoration of polluted sites. Moreover, bacterial association with roots of the plants has shown beneficial effect on the host plant for phytoremediation of contaminated ecosystem (2).

Serpentine or ultramafic soils are naturally occurring metalliferrous outcrops distributed all over the world and characterized by high levels of nickel, chromium, and cobalt, in addition to iron and magnesium. The soil is characteristically porous, relatively infertile due to low level of calcium than magnesium, has low Ca/Mg ratio, and is poor in essential plant nutrients like nitrogen, phosphorus, and potassium [3]. The unusual chemistry of serpentine soil and the underlying bedrocks has given rise to sparse vegetation cover or barren land devoid of plants in some areas, resulting in a sharp contrast from the lustrous foliage of neighboring non-serpentine areas, thus delineating the presence of edaphic discontinuity. Interestingly, these outcrops support a unique assemblage of endemic serpentine-tolerant flora termed "serpentinophytes," which develops morphological and physiological adaptability only to serpentine habitat. These serpentine-associated plants have led to the evolution of a distinctive structure of serpentine communities that are unable to compete in non-serpentine environments. A vast majority of serpentinophytes are metal hyperaccumulators that concentrate metals in their above-ground shoot system to levels higher than that present in soil.

This unique group of endemic serpentinophyte metal hyperaccumulator plants showed evolution of diverse population of metal-resistant microbial community in the plant rhizosphere and endosphere [4]. Majority of these plants hyperaccumulate nickel in their leaves and drive a hypothetical nickel cycle to concentrate metal in the rhizosphere zone for potential adaptability of the microbiota to heavy metals [5–7]. The plant-associated microbiome of serpentinophytes have also shown potential benefits by functioning as plant growth-promoting traits following synthesis of plant hormones, siderophores, and growth associated enzymes. Moreover mycorrhizal population in the root helped the plants in nutrient acquisition as well as metal hyperaccumulation [8, 9]. Serpetinophytes are unique examples of plant–microbe association to study the mutualistic role of microbiome that helps plants not only to survive under metal stress but also aids in metal hyperaccumulation [10, 11]. The higher proportion of metal-resistant microorganisms found in proximity to the metal hyperaccumulating serpentinophytes has driven scientists across the world to explore a wide diversity of endemic serpentinophytes in different parts of the world and also investigate the plant-associated microbiome for their exploitation in heavy metal bioremediation. Since the plants are thriving under naturally metal percolated ecosystem, the microbiome diversity may be unique and also shows novel strategies for metal bioremediation. Assessment of metal tolerance mechanisms in serpentinophytes associated microbiota may find potential in phytoextraction and biomining of metal values. This review focuses attention on assessing the plant–microbe interaction pertaining especially to the serpentine soil-associated plants and explore their potential in remediation and restoration of metal-contaminated ecosystem.

9.2 Serpentinophytes and serpentine syndrome

"Serpentinite" rocks undergo weathering and pedogenesis producing a naturally occurring metalliferous ultramafic / serpentine soil that is poor in essential plant nutrients but enriched with metals like iron, nickel, chromium, and cobalt. This condition of serpentine soils, commonly referred to as the "serpentine syndrome", results in poor plant productivity and high rates of endemism leading to the development of a rare vegetation pattern that contrasts them from the neighboring non-serpentine areas [12, 13]. The initial phases of soil weathering are slow, leading to sparse plant occupancy on the shallow and fragmented soil, resulting in formation of serpentine "barrens". As weathering proceeds, soil becomes deeper and fine-textured to support development of productive plant communities. Serpentine habitats are highly selective for serpentine-loving or serpentinophilous plants, called "serpentinophytes" that are adapted to grow on such metalliferous habitat [14]. These serpentinophilous plants develop unique adaptive features as manifested in their morphology, physio-biochemical and cytological alterations for metal tolerance or avoiding metal toxicity.

Serpentine plants show some morphological characteristics deviating from the typical species growing on non-serpentine soil. These changes, commonly designated as "serpentinomorphoses," have arisen in different genera independently and include: stenophyllism, glabrescence, plagiotropism, nanism, macrorrhyzy, and glaucescence. Serpentinomorphoses are due to adaptations to intense light, dry microclimate, and low nutrient content, which include both hereditarily fixed deviations that persist under cultivation in normal soil and non-hereditary modifications. There are several reports on possible mechanisms of metal detoxification by chelation, sequestration in vacuoles, or even exclusion from the protoplast in selected serpentinophytes [15, 16]. The phenomenon of metal hyperaccumulation amongst serpentinophytes is not only associated with enhanced ability of plants to survive in metalliferous toxic environment but also benefits the plant by allelopathy, defense against herbivory, and resistance to pathogens of diverse types.

Serpentine plants share several common features, which may be summarized as: (i) the flora is relatively poor both in individuals and in species particularly in open vegetation on skeletal soil; (ii) on serpentine outcrops, several species are represented by particular races, which differ ecologically and sometimes morphologically from the typical representative of the species; (iii) many of these plants occur very disjunctively (disjunction of 80 km or more); (iv) the flora contains both basophilous (e.g., *Asplenium viride* and *Silene acaulis)* and acidophilous *(Calluna vulgaris* and *Deschampsia flexuosa)* plants; (v) some species on ultramafic substrata have relatively xeromorphic characters; and (vi) the flora is often dominated by certain families such as Caryophyllaceae, Fabaceae, Asteraceae, Poaceae, and Ericaceae [3]. In general, the serpentine plant communities vary remarkably in different regions of the world, forming distinctive variants of conifer or mixed conifer-hardwood forest, chaparral, or grassland and is distinguished from the neighboring non-serpentine, both by species composition and by physiognomy.

9.2.1 Diversity and endemism in serpentinophytes

Earlier plants growing on serpentine outcrops were classified as (i) magnesium restrictors (plants which tolerate high Mg by restricting uptake); (ii) calcium extractors (plants able to take up calcium more than races of the same species native to calcium sufficient sites); (iii) pathogen escapees (species thrive on serpentine soils and escape pathogens that cannot tolerate this unusual environment); (iv) metal-resistant species (plants exclude heavy metals from uptake); and (v) hyperaccumulators (plants retain enormous amount of toxic metals and are referred to as metallophytes). Kruckeberg [3] have divided serpentinophytes into three categories: (i) endemic taxa, linked exclusively to peridotites; (ii) preferential taxa, distribution mainly associated with peridotite but occasionally found on other substrata; and (iii) taxa living on a variety of substrata but also on peridotite (bodenvag taxa). The endemic serpentine species, often referred to as obligate serpentinophytes, are those that grow exclusively on serpentine soils and are not found on other types of substrate. Moreover, they are characterized by having a peculiar metabolism for avoiding the toxic effect of heavy metals [12]. Serpentine-obligate plants are mostly endemic, and their habitats are sites of floristic differentiation and speciation. The serpentine facultative species, on the contrary, show a serpentine-morph syndrome, a complex of morpho-functional adaptations (stenophyllous, glabrous, glaucous, macro-rooted plants, plagiotropism, and nanism) that distinguish the serpentine populations from those living on other substrates. Reeves et al. [17] recently reviewed the global distribution, ecology, and conservation status of serpentinophyte hyperaccumulators and also discussed their potential applications in phytoremediation and agromining.

Serpentinophytes are distributed worldwide from the American subcontinent, through Europe, Africa and, to some extent, in Asia and Australia. The state of California, USA has been reported to have a serpentine coverage of 6,000 km^2, representing only 1.5% of the total land area. Initially, 450 taxa were reported associated with serpentine in California, but later about 670 serpentine-associated taxa with endemics representing approximately 12.5% of the plant species endemic to California have been described. These include six families – Asteraceae, Liliaceae, Brassicaceae, Polygonace, Scrophulariaceae, and Apiaceae, while the most dominant and diverse genera were *Streptanthus*, *Erigonum*, followed by *Hesperolinon* and *Arctostaphylos*. The most important area for obligate serpentinophytes in Europe is situated in the Balkans, which is characteristically heterogeneous. The number of obligate serpentinophytes occurring in the Balkan Peninsula is smaller and represented mostly by *Alyssum markgrafii, Asplenium adulterinum, A. balcanicum, A. cuneifolium, A. smolikanum, Bornmuellera dieckii, Fumana bonapartei, Gypsophila spergulifolia, Halacsya sendtneri, Haplophyllum boissierianum, Onosma elegantissima, Onosma euboica, Polygala dorfleri, Potentilla visianii, Saponaria sicula, Sesleria serbica,* and *Sedum serpentinii* [17].

A complete eco-floristic checklist of serpentinophytes of south Iberian Peninsula (Andalusia, Spain) and its comparison with other serpentine-endemic floras in the Mediterranean Basin has been made [18]. The list comprises taxa with a variable degree of serpentinophily and is categorized as: obligate serpentinophytes (obligate endemics), preferential serpentinophytes (populations located mainly on serpentine), and sub-serpentinophytes (with some populations located on magnesium-rich substrata). Obligate serpentinophytes increase with the area of the outcrop and the genera *Alyssum*, *Arenaria*, *Armeria*, *Centaurea*, and *Silene* were the most frequent. The serpentinized area of northeast Portugal has 135 serpentinophytes belonging to 39 families, and the nickel accumulating taxa so far identified include *Alyssum serpyllifolium*; *Bromus hordeaceus*; *Linaria spartea*; *Plantago radicata*; *Lavandula stoechas*, and *Cistus salvifolius*. Obligate serpentinophytes, *Fumana bonapartei*, and *Stachys recta*, and the facultative serpentinophyte *Seseli rigidum* have been distinguished in Serbia. Ni-accumulating serpentinophytes, *Alyssum markgrajii*, *Alyssum murale*, and species of *Filago, Znula, Zicris, Galamintha, Marrubium, Teucrium, Lotus, Ononis,* and *Xeranthemum* have been reported from Albania. In Italy, serpentines range from the central-western Alps to the northern Apennines, and serpentine plants are most represented by families of Compositae, Graminaceae, Leguminosae, Caryophyllaceae, Rosaceae, Labiatae, Umbelliferae, Cruciferae, Orchidaceae, Cyperaceae, Liliaceae, Scrophulariaceae, and Ranunculaceae. About 11 serpentine endemic taxa were reported, which include *Alyssum bertolonii*, *Armeria denticulata*, *Asplenium cuneifolium*, *Biscutella pichiana*, *Centaurea aplolepa*, *Euphorbia nicaeensisi*, *Leucanthemum pachyphyllum*, *Minuartia laricifolia*, *Stachys recta*, *Stipa etrusca*, and *Thymus acicularis* [17]. The serpentine flora of the New Caledonian ultramafics in South Pacific Island show a high level of endemism, and the greatest number of Ni-hyperaccumulator taxa has been reported from this island [19]. Extremely high Ni concentrations were observed in some hyper-nickelophores (*Hybanthus austrocaledonicus* and *Sebertia acuminata*) along with extremely high concentrations of Mn in *Garcinia amplexicaulis* and *Grevillea gillivrayi*. Likewise, Co and Cr in high concentrations were recorded in *Homalium kanaliense* and *Agatea longipedicillata*, respectively.

Ni-hyperaccumulating endemic serpentinophytes belonging to Brassicaceae, Asteraceae, and the Caryophyllaceae have been reported from Turkey [20]. The notable endemic serpentine plants include *Alyssum, Thlaspi, Bornmuellera, Pseudosempervivum,* and *Centaurea*. The Republic of Cuba, located in the Gulf of Mexico is characterized by serpentine-loving plants, which display unique adaptive features like low metabolic rate, low biomass productivity, and extended longevity. Endemic and obligate ultramafic species that have the ability to hyperaccumulate Ni have been exhibited in the families Asteraceae (*Koanophyllum, Pentacalia, Senecio*), Buxaceae (*Buxus*), Euphorbiaceae (*Leucocroton, Phyllanthus*), Rubiaceae (*Ariadne, Psychotria, Rondeletia*), and Tiliaceae (*Tetralix*). Diversity of Sri Lankan serpentinophytes, as established, encompasses species belonging to Asteraceae, Euphorbiaceae, Fabaceae, Poaceae, and Rubiaceae. Of particular interest, was the discovery of Ni-hyperaccumulators that include *Evolvulus alsinoides, Hybanthus enneaspermus,* and *Crotalaria biflora*, while species like *Geniosporum tenuiflorum, Clerodendrum infortunatum, Croton bonplandianus, Waltheria indica,* and *Tephrosia villosa* were recorded as Cu hyperaccumulators [17]. The serpentine plants in Andaman and Nicobar Islands, India, mostly belonged to Rubiaceae, Euphorbiaceae, Memecyclaceae, Clussiaceae, and Arecaceae, and 32.5% of them were found to be endemic. Maximum species diversity was recorded in Rubiaceae and Euphorbiaceae. Species like *Rinorea bengalensis, Dichapetalum gelenioides, Strychnos wallichiana, S. minor, S. andamanensis, Orophea hexandra,* and *Anoxogorea luzoniensis* were identified as potent nickel hyperaccumulators [21]. The Australian serpentinite-endemic plants, particularly those from Central Queensland, include *Brachychiton bidwillii, Bursaria reevesii, Capparis thozetiana, Corymbia xanthope, Hakea trineura, Neoroepera buxifolia, Pimelea leptospermoides, Pultenaea setulosa, Scleria tryonii,* and *Stackhousia tryonii*. The Ni-hyperaccumulating plants are *Hybanthus floribundus* from Western Australia and *Stackhousia tryonii* as well as *Pimelea leptospermoides*, both from Central Queensland [17].

9.3 Microbial communities associated with serpentinophytes

Serpentine soil chemistry is known to influence not only the plant distribution and species endemism but also plant adaptability to extreme environment. Microbial associations with plants in such extreme habitat were hypothesized to significantly influence metal hyperaccumulation and / or exclusion as a means for survival in these serpentinophytes. Hence plant–microbial interactions in serpentine ecosystem have been a unique attribute where microorganisms play a pivotal role in endurance of host plants in edaphic conditions, either by providing plant growth promoting factors or aiding in metal detoxification [8, 9]. Several research findings have established the presence of rhizospheric and endophytic microbial association, including arbuscular mycorrhizal assemblages in serpentinophytes all over the globe and have concluded that serpentine tolerant species harbor diverse microbial communities, which show unique traits of metal resistance. The same microbial members may be present on adjacent non-serpentine sites or in association with non-serpentinophytes, but their adaptive features show distinct variations with respect to metal tolerance and bioavailability of metals in soil [22, 23]. Stochastic as well as deterministic processes including soil characteristics, host plant life stage, neighboring plant communities were studied in roots of *Knautia arvensis* (Dipsacaceae) from three serpentine grasslands and adjacent non-serpentine sites [24]. Around 81 molecular taxonomical operational units of arbuscular fungal mycorrhiza (AFM MOTU) belonging to Glomeromycota were detected, following 454-sequencing of ITS region of rDNA. The researchers found a strong positive correlation between the AFM MOTUs and neighboring plant community composition, whereas serpentine soil characteristics like K and Cr concentration negatively influenced AFM root colonization.

Several researchers tried to explore below-ground diversity of fungi in association with serpentine plants and correlate soil characteristics and plant diversity with the mycofloral assemblage. Although, the serpentine plant-associated ectomycorrhizal and arbuscular mycorrhizal associations in the plant roots as mutualistic symbionts were found critical for nutrient cycling and energy exchange below the soil surface, reports on serpentine-endemic fungi are lacking. Fruit bodies of hypogeous fungi were reduced in numbers, but distinct fungal sporocarps of serpentine isolates of *Cenococcum geophilum* and *Aculospora* were found to have adapted in serpentine environment [22]. Fungal colonization was found to be important in revegetation of barren serpentine areas as well as acceleration of Ni-hyperaccumulation in the plant [22, 25]. However, adaptation of *Avenula sulcata*, herbaceous serpentinophyte from New Caledonia, to serpentine syndrome did not involve mutualistic association with community of soil fungi [9].

Microorganisms, by virtue of their adaptability to metal-rich habitat and rhizosphere chemistry, occupy the serpentine niche and remain associated with the metal hyperaccumulators, although a rationale existed that due to metalliferrous nature of the ecosystem, less diverse forms of microorganisms may exist. Moreover, nonexistence of diazotrophs in soil due to lack of ammonifiable nitrogen and decomposable carbon had also been reported much earlier. However, elaborate research on microbiota associated with Ni-hyperaccumulating tree *Sebertia acuminita* from New Caledonian ultramafics documented accumulation of nearly 25,000 mg Ni / kg dry leaves. Metal absorption by roots took place from the deep soil layers, followed by the translocation of nickel to leaves, and after the abscission of the leaves, the metal was released in the litter. As a consequence, top soil layers in the plant rhizosphere contain high nickel concentrations that exerted a stronger selective pressure for evolution Ni-resistance in soil bacterial communities. This proposition was proved to be true for most the microfloral assemblage underneath the metal hyperaccumulators *Alyssum bertolonii, Alyssum murale, Berkhyyea coddii, Thlaspi geosingense, T dasycarpum, Rinorea bengalensis, Dichapetelum gelonioides,* and *Isatis pinnatiloba* [5, 6, 10, 26–29].

In recent years, scientists have revealed the presence of metal-resistant microbial communities endogenous to the Ni-hyperaccumulating serpentinophytes like *Alyssum murale, A. serpyllifolium,* and *Noccaea caerulescens.* It has been observed that abiotic factors in ultramafic ecosystem modulate the soil microbial resources as well as the diversity of rhizospheric and endophytic microbial communities associated with serpentinophytes [30]. Metagenomic studies conducted in New Caledonia revealed that vegetation cover and soil properties significantly shaped the large phylogenetic distribution of operational taxonomic unit within microbial populations, which proved that the above-ground vegetation had a strong correlation with microbial species richness and distribution pattern over ultramafic sites [31]. Various techniques including phenetic, genetic, and metagenomic approaches have been used by several researchers to elucidate the microbial community structure as well as endophytic assemblage of endemic serpentinophytes. Moreover, these plant-associated microorganisms have been shown to play significant role in metal bioavailability and uptake for the plants and also in microbe-assisted phytomining [30].

Although several researchers have reported genetic diversity of serpentine microbiota associated with hyperaccumulator plants as well as in soil microbiome, a thorough search for metal responsive genes amongst serpentinophytes-associated bacteria and fungi is lacking. DNA-DNA hybridization experiments have showed strong homologies in metal-resistant determinants between rhizospheric nickel-resistant strains of *Sebertia acuminata* from New Caledonian ultramafics and microorganisms obtained from anthropogenically polluted sites. Moreover, nickel and cobalt resistance operons, *ncc* and *cnr,* occurred abundantly or in association with *nre* operon amongst the serpentine bacteria where metal responsive genes were mostly borne on the chromosomes and showed strict inducible nickel resistance [4]. Partial amplification of the metal-resistant loci, *czc, chr, mer,* and *ncc* from gram-positive and gram-negative rhizobacteria (*Arthrobacter rhombi, Microbacterium arabionogalactanolyticum, Clavibacter xyli, Rhizobium mongolense, Variovorax paradoxus*) of Ni-hyperaccumulator *A. murale* growing over serpentine areas of Oregon, USA showed that these genes confer resistance to mercury, nickel, cobalt, copper, chromium, and zinc [27]. Moreover, metal resistance genes from *Microbacterium arabinogalactanolyticum* AY509226 showed homology to those previously reported in the metallophilic bacterium, *Cupriavidus metallidurans,* CH 34, obtained from polluted environment.

9.4 Bacterial and fungal rhizobiome of hyperaccumulator plants

Plant colonization of serpentine habitat supported higher microbial population, and the number of bacteria and fungi decreased gradually on transect from a cluster of plants to a bare site. The rhizosphere of *Grevillea gillivrayi* and *Costularia comosa* appeared to have a strong positive influence on microbial density. Gram-negative bacteria were found to be predominant in the New Caledonian ferralitic oxidic soil, which contained only 28% gram-positive bacteria. The actinomycetes were dominated by the genera, *Streptomyces* and *Streptoverticillum*, representing more than 70% of the actinomycetal flora. A high genetic diversity and heterogeneity occurred amongst the rhizobacterial communities of serpentine ecosystem, and the presence of endemic Ni-hyperaccumulating plants *A. bertolonii, A. murale, Isatis pinnatiloba, T. geosingense, T. dasycarpum*, etc., shaped the microbial community composition [32].

9.4.1 Rhizobacteria

Rhizobacterial assemblage, their characteristics, and significance in serpentine metal-hyperaccumulating plants worldwide has been tabulated in Table 9.1. Both cultivation-dependent and cultivation-independent approaches were used to assess rhizobacterial communities associated with the Ni-hyperaccumulator plant, *Thalspi geosingense* in eastern Austria [10], and a total of 47 nonchimeric 16S rRNA gene sequences were obtained. Comparison of data with terminal restriction length polymorphism (T-RFLP) profiles confirmed the presence of *Halophaga / Acidobacterium* division (27% of the clones), α-*Proteobacteria* (22%), high G + C gram-positive clones (16%), *Cytophaga / Flexibacter / Bacteroides* division (10%), and the rest as γ- or β-Proteobacteria. The rhizosphere bacteria were identified as members of *Methylobacterium, Rhodococcus, Okibacterium, Verrucomicrobia, Burkholderia, Rhizobium, Microbacterium*, etc. However, low G + C gram-positive clones belonging to *Bacillus* and *Paenibacillus* were found to be associated with the shoot instead of the rhizosphere [10]. Genetic diversity of heterotrophic bacterial communities in Tuscan serpentines in Italy were studied with respect to the Ni-hyperacumulator *A. bertolonii* following amplified ribosomal DNA restriction analysis (ARDRA), T-RFLP studies, and phylogenetic analysis of 16S rDNA sequences [5]. Taxonomic interpretation showed that most of the sequenced TRFs showed similarities with members of Proteobacteria, *Acidobacteria*, Verrucomicrobia, Chlorobi, *Bacillus, Paenibacillus*, Actinobacteria, *Metylosinus*, etc. Several species of *Pseudomonas* (*P. syringae, P. brassicacearum, P. corrugate, P. synxantha, P. putida, P. mandelii*, etc.) and *Chryseobacterium meningosepticum* dominated the rhizosphere, while *Streptomyces* were mainly present in the soil. The rhizosphere microbial communities of

Turkish Ni-hyperaccumulators, *Isatis pinnatiloba* and *A. dasycarpum* were assessed following fatty acid methyl ester (FAME) analysis and identified as members of *Arthrobacter, Bacillus, Microbacterium, Staphylococcus, Xanthobacter, Novosphingobium, Acinetobacter*, etc. [29].

Bacterial diversity, richness, and composition were assessed from rhizosphere of ten woody nickel-hyperaccumulators from ultramafics of Halmahera Island, Indonesia, and it was observed that the obligate hyperaccumulator, *Rinorea* aff. *Bengalensis*, is the main factor driving the bacterial diversity in the rhizosphere, while the impact of soil conditions was important for other hyperaccumulator species like *Ficus trachypison* and *Trichospermum morotaiense* [39]. Rhizobacterial community diversity from underneath Ni-hyperaccumulators *Noccaea ochroleuca, Odontarrhena smolikana, O. rigida* and *O. chalcidica* growing on ultramafic regions of Librazhd and Pogradec in eastern Albania had a strong influence on soil cation exchange capacity. Proteobacteria was found to be the dominant phylum and based on metagenomes, genes for aminoacids, lipid and carbohydrate metabolism were found to be the major gene families [36].

Table 9.1: Rhizosphere bacterial resources associated with serpentinophytes, their characteristics, and significance.

Serpentinophyte	Origin	Associated rhizobacteria	Characteristics / Significance	References
Alyssoides utriculata	Italy, Europe	*Arthrobacter, Bacillus, Streptomyces*	Nickel tolerance and PGPR activity	[23]
Alyssum bertolonii	Italy, Europe	*Pseudomonas, Metylosinus, Acidobacteria, Bacillus, Paenibacillus, etc.*	Resistance to Ni, Co, Cr, Zn, Cu, etc.; production of siderophore	[5]
Alyssum murale	Oregon, USA	*Sphingomonas, Microbacterium, Arthrobacter, Rhizobium, Variovorax*	Resistance to heavy metal, solubilization of PO_4; production of siderophore	[27]
Alyssum serpyllifolium ssp. lusitanicum, A. serpyllifolium ssp. malacitanum	Iberian Peninsula	*Arthrobacter, Streptomyces, Rhodococcus, Microbacterium*	Resistance to Ni; solubilization of PO_4; production of IAA	[33]
Alyssum serpyllifolium, Dactylis glomerata	Iberian Peninsula, Spain	*Blastococcus, Geodermatophilus, Modestobacter, Rhodococcus, Pseudonocardia, Devosia* and *Hyphomicrobium*	Alphaproteobateria mediated assistance in trace metal tolerance by plants	[34]

Table 9.1 (continued)

Serpentinophyte	Origin	Associated rhizobacteria	Characteristics / Significance	References
Costularia sp.	New Caledonia, South Pacific	*Caballeronia novacaledonica* STM10272, *Paraburkholderia ultramafica* STM10279	High Ni tolerance related to biofilm; production of acidic polysaccharide	[35]
Isatis pinnatiloba	Turkey, Europe	*Staphylococcus, Xanthobacter, Novosphingobium, Acinetobacter*, etc.	Tolerance and resistance to Ni, Co, etc.	[29]
Noccaea ochroleuca, Odontarrhena smolikana, O. rigida, O. chalcidica	Librazhd and Pogradec, Albania	Proteobacteria, Actinobacteria, Acidobacteria, Gemmatimonadetes, Chloroflexi, Bacterioidetes, Nitrospirae	Influence on rhizospheric soil bacterial diversity catalyzed by cation exchange capacity	[36]
Rinorea bengalensis, Dichapetelum gelonioides ssp. *andamanicum*	Andaman Islands, India	*Bacillus, Streptomyces, Pseudomonas, Cupriavidus*	Resistance to Ni, Cr, Co, Zn, Mn; intracellular accumulation of PHA	[6, 37]
Rinorea aff. *bengalensis*	Borneo Island, Sabah, Malaysia	Proteobacteria, Acidobacteria, Actinobacteria	Assistance in concentrating nickel in plant rhizosphere; resistance to high Ni levels	[38]
R. aff. *bengalensis, Ficus trachypison, Trichospermum morotaiense*	Halmahera Island, Indonesia	Proteobacteria, Acidobacteria, Actinobacteria, Chloroflexi	Influence on bacterial composition, richness and diversity	[39]
Sebertia acuminata	New Caledonia, South Pacific	*Hafnia, Acinetobacter, Commamonas, Alcaligenes, Burkholderia, Streptomyces, Streptoverticillum*	Occurrence of Ni, Co, Cu, and Zn resistant plasmids	[4]
Tetraria arundinacea, Chamaedendron. nervosa, T. pubescens	New Caledonia, South Pacific	*Curtobacterium citreum* BE, BB, AM	Ni tolerance and bioaccumulation; Ni chelation by exopolysaccharides	[11]

Table 9.1 (continued)

Serpentinophyte	Origin	Associated rhizobacteria	Characteristics / Significance	References
Thlaspi geosingense	Austria, Europe	*Halophaga, Acidobacterium, Verrucomicrobia, Cytophaga, Flexibacter, Methylobacterium, Okibacterium, Rhodococcus*	Resistance to Ni, production of siderophore, ACC deaminase activity	[10]

9.4.2 Fungi and mycorrhiza

The first report on the mycoecological survey of serpentine ecosystem was conducted in Cascade Mountains, Washington, USA and based on sporocarp studies, less diverse fungal communities were reported. In addition, a higher proportion of mycorrhizal symbionts occurred in serpentine than non-serpentine areas. The fungal communities in New Caledonian ultramafics mostly belonged to the family Moniliaceae and the genera *Aspergillus,* and *Penicillium* accounted for > 50% of the total mycoflora. *Trichoderma*, *Paecilomyces*, *Scropulariopsis,* and *Cephalosporium* were also present, in addition to small numbers of Phycomycetes, especially *Pythium*, *Mucor*, and *Mortierella*. Common soil fungi like *Fusarium* (Tuberculariaceae) and *Alternaria* (Dermatiaceae) were not detected, which might be due to their sensitivity to metal ions. Similar observations on the dominance of *Aspergillus, Mortierella, Pythium, Penicillium, Rhizopus, Paecilomyces,* and *Trichoderma* were reported from serpentine sites of Andaman Islands, India [6]. Rhizoplane and rhizospheric fungi have been reported in association with Zn-hyperaccumulating plant, *Thlaspi caerulescens* from serpentine sites, and the fungi were able to solubilize zinc oxide and zinc phosphate under laboratory conditions. Table 9.2 enumerates fungal and mycorrhizal resources associated with metal-hyperaccumulating serpentinophytes and their significance.

The occurrence of vesicular arbuscular mycorrhizae in the grassland community was first reported from serpentines of California, and the common fungal symbionts were *Glomus fasciculatum* and *Glomus tenue*. Colonization of arbuscular mycorrhizae with the abundant formation of arbuscules was observed in Ni-hyperaccumulators of South Africa like *Berkheya coddii*, *B. zeyheri*, *Senecio coronatus,* and *S. anomalochrous* of Asteraceae growing in South Africa, and all plants showed obligatory mycorrhizal symbiosis [26]. Mycorrhizal frequency and relative arbuscular richness with respect to spores and auxiliary cells of *Gigaspora* sp., small vesicles of *Glomus tenue*, etc. were illustrated. The genetic divergence amongst ectomycorrhizal fungi reflects adaptation to serpentine soil factors. Amplified fragment length polymorphism (AFLP) and RFLP

Table 9.2: Fungal and mycorrhizal resources associated with serpentinophytes, their characteristics, and significance.

Serpentinophyte	Origin	Associated fungi	Characteristics / Significance	References
Alyssoides utriculata	Italy, Europe	*Penicillium, Mucor, Trichoderma harzianum, Penicillium ochrochloron*	Tolerance to Ni	[23]
Avenula sulcata ssp. *sulcata*	Portugal, Europe	Arbuscular mycorrhizal fungi: *Glomus, Gigaspora, Scutellospora, Aculospora, Entrophospora*	Promote adaptation of host to edaphic conditions; little influence on microbial symbiosis	[9]
Berkheya coddii	Mpumalanga, South Africa	AMF: *Glomus, Gigaspora* Endophytic fungi: *Alternaria, Aspergillus, Bipolaris, Cladosporium, Epicoccum, Fusarium* Yeasts: *Cryptococcus, Meyerozyma, Rhodotorula*	Resistance to Ni	[26]
Costularia spp.	New Caledonia	Arbuscular mycorrhizal fungi: *Glomus*	Enhancement of plant growth, reduction of Ni content in roots	[40]
Hakea verrucosa	Bandalup Hill, Western Australia	AFM hyphae, vesicles, intracellular arbuscules and hyphal coils	Reduced Ni uptake in host increased shoot P and K concentration	[41]
Knautia arvensis	Temperate Grassland, Europe	Arbuscular mycorrhizal fungal colonization, 81 molecular OTUs belonging to the Glomeromycota	Enhancement of host phosphate nutrition under greenhouse condition	[24]
Nothofagus aequilateralis, Arillastrum gummiferum	Tropical rainforest, New Caledonia	311 OTUs, dominated by *Cortinarius* spp.	Plays a central role in ultramafic ecosystems functioning	[42]
Onosma echioides	Tuscany, Italy	*Ceratobasidium, Coprinellus* (Basidiomycota)	Enhanced Ni tolerance in host plant	[43]
Pinus sylvestris, Quercus petraea	Eastern Austria, Europe	Basidiomycetous fungi; dominance of Russulaceae and *Phialocephala fortinii*	Colonization and tolerance to extreme edaphic conditions	[44]
Psychotria douarrei	New Caledonia	Arbuscular mycorrhizal fungi: *Glomus*	Tolerance and adaptation to Ni	[7]

Table 9.2 (continued)

Serpentinophyte	Origin	Associated fungi	Characteristics / Significance	References
Rinorea bengalensis	Andaman Islands, India	*Aspergillus, Mortierella,*	Resistance to Ni, Cr, Co, Zn, Mn, etc.	[6]
Tristaniopsis spp.	Koniambo massif, New Caledonia	*Cortinarius, Pisolithus, Russula, Heliotales, Boletellus*	Ectomycorrhizal symbiosis mediated improvement of plant adaptation	[45]

analysis revealed a high degree of genetic diversity in ectomycorrhizal fungus, *Cenococcum geophilum,* from serpentine and non-serpentine soils in Portugal, where serpentine strains lacked a group I intron frequently found within the 18S rRNA gene of those species.

The community of ectomycorrhizal and co-associated fungi from serpentine site in Austria forested with *Pinus sylvestris* and *Quercus petraea* was explored [44] following a combinational approach of morphotyping and direct sequencing of the nuclear ribosomal internal transcribed spacer region 2 and of a part of the large-subunit region. Co-associated fungal members from selected ectomycorrhizal strains were identified by RFLP and sequencing of the representative clones from the library. A successful colonization of tree roots with mycorrhizal members showed intra-fungal association in serpentine edaphic conditions. Aphyllophoralean fungi representing basidiomycete order and Rusulaceae were found to be dominant. *Phialocephala fortinii* was the most frequently encountered taxon from the diverse assemblage of ectomycorrhiza co-associated fungi. Soil metal content, especially Ni and the Mg/Ca ratio, along with vegetation pattern and endemic hyperaccumulators on serpentine ecosystem have been found to influence the diversity of AFM. Moreover, in vitro studies have shown that inoculation of AFM not only enhanced nickel uptake in the hyperaccumulating plants but also improved phosphorus uptake and biomass yield of metal-sensitive plants [25]. Fungal mutualists, thus, find great potential for development as a biological fertilizer to support reclamation of nickel post-mining land on serpentine habitat.

9.5 Endophytic microbial population of serpentinophytes

The potential applications of endophytic bacteria are manyfold, and the exploitation of bacterial communities in metal tolerant and metal-hyperaccumulating plants could have a direct impact on plant performance. One of the most basic questions concerns the relationships existing between endosphere (endophytic) and rhizosphere bacteria in hyperaccumulators. Endophytic bacteria are thought to enter plant tissue mainly from roots or through the sites of wounding [46], but some endophytes are transferred from one plant generation to the next through the seeds. Consequently, bacteria of the endosphere are, at least partly, a subset of rhizosphere and of phyllosphere community. Endospheric bacterial resources associated with serpentine plants, their characteristics, and significance are presented in Table 9.3.

Table 9.3: Endospheric bacterial resources associated with serpentinophytes, their characteristics, and significance.

Serpentinophyte host/organ	Origin	Associated endophyte	Characteristics/ significance	References
Alyssum bertolonii (Root, stem, leaf)	Italy, Europe	Alpha and Gamma proteobacteria, Clostridia	Plant by plant variations in leaf associated microbial community, resistance to Ni	[47]
Alyssum serpyllifolium (Leaf)	Spain, Europe	*Pseudomonas azotoformans* ASS 1	Promotion of phytoremediation efficiency in *Trifolium arvense* growing on multi-metal contaminated soils	[48]
Berkheya coddii (Leaf)	Mpumalanga, South Africa	*Bacillus, Lysinibacillus*	Resistance to Ni	[49]
Noccaea caerulescens (Root)	Italy, Europe	*Microbacterium, Arthrobacter, Agreia, Bacillus, Sthenotrophomonas, Kocuria, Variovorax*	Resistance to Ni; plant growth promoting activities, enhancement of Ni translocation in the host plant	[50]
Noccaea caerulescens (Seed)	France, Europe	γ-Proteobacteria, α-Proteobacteria, Actinobacteria, Firmicutes	Plant growth promoting activities	[51]

Ni-hyperaccumulating plants growing on metalliferous outcrops provide a unique niche for colonization of bacterial endophytes in the internal tissues of the host plant and were adapted to survive in high concentrations of heavy metals. There is ample evidence of beneficial role of endophytic microorganisms on plants by promoting plant growth or conferring the host tolerance to biotic and abiotic stresses. Culture-independent analysis of endophytic bacterial communities of Ni-hyperaccumulator, *T. geosingense*, following T-RFLP analysis indicated presence of novel endophytes belonging to proteobacteria and high G + C gram-positive groups [10]. Cloning and sequencing of endophytic 16S rRNA genes categorized the bacterial strains into six clone libraries with 67% of clones clustering in α, ß, and γ-proteobacteria. The occurrences of *Sphingomonas, Methylobacterium, Rhodococcus, Curtobacteria*, etc., in the stem of *T. geosingense* were in concordance with those isolated from the root and stem of Zn-hyperaccumulator, *T. caerulescens* [46].

Several molecular techniques including denaturing gradient gel electrophoresis, temperature gradient gel electrophoresis, etc. have been applied to study the population dynamics of endophytic microorganisms, and results indicated that individual plants host a broad phylogenetic range of bacterial endophytes. Eighty-three endophytic bacteria were isolated from roots, stems, and leaves of a nickel-hyperaccumulator plant, *A. bertolonii* (Brassicaceae) endemic to serpentine outcrops of Central Italy and classified by restriction analysis of 16S rDNA (ARDRA) and partial 16S rDNA sequencing into 23 different taxonomic groups [47]. The aerobic heterotrophic culturable bacteria mostly colonizing inside lower roots belonged to the genera *Staphylococcus, Curtobacterium, Bacillus, Arthrobacter, Pseudomonas, Microbacterium, Leifsonia*, and *Paenibacillus*.

9.6 Microbial association and phenomenon of hyperaccumulation

The phenomenon of metal hyperaccumulation in endemic serpentinophytes has found a positive impact on plant-associated microbiota that has opened an important arena of research on environmental microbiology. Microorganisms can have a great impact on the performances of revegetation of polluted soils, and investigations pertaining to serpentine soils may provide new insight into bacterial diversity under unfavorable conditions, new isolates, and probably new genetic information on heavy-metal resistance, which could be exploited. Most of the research on metal hyperaccumulators has focused attention on the physiological mechanisms of metal transport in plants, and the plants' storage and tolerance of the accumulated metals. Considerably less information is available on the rhizosphere processes of hyperaccumulators, where root proliferation and effective uptake mechanisms constitute the key processes that distinguish metal hyperaccumulators from non-accumulating plants. The roles of root exudates and microbial activity in the soil are largely unknown, although it has been suggested that root exudates of the members of the genus *Thlaspi* normally do not enhance mobilization of metals in soil.

The metal-resistant microbes in rhizosphere can affect trace metal mobility and availability to the plants through release of chelating agents, acidification, phosphate solubilization, and redox changes [10, 27]. Chemolithotrophic bacteria have been shown to enhance environmental mobility of metal contaminants via soil acidification, or, in contrast, to decrease their solubility due to precipitation as sulfides.

Moreover, the naturally occurring bacteria in the metal-percolated ultramafic soils have acquired resistance against nickel, chromium, cobalt, and other heavy metals [6]. Furthermore, these metal-resistant bacteria have exceptional ability to promote the growth of the host plant by various mechanisms, namely fixation of atmospheric nitrogen, production of siderophores, solubilization of phosphate, or production of plant growth regulators [27]. Exploitation of specific traits related to metal–microbe interactions appears to be an essential requirement in order to utilize the microbial diversity for bioremediation of environmental pollutants. Remediation options currently available for contaminated soils and groundwater are frequently expensive, environmentally invasive, and do not make efficient use of existing biological resources, and, as a consequence, they cannot be employed for the treatment of large contaminated areas with diffuse pollution problems.

The indigenous microbiota from the rhizosphere or endosphere of serpentinophytes is endowed with metal mobilizing ability that is significant for efficient phytoremediation of metal-contaminated environment. In general, a majority of the rhizospheric and endospheric bacteria associated with endemic hyperaccumulators worldwide show the phenomenon of metal tolerance, especially to nickel (Table 9.1, 9.3). Bacteria associated with *Rinorea bengalensis* in Andaman serpentines, India, showed the presence of constitutive Cr(VI)-reductase activity, Ni uptake in aqueous solution, accumulation of intracellular polyhydroxyalkanoates, as well as production of extracellular polysaccharide during growth under metal-stressed condition [37]. Two ultramafic bacteria, *Caballeronia novacaledonica* STM10272 and *Paraburkholderia ultramafica* STM10279 showed high levels of Ni resistance, and the mechanism of such metal tolerance was attributed to biofilm formation and production of acidic EPS [35]. Both species allowed metal uptake in *Tetraria comosa*, an endemic herbaceous species growing on New Caledonian ultramafic soils. Ni-resistant strains of *Curtobacterium citreum* isolated from the same sites showed production of exopolysaccharides, which was involved in Ni chelation. The bacterial strains also showed plant growth promotion and when inoculated with sorghum, exhibited reduced metal uptake in the plant [35].

The much diverse fungal population associated with hyperaccumulators on ultramafic sites have also shown metal resistance and metal mobilization potentials. Fungi-colonizing rhizosphere of *Rinorea bengalensis* (India) showed high tolerance efficiency of cobalt in comparison to Ni and Cr, and few of them showed cobalt biosorption potential [6]. The Zn-solubilizing fungi associated with *Thlaspi caerulescens* did not show any adaptive tolerance to metals. Arbuscular mycorrhizal fungi were most commonly found associated with serpentine hyperaccumulators, and they not only enhanced plant nutrition under edaphic stress but also helped in Ni accumulation by the host plants [25]. A non-mycorrhizal plant, *Hakea verrucosa* from Australian ultramafics, sustained hostile environment through association of cluster roots to access phosphorus. The roots showed fungal hyphae, vesicles, as well as intracellular arbuscules and hyphal coils [41].

The diverse endophytic bacterial community reported from Ni-hyperaccumulators showed a high level of resistance to heavy metals that might influence plant growth and phytoremediation [8]. Moreover, endophytic bacteria may be genetically modified to confer new abilities to the host plant for phytoextraction purposes. The bacterial endophytes of *A. bertolonii* showed resistance to heavy metals Ni, Cr, Co, Zn, and Cu and were segregated into 42 metal-resistant phenotypes [47]. About 80% of the isolates (*Staphylococcus, Pseudomonas, Microbacterium, Curtobacterium*) from leaves and stem were, in fact, resistant to high chromate concentrations, whereas around 30% of root isolates showed resistance to the same metal concentrations. An endophytic bacterium, *Pseudomonas azotoformans* ASS1, from leaves of *A. serpyllifolium*, possessed resistance to abiotic stress including heavy metals of Cu, Zn, and Ni as well as showed PGPR properties [48]. The bacterium promoted phytoremediation efficiency in *Trifolium arvense* by protecting the host against draught and metal stress on semi-arid ecosystem. The bacterium enhanced chlorophyll content, catalase, peroxidase, superoxide dismutase activities, and proline content (only in the absence of drought) in plant leaves, whereas there was a decrease in the concentrations of malondialdehyde.

Majority of the endophytes showed siderophore production, which seemed to be a general phenotype, because endophytic population thrives at extremely low levels of free iron ions within plant tissues, and this may be related to formation of Ni-siderophore complexes in the apoplast, causing decreased Ni toxicity. The high percentage of 1-aminocyclopropane-1-carboxylic acid (ACC) deaminase producing endophytes (36%) in *Thalspi geosingense* explained the intimate relationship of endophytes with the Ni-hyperaccumulator, as they not only showed metal tolerance but also exhibited beneficial effect on plant growth in metal percolated serpentine habitat. Culturable endophytes belonging to *Microbacterium, Arthrobacter, Agreia, Bacillus, Sthenotrophomonas, Kocuria,* and *Variovorax* from roots of *N. caerulescens* tolerate 6 mM Ni and promoted plant growth. Seed inoculation with these metal tolerant endophytes under in vitro experiments showed root colonization and root-to-shoot Ni translocation in metal hyperaccumulator plants, therefore possessing a significant role in phytoextraction technologies [50].

9.7 Role of microbiota in phytoextraction and phytoremediation

Metal accumulation in hyperaccumulators depends mostly on rhizosphere processes of serpentinophytes, including a high population of bacteria around the roots that significantly influence the growth of plants as well as bioavailability of soluble metals. Research on rhizobacteria associated with serpentine hyperaccumulating plants is progressing towards development of new strategies for improving current phytoremediation technologies. Technological improvements have been achieved with genetically modified bacterial strains developed by transferring Ni-resistance determinants from *Cupriavidus metallidurans* or phytochelatin synthase from *Arabidopsis thaliana* into endophytic *Burkholderia cepacia, Herbaspirillum seropedicae,* and *Escherichia coli.* However, biosafety issues have limited the use of recombinant strains, unless natural gene transfer has taken place. Naturally occurring metal-resistant bacteria associated with serpentinophytes may be an ideal tool for development of species-specific inoculants to increase the yields of metal uptake and phytoremediation in metal-contaminated sites [52]. The plant growth promoting bacteria (PGPB) have an exceptional ability to promote the growth of the host plant by various mechanisms, namely fixation of atmospheric nitrogen, production of ACC-deaminase, siderophores, solubilization of phosphate, or production of plant growth hormones [8, 27, 48]. Further, a number of different heavy metal tolerance mechanisms have been discovered in these PGPB [8], which involve exclusion, active removal, biosorption, precipitation, or bioaccumulation of metals. These processes can influence solubility and bioavailability of metal to host plant and thereby improve the efficiency of phytoextraction process. In addition, these PGPR also contribute to indirect phytoextraction by promoting the shoot and root biomass of hyperaccumulators. Production of ACC-deaminase by rhizobacteria is beneficial for modulating ethylene level in the hyperaccumulator plant [52].

Rhizobacterial strains of *Alyssum murale* have displayed plant growth promoting activities and metal accumulation. It has been observed that root architecture, effective uptake systems, and partial depletion of metals in the rhizosphere are relevant systems involved in hyperaccumulation of metal ions. However, the role of root exudates as well as rhizosphere acidification in metal mobilization has not been proved. Moreover, it has been found that rhizosphere microorganisms increase the availability of heavy metals for plant uptake [53]. Ni-resistant rhizobacteria may improve metal availability and solubility either through production of metal chelating substances or siderophores or decrease of soil pH. Siderophore production not only prevents iron deficiency in the producer bacterium and the plant but also aids in uptake of other metals. Seeds of Ni-hyperaccumulators, *Isatis pinnatiloba* and *Alyssum dasycarpum,* germinate and grow in varying concentrations of nickel solution in greenhouse conditions under promotion of PGPR. The plants showed a better stimulatory effect on the rhizobacteria, which showed resistance to Ni, Pb, Cd, and Zn, and thereby influenced phytoremediation processes [29]. Rhizobacteria from Ni-hyperaccumulator *A. pintodasilvae* selected on the basis of metal tolerance, plant growth promoting traits, and Ni-solubilizing capacity were used in re-inoculation experiments in Ni-rich soils as well as sewage–sludge-affected agricultural soil. The inoculants influenced plant biomass yield and metal accumulation in shoot. The bacterium, *Arthrobacter nicotinovorans* SA40, was able to promote plant growth and phytoextraction of Ni in both soil types and was claimed to be an ideal candidate for future field-based trials [54].

Investigations pertaining to illustrate the relationship between extractable metals and microbiological characteristics of several ultramafic soils in New Caledonia have reported release of Ni and Co in serpentine soils due to microbial processes involving organotrophic microorganisms. The release of metals in these soils constitutes a selective pressure that causes the microflora to evolve, by natural selection, higher metal tolerance, enabling an adequate degree of activity to be maintained in an increasingly toxic environment. There are reports on the weathering of oxides of iron and manganese together with release of trace metals by bacterial reduction in New Caledonian ultramafic soils [7, 40]. The impact of fungi on geological processes involves mineral bio-weathering following the alteration of rocks and minerals mediated by biomechanical and biochemical activities of saprotrophic and mycorrhizal fungi. *Verticillium leptobactrum, Paecilomyces lilacinum, Aspergillus fumigates*, etc. were abundant in several serpentine sites in the Western Alps and were able to extract different amounts of iron and magnesium from asbestos, in vitro.

The serpentine bacteria have been shown to possess several traits that can alter heavy metal mobility and availability to the plants. Investigation reveals that organic molecules excreted by the plant are used as substrate by the microorganisms leading to their enhanced population density in the zone close to the roots than in the bulk soil or soil without vegetation. Plants also produce humus, which increases the microbial life in the top soil layer. In return, the rhizobacterial population may produce iron chelators and siderophores for ensuring iron availability, reduce soil pH, and/or solubilize metal-phosphates [10, 27]. The rhizosphere provides a complex and dynamic niche, where microorganisms, in association with roots, form unique complex that have considerable potential for the detoxification of hazardous materials. The chemolithotrophic bacteria have been shown to enhance environmental mobility of metal contaminants via soil acidification, or in contrast, to decrease their solubility due to precipitation as sulfides. Some are also known to influence root parameters, such as root morphology and growth. An increase in root exudation of organic solutes could affect the rate of phytosiderophore release. In turn, rhizosphere microorganisms may interact symbiotically with roots to enhance the potential for metal uptake [55].

9.8 Contribution of associated microbes in phytomining

Biomining using metal-resistant microorganisms is a well-established phenomenon whereby acidophilic, chemolithotropic, and iron- and sulphur oxidizing bacteria have been used to recover metals from insoluble metal ores. Metals in soluble state are easily absorbed by the plants root system, leading to phytomining. Enhanced uptake of Ni by hyperaccumulators is important to the overall process of phyroextraction of metal from serpentine soils. It has been found that inoculation of rhizobacteria *M. oxydans* AY509223 significantly increased Ni extraction as well as uptake by *A. murale,* compared to uninoculated seeds in addition to insignificant increase of fresh and dry weight of *A. murale* shoots grown in Ni-amended or unamended soils. These suggest that Ni-hyperaccumulation by rhizobacteria is an important development of inoculum for enhancing uptake during commercial phytomining of Ni. Very recently, Benrizi et al. [30] have documented a list of microorganisms associated with metal hyperaccumulators, which are significant for phytoextraction and phytomining. Interestingly, these organisms were endowed with plant growth promotion traits like production of phytohormones, siderophores, glutathione synthatase, ACC deaminase, and solubilization of phosphate. Further, they have led to increase of rooting, biomass augmentation as well as metal removal / uptake in the shoot along with certain degree of tolerance to abiotic stress.

The inoculation of Ni-mobilizing PGPB, *Pseudomonas* sp. SRI2, *Psychrobacter* sp. SRS8, and *Bacillus* sp. SN9 were effective in protecting plants from growth inhibition caused by Ni. Bioinoculation of metal-resistant PGPB, significantly increased the growth of Brassica *juncea, B. oxyrrhina,* and *Ricinus communis* in Ni contaminated soils, thus increasing the efficiency of phytoextraction directly by enhancing Ni accumulation in plant tissues and indirectly by promoting the shoot and root biomass. Further, a significantly positive correlation was observed between the bacterial Ni mobilization in soil and the total Ni uptake in both plant species [8]. Similarly, the Ni-resistant PGPB *Psychrobacter* sp. SRS8 from rhizosphere of *A. serpyllifolium* had significant potential in increasing biomass production and Ni uptake by energy crops, *Ricinum communis* (castor) and *Helianthus annus* (sunflower), grown in Ni-contaminated sites [55]. The inoculation of PGPB not only protects the host against inhibitory effect of toxic nickel but also causes changes in the growth, protein, leaf chlorophyll content, and nutrient status of the host plant. Production of siderophore and enhancement of antioxidant enzymes by inoculation of SRS8 suggested that PGPB might have contributed in limiting the oxidative damage to biomolecules in response to Ni-stress and improving phytoextraction as well as phytomining potential of the host plants.

Moreover, the inoculation of plants with arbuscular mycorrhizal fungi has led to increased plant growth and better adaptation in metal-enriched ecosystem, and Benririzi et al. [30] have suggested that AFM may aid in nickel phytomining technologies. Although presence of metal reduced fungal colonization and sporulation, mycorrhizal symbionts seemed to act as a barrier to the absorption of Ni by the plant and reduced root-to-shoot Ni translocation. This was observed by Amir et al. [7] in case of endemic nickel hyperaccumulators, *Alphitonia neocaledonica* and *Cloezia artensis* from ultramafic shrub lands of New Caledonia. The fungus, *Glomus etunicatum,* showed a stronger positive effect on plant biomass and phosphorus uptake, and a greater reduction in toxicity symptoms and Ni concentration in roots and shoots.

Several serpentine hyperaccumulators were initially reported to be non-mycorrhizal; however, studies showed that Zn- and Cd-hyperaccumulator *Thlaspi praecox* from metal-rich sites of Slovenia have mycorrhizal colonization like *Berkheya coddii* from South African serpentines. In the latter, mycorrhizal symbiosis effected a 20-fold increase of biomass and better Ni yield, although AFM inoculation reduced Ni uptake in root and shoot. Thus non-mycorrhizal plants can also aid in metal phytomining, with improved metal uptake efficiency and high plant biomass, due to mutualistic associations of AFM [26].

9.9 Present status of applications of microbiota in bioremediation

Serpentinophytes-associated microbiota are characterized by site-specific adaptability and a high level of genetic diversity, independent of the metal concentrations present in the soil. Bacteria and fungi associated with metal hyperaccumulators have shown high degree of tolerance to several heavy metals, by virtue of their ability of metal uptake and / or efflux mechanisms controlled by specific genes on the chromosome. However, the degree of metal resistance varied amongst the different phenotypes, and the rhizosphere chemistry of endemic hyperaccumulators determines the metal-resistant population of microbial communities. The culturable metal-resistant phenotypes also showed wide diversification in metal tolerance strategies of bioaccumulation, biosorption, production of expolysaccharides, and enzymatic bioreduction of toxic metals under in vitro systems. The rhizomicrobiota and endospheric organisms were also found to produce plant growth promoting attributes, which helps the plant to increase root and shoot biomass under metal-stressed environment. The rhizospheric processes including production of carboxylic acid, abscisic acid, siderophores, and solubilization of phosphates helped in mobilizing metals for plant uptake and phytoremediation. In recent years, metagenomics and metaproteomics data can be useful in assessing the microbial community and complex microbial consortium present in such extreme metalliferrous niche. Microbial proteins may serve as bioindicators of positive interactions between plant roots and soil microcosms, which might, in turn, lead to prediction of hyperaccumulator phenotypes [30, 50]. Omics-based analysis of plant-microbial interactions, coupled with metabolic profiling of microorganisms associated with serpentinophytes by in situ mass spectrometry studies, will be a potential tool for phytotechnologies and future bioremediation strategies.

Uptake of trace elements or heavy metals by plants in association with microorganisms has been attributed to changes made in the absorptive properties of the roots, increase of root length, root hairs, and surface area by the plant-specific rhizomicrobiome. Translocation of metal ions into the shoot was possible through beneficial effects of PGPR, which cause trace metal complexation and mitigation of metal ion phytotoxicity. However, use of these specific microbial species in bioremediation and phytoextraction studies is still inconsistent [55]. In particular, the use of potent microbial genera as bioinoculants under in vitro studies for metal uptake in greenhouse conditions have faced several complexities like plant-specificity, soil properties, metal concentrations, availability of mono- or polymetallic conditions, survival of inoculums, etc. These issues have not only minimized the possibility of using such microorganisms in phytoremediation or phytoextraction technologies, but also made the transfer of research technology from laboratory to land scale a more difficult one. Moreover, the use of genetically modified serpentine-associated microbiota may serve as a promising approach for phytoremediation processes. Although, their performances under natural environmental conditions have been exploited to a certain extent, they face competitions with the natural microbiota having better prospects of metal bioremediation.

A strong antagonism may be encountered in the soil due to introduction of a newly selected microbe or host plant for bioremediation of metal-polluted sites. The microbial inoculants of either bacteria or fungal communities may not be compatible with the host plant or the mode of application of bioinoculants may not be suitable, since these are not naturally selected by the host. Effective root colonization or rhizosphere competence is advantageous in plant-microbe interactions that ultimately contribute to the success of bioaugmentation. However, use of endophytes may be a better replacement that will increase the chances of establishment of an active and stable population, without competing with diverse microbial assemblage in soil. Co-inoculation of root endophytes, *Arthrobacter* and *Microbacterium,* with Ni-hyperaccumulator, *Noccaea caerulescens* on serpentine soil had a comparatively better effect on bioremediation and Ni-mining, along with translocation of the metal than individual inoculation [50, 51]. Combined inoculation of AFM and PGPR on growth of the "energy crop" maize or model plant, *Sorghum vulgare,* on metal polluted soils is also one of the important phytoremediation strategies in present years [7, 30].

9.10 Conclusion and future perspectives

The vast majority of metal-hyperaccumulating plants thriving on edaphic metalliferrous serpentine outcrops provide an interesting insight for studying metal accumulation by plants and their utilization in phytoremediation technologies. The associations of soil rhizospheric processes as well as plant rhizobiome with these serpentinophytes have shown to process metal mobilization and solubilization for uptake by root system. Moreover, several mechanisms of metal translocation and chelation are involved within the plants, which require support from the plant rhizobiome. Plant growth-promoting traits from endophytic communities as well as arbuscular mycorrhizal populations have shown to stimulate the serpentinophytes and even non-serpentinophytes for metal translocation, and they can serve as potential bioinoculants for restoration of metal-contaminated sites and barren lands. Although, several culture-dependent and culture-independent studies have been conducted to evaluate the diversity of microbiota associated with endemic serpentinophytes all over the globe, very few researchers have tried to establish their metal tolerance mechanism and genetic determinants for heavy metal tolerance. The sustenance of the program underlying the use of this rare and unique microbial assemblage of serpentinophytes for biomining and phytoremediation has just started and field scale evaluation of such microorganisms has received much attention. The complex and dynamic micro-environmental niche present in the rhizosphere of serpentinophytes has enabled adaptation of bacterial and fungal species that show resistance to extreme levels of heavy metals. Metal resistance potentials of these microorganisms need to be expressed in detail for their future exploitation in the field of environmental biotechnology.

References

[1] Berg, G. Plant-microbe interactions promoting plant growth and health: Perspectives for controlled use of microorganisms in agriculture. Appl Microbiol Biotechnol. 2009, 84, 11–18.

[2] Bulgarelli, D., Schlaeppi, K., Spaepen, S., van Themaat, E.V.L., Schulze-Lefert, P. Structure and functions of the bacterial microbiota of plants. Ann Rev Plant Biol. 2013, 64, 807–838.

[3] Kruckeberg, A.R. The influence of lithology on plant life. In: Kruckeberg, A.R., ed. Geology and plant life: The effects of landforms and rock type on plants. University of Washington, Seattle, USA, University of Washington Press, 2002, 160–181.

[4] Stoppel, R.D., Schlegel, H.G. Nickel resistant bacteria from anthropogenically nickel polluted and naturally nickel percolated ecosystems. Appl Environ Microbiol. 1995, 61, 2276–2285.

[5] Mengoni, A., Barzanti, R., Gonnelli, C., Gabbrielli, R., Bazzicalupo, M. Characterization of nickel-resistant bacteria isolated from serpentine soil. Environ Microbiol. 2001, 3, 691–708.

[6] Pal, A., Dutta, S., Mukherjee, P.K., Paul, A.K. Occurrence of heavy metal-resistance in microflora from serpentine soil of Andaman. J Basic Microbiol. 2005, 45, 207–218.

[7] Amir, H., Lagrange, A., Hassaïne, N., Cavaloc, Y. Arbuscular mycorrhizal fungi from New Caledonian ultramafic soils improve tolerance to nickel of endemic plant species. Mycorrhiza. 2013, 23, 585–595.

[8] Rajkumar, M., Prasad, M.N.V., Freitas, H., Ae, N. Biotechnological applications of serpentine soil bacteria for phytoremediation of trace metals. Crit Rev Biotechnol. 2009, 29, 120–130.

[9] Fitzsimons, M.S., Miller, R.M. Serpentine soil has little influence on the root-associated microbial community composition of the serpentine tolerant grass species *Avenula sulcata*. Plant Soil. 2010, 330, 393–405.

[10] Idris, R., Trifonova, R., Puschenreiter, M., Wenzel, W.W., Sessitsch, A. Bacterial communities associated with flowering plants of the Ni-hyperaccumulator *Thlaspi goesingense*. Appl Environ Microbiol. 2004, 70, 2667–2677.

[11] Bourles, A., Guentas, L., Chalkiadakis, E., et al. New Caledonian ultramafic conditions structure the features of *Curtobacterium citreum* strains that play a role in plant adaptation. Can J Microbiol. 2019, 65, 880–894.
[12] Brady, K.U., Kruckeberg, A.R., Bradshaw, H.D. Jr. Evolutionary ecology of plant adaptation to serpentine soils. Ann Rev Ecol Evol Sys. 2005, 36, 243–266.
[13] Bini, C., Maleci, L. The serpentine syndrome (H. Jenny, 1980): A proxy for soil remediation. Int J Environ Qual. 2014, 15, 1–13.
[14] Selvi, P. Diversity, geographic variation and conservation of the serpentine flora of Tuscany (Italy). Biodivers Conserv. 2007, 16, 1423–1439.
[15] Wójcik, M., Gonnelli, C., Selvi, F., Dresler, S., Rostański, A., Vangronsveld, J. Metallophytes of serpentine and calamine soils – Their unique ecophysiology and potential for phytoremediation. Adv Bot Res. 2017, 83, 1–42.
[16] D'Amico, M.E., Freppaz, M., Zanini, E., Bonifacio, E. Primary vegetation succession and the serpentine syndrome: The proglacial area of the Verra Grande glacier, North-Western Italian Alps. Plant Soil. 2017, 415, 283–298.
[17] Reeves, R.D., van der Ent, A., Echevarria, G., Isnard, S., Baker, A.J.M. Global distribution and ecology of hyperaccumulator plants. In: van der Ent, A., Baker, A.J.M., Echevarria, G., Simonnot, M.O., Morel, J.L., ed. Agromining: Farming for Metals. Switzerland, Springer Nature AG, 2021, 133–154.
[18] Perez-Latorre, A.V., Hidalgo-Triana, N., Cabezudo, B. Composition, ecology and conservation of the south-Iberian serpentine flora in the context of the Mediterranean basin. Anales Jard Bot. 2013, 70, 62–71.
[19] Boyd, R.S., Jaffré, T. Elemental concentrations of eleven New Caledonian plant species from serpentine soils: Elemental correlations and leaf-age effects. Northeast Nat. 2009, 16, 93–110.
[20] Adıgüzel, N., Reeves, R.D. Important serpentine areas of Turkey and distribution patterns of serpentine endemics and nickel accumulators. Bocconea. 2012, 24, 7–17.
[21] Pal, A., Paul, A.K. Serpentinophytes: Resource survey and conservation for potential applications in biotechnology. In: Kapoor, R., Kaur, I., Koul, M., ed. Plant reproductive biology and conservation. New Delhi, India, IK International Publishing House Pvt. Ltd, 2015, 354–388.
[22] Southworth, D., Tackaberry, L.E., Massicotte, H.B. Mycorrhizal ecology on serpentine soils. Plant Ecol Divers. 2014, 7, 445–455.
[23] Rosatto, S., Roccotiello, E., Di Piazza, S., et al. Rhizosphere response to nickel in a facultative hyperaccumulator. Chemosphere. 2019, 232, 243–253.
[24] Kohout, P., Doubková, P., Bahram, M., et al. Niche partitioning in arbuscular mycorrhizal communities in temperate grasslands: A lesson from adjacent serpentine and nonserpentine habitats. Mol Ecol. 2015, 24, 1831–1843.
[25] Husna,, Tuheteru, F.D., Arif, A. Arbuscular mycorrhizal fungi and plant growth on serpentine soils. In: Wu, Q.S. ed. Arbuscular Mycorrhizas and Stress Tolerance of Plants. Singapore, Springer Nature, 2017, 293–303.
[26] Mesjasz-Przybyłowicz, J., Przybyłowicz, W.J. Ecophysiology of nickel hyperaccumulating plants from South Africa–from ultramafic soil and mycorrhiza to plants and insects. Metallomics. 2020, 12, 1018–1035.
[27] Abou-Shanab, R.A., van Berkum, P., Angle, J.S. Heavy metal resistance and genotypic analysis of metal resistance genes in gram-positive and gram-negative bacteria present in Ni-rich serpentine soil and in the rhizosphere of *Alyssum murale*. Chemosphere. 2007, 68, 360–367.

[28] Pal, A., Wauters, G., Paul, A.K. Nickel tolerance and accumulation by bacteria from rhizosphere of nickel hyperaccumulators from serpentine soil of Andaman, India. Plant Soil. 2007, 293, 37–48.
[29] Turgay, O.C., Gormez, A., Bilen, S. Isolation and characterization of metal resistant-tolerant rhizosphere bacteria from the serpentine soils in Turkey. Environ Monit Assess. 2012, 184, 515–526.
[30] Benizri, E., Lopez, S., Durand, A., Kidd, P.S. Diversity and role of endophytic and rhizosphere microbes associated with hyperaccumulator plants during metal accumulation. In: van der Ent, A., Baker, A.J.M., Echevarria, G., Simonnot, M.O., Morel, J.L., ed. Agromining: Farming for metals. Switzerland, Springer Nature AG, 2021, 239–279.
[31] Bordez, L., Jourand, P., Ducousso, M., et al. H. Distribution patterns of microbial communities in ultramafic landscape: A metagenetic approach highlights the strong relationships between diversity and environmental traits. Mol Ecol. 2016, 25, 2258–2272.
[32] Mengoni, A., Schat, H., Vangronsveld, J. Plants as extreme environments? Ni-resistant bacteria and Ni-hyperaccumulators of serpentine flora. Plant Soil. 2010, 331, 5–16.
[33] Álvarez-López, V., Prieto-Fernández, Á., Becerra-Castro, C., Monterroso, C., Kidd, P.S. Rhizobacterial communities associated with the flora of three serpentine outcrops of the Iberian Peninsula. Plant Soil. 2016, 403, 233–252.
[34] Touceda-González, M., Kidd, P.S., Smalla, K., Prieto-Fernández, A. Bacterial communities in the rhizosphere of different populations of the Ni-hyperaccumulator *Alyssum serpyllifolium* and the metal-excluder *Dactylis glomerata* growing in ultramafic soils. Plant Soil. 2018, 431, 317–332.
[35] Bourles, A., Amir, H., Gensous, S., et al. Investigating some mechanisms underlying stress metal adaptations of two *Burkholderia* sensu lato species isolated from New Caledonian ultramafic soils. Eur J Soil Biol. 2020, 97. doi: https://doi.org/10.1016/j.ejsobi.2020.103166.
[36] Lopez, S., Goux, X., Echevarria, G., Calusinska, M., Morel, J.L., Benizri, E. Community diversity and potential functions of rhizosphere-associated bacteria of nickel hyperaccumulators found in Albania. Sci Total Environ. 2019a, 654, 237–249.
[37] Pal, A., Paul, A.K. Accumulation of polyhydroxyalkanoates by rhizobacteria underneath nickel-hyperaccumulators from serpentine ecosystem. J Pol Environ. 2012, 20, 10–16.
[38] Lopez, S., van der Ent, A., Sumail, S., et al. Bacterial community diversity in the rhizosphere of nickel hyperaccumulator plant species from Borneo Island (Malaysia). Environ Microbiol. 2020, 22, 1649–1665.
[39] Lopez, S., Goux, X., van der Ent, A., Erskine, P.D., Echevarria, G., Calusinska, M., Morel, J.L., Benizri, E. Bacterial community diversity in the rhizosphere of nickel hyperaccumulator species of Halmahera Island (Indonesia). Appl Soil Ecol. 2019b, 133, 70–80.
[40] Lagrange, A., Ducousso, M., Jourand, P., Majorel, C., Amir, H. New insights into the mycorrhizal status of Cyperaceae from ultramafic soils in New Caledonia. Can J Microbiol. 2011, 57, 21–28.
[41] Boulet, F.M., Lambers, H. Characterization of arbuscular mycorrhizal fungi colonization in cluster roots of shape *Hakea verrucosa* F. Muell (Proteaceae) and its effect on growth and nutrient acquisition in ultramafic soil. Plant Soil. 2005, 269, 357–367.
[42] Carriconde, F., Gardes, M., Bellanger, J.M., et al. Host effects in high ectomycorrhizal diversity tropical rainforests on ultramafic soils in New Caledonia. Fungal Ecol. 2019, 39, 201–212.
[43] Ludo, A., Muller, H., Hilger, H.H. Insights into the effects of serpentine soil conditions on the community composition of fungal symbionts in the roots of *Onosma echioides*. Soil Biol Biochem. 2015, 81, 1–8.

[44] Urban, A., Puschenreiter, M., Strauss, J., Grofer, M. Diversity and structure of ectomycorrhizal and co-associated fungal communities in a serpentine soil. Mycorrhiza. 2008, 18, 339–354.
[45] Waseem, M., Ducousso, M., Prin, Y., et al. Ectomycorrhizal fungal diversity associated with endemic *Tristaniopsis* spp. (Myrtaceae) in ultramafic and volcano-sedimentary soils in New Caledonia. Mycorrhiza. 2017, 27, 407–413.
[46] Lodewyckx, C., Vangronsveld, J., Porteous, F., et al. Endophytic bacteria and their potential applications. Crit Rev Plant Sci. 2002, 21, 583–606.
[47] Barzanti, R., Ozino, F., Bazzicalupo, M., et al. Isolation and characterization of endophytic bacteria from the nickel hyperaccumulator plant *Alyssum bertolonii*. Microb Ecol. 2007, 53, 306–316.
[48] Ma, Y., Rajkumar, M., Moreno, A., Zhang, C., Freitas, H. Serpentine endophytic bacterium *Pseudomonas azotoformans* ASS1 accelerates phytoremediation of soil metals under drought stress. Chemosphere. 2017, 185, 75–85.
[49] Postma, F., Mesjasz-Przybyłowicz, J., Przybyłowicz, W., Stone, W., Mouton, M., Botha, A. Symbiotic interactions of culturable microbes with the nickel hyperaccumulator *Berkheya coddii* and the herbivorous insect *Chrysolina clathrata*. Symbiosis. 2012, 58, 209–220.
[50] Visioli, G., D'Egidio, S., Vamerali, T., Mattarozzi, M., Sanangelantoni, A.M. Culturable endophytic bacteria enhance Ni translocation in the hyperaccumulator *Noccaea caerulescens*. Chemosphere. 2014, 117, 538–544.
[51] Durand, A., Sterckeman, T., Gonnelli, C., et al. A core seed endophytic bacterial community in the hyperaccumulator *Noccaea caerulescens* across 14 sites in France. Plant Soil. 2021, 459, 203–216.
[52] Sessitsch, A., Puschenreiter, M. Endophytes and rhizosphere bacteria of plants growing in heavy metal-containing soils. In: Dion, P., Nautiyal, C.S., ed. Microbiology of extreme soils. Berlin Heidelberg, Springer-Verlag, 2008, 317–332.
[53] Alford, E.R., Pilon-Smits, E.A.H., Paschke, M.W. Metallophytes – A view from the rhizosphere. Plant Soil. 2010, 337, 33–50.
[54] Cabello-Conejo, M.I., Becerra-Castro, C., Prieto-Fernández, A., et al. Rhizobacterial inoculants can improve nickel phytoextraction by the hyperaccumulator *Alyssum pintodasilvae*. Plant Soil. 2014, 379, 35–50.
[55] Sessitsch, A., Kuffner, M., Kidd, P., et al. The role of plant-associated bacteria in the mobilization and phytoextraction of trace elements in contaminated soils. Soil Biol Biochem. 2013, 60, 182–194.

Komal Agrawal and Pradeep Verma

10 Microbial desalination cell: a versatile insight towards sustainable contribution

Abstract: As the name microbial desalination cell indicates, it involves the use of two approaches, namely, microbial fuel cell and electrodialysis, for the treatment and desalination of wastewater (WW), along with the production of renewable energy. Since it involves two approaches and enables multiple and simultaneous benefits, the system has garnered tremendous attention of the scientific community. In the case of the microbial desalination system, the charged particles (anions and cations) in the desalination chamber (Dc), between the anode and cathode, move to the anodic and the cathodic chamber, respectively, under the influence of the electric fields. These electric field is created by the activity of the exoelectrogens (e.g., bacteria) in the anode which acts on the organic matter (e.g., WW) and generates electricity by oxidation. As it is an emerging technology and due to the focus on the above aspect, the present chapter would deal with the basic concepts of microbial desalination cell, its designing, operational parameters, and various types of microbial desalination cells. Further, the chapter will also deal with the limitations in the use of microbial desalination cell as well as its future prospects.

Keywords: microbial desalination cell, wastewater, current, desalination, treatment

10.1 Introduction

The rapid decline in the availability of the freshwater resources and the unregulated use of the existing resources has resulted in the development of technologies that favor the recovery and reuse of the treated WW. Among all the technologies, microbial desalination has been largely adopted for the recovery of WW. It is estimated that the demand for freshwater has increased and reached 4.6 trillion m^3/year [1]. The desalination system has enabled efficient treatment of the seawater in the Arabic Gulf countries [2]. The desalination process uses seawater (97%), which constitutes a majority of the Earth's water. The process then separates the sea water into freshwater and highly concentrated salt water [3, 4]. The process of desalination has been used along with the various energy resources, such as wind, biomass,

Komal Agrawal, Pradeep Verma, Bioprocess and Bioenergy Laboratory, Department of Microbiology, Central University of Rajasthan, Bandarsindri, Kishangarh, Ajmer, Rajasthan 305817, India

https://doi.org/10.1515/9783110727227-010

geothermal energy, solar, and hydropower, to effectively run the desalination power plants and also to make them sustainable and energy-efficient [5, 6].

In the case of desalination, two approaches have been used – one approach uses thermal techniques (there are reports of the use of hybrid technology as well, i.e., thermal-solar process) and the other approach uses membrane separation for desalination (e.g., reverse osmosis, electrodialysis, and microbial desalination cells (MDCs)) [7–9]. As with other technologies that have been employed for the treatment of water, the major issues are the high consumption of energy and the revenue generated after the process. Among the various desalination processes, MDC is the most feasible and energy-saving process. In the MDC, the Dc is added between the anode and the cathode chambers. In the general mechanism of microbial fuel cell (MFC), the metabolic potential of the microorganism is used to induce current. It consists of two chambers (anode and cathode) that are linked by an electrical device/resistance to complete the circuit. Bacteria are placed in the anode chamber, where they act on the organic matter and eventually release electrons (e^-). These e^- are transferred to the cathode via the circuit, which is usually visible to air and is the place where reduction occurs –O_2 and H react to form H_2O [10–16]. In the case of MDC, the organic matter is degraded at the anode and e^- flow via the circuit, creating positive charge at the anode and negative charge at the cathode. The potential difference so created flows the Cl^- from the Dc through the anion exchange membrane to the anode and Na^+ from the cation exchange membrane to the cathode, thus resulting in desalination of the water [17, 18]. Thus, the present chapter would discuss the basic concept of the MDC, the gradual development of MDC as reported in literature, its limitations, and its future prospects will also be elaborated.

10.2 Concept of MDC

As mentioned above, the MDC consists of three chambers, the anode, the cathode, and the Dc (Figure 10.1). The Dc is connected to the electrodes by the anion and cation exchange membrane that faces the anode and cathode, respectively. The activity at the anode basically depends on the exoelectrogens (e.g., bacteria). These exoelectrogens act on the organic matter (e.g., WW) and generate electricity by oxidation. Also, the e^- are transferred from the anode, outside their cells, by the exoelectrogens, which are predominantly found in WW and can be used as pure/mixed culture [19–24]. The survival and maintenance of the exoelectrogens are very crucial for the functioning of the MDC. The selection and optimization of the exoelectrogens can significantly enhance the efficiency of the MDC. The most commonly used exoelectrogens include nitrate-reducing bacteria, which are dissimilatory metal-reducing [25, 26]. The MDC has been extensively studied but the biological driving force has not been extensively studied and reported in literature. The detailed mechanism and the role of exoelectrogen are yet to be studied in detail [19–24].

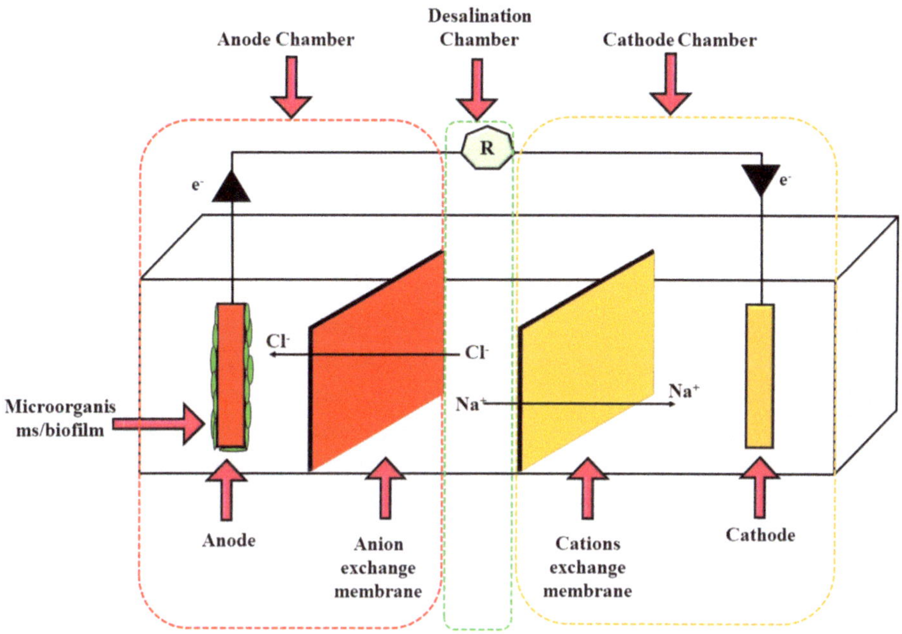

Figure 10.1: The schematic representation of the basic microbial desalination cell (MDC).

10.3 Types of MDC

The various types of MDC reported in literature are discussed as follows (Figure 10.2):

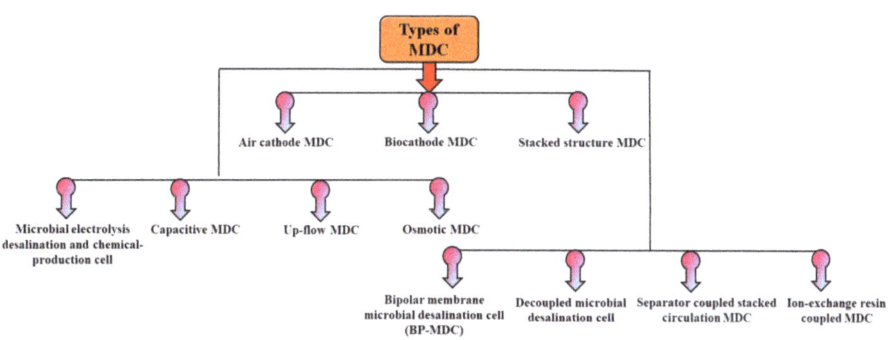

Figure 10.2: Schematic representation of the various types of MDC reported in literature.

10.3.1 Air cathode MDC (AC-MDC)

In the case of AC-MDC, oxygen is used as the terminal e^- acceptor. The three major benefits include economic feasibility, non-toxic, and high reduction potential [27]. In the study by [28], efficient salt removal was attained in a continuously operated up-flow MDC with an air cathode. It was reported that current production and salt removal was ~62 mA and 99% (initial TDS- 30 g/L), respectively. Despite the advantages, there is a limitation with AC-MDC, i.e., under ambient condition, the redox kinetics are slow over other reported catholytes, and, thus, would require the use of more expensive materials, such as platinum, as catalytic materials, which would reduce the activation over the potential for oxygen reduction. The other drawback would be maintaining the dissolved oxygen concentration in the cathode, which can be overcome by exposing the surface to the atmospheric air, oxygen transfer by using passive methods, and the use of high surface area carbon substrates that would minimize the use of platinum (an expensive catalyst) [27, 29] (Table 10.1).

Table 10.1: The various types of microbial desalination cell (MDC) reported in literature.

Sl. No	Type of MDC	Application	Removal	Reference
1.	Air Cathode MDC	WW	Salinity removal efficiency was 93.6%, chemical oxygen demand (COD) was 97.3%, and conductivity was 68 ± 12 µS/cm.	[30]
2.	Three-chamber air–cathode microbial desalination cell	Aerobic sludge was collected from second sedimentation from the local WW treatment plant (WWTP).	Of the three systems used, the desalination rates and Coulomb efficiencies of the system with 16, 12, and 8 cm electrode spacing were 84.86, 87.71, 83.43%, and 17.29, 18.60, and 17.95%, respectively.	[31]
3.	Two systems of MDC were used where the 1st was an air cathode for efficient oxygen reduction, and the 2nd was liquid catholyte, with Fe^{3+}/Fe^{2+} solution (a ferro ferricyanide complex)	The two types of samples selected were brackish water and sea water.	Both systems had a desalination efficiency of over 90% for brackish water. Nominal desalination rates for brackish and seawater were in the range of 0.17–0.14 $L \cdot m^{-2} \cdot h^{-1}$ with the 1st system and 1.5–0.7 $L \cdot m^{-2} \cdot h^{-1}$	[32]

Table 10.1 (continued)

Sl. No	Type of MDC	Application	Removal	Reference
			with the 2nd system. Organic matter present in WW was effectively removed at 0.9 and 1.1 kg $COD \cdot m^{-3} \cdot day^{-1}$ using the 1st system and 7.1 and 19.7 kg $COD \cdot m^{-3} \cdot day^{-1}$ with the 2nd system for brackish and sea water.	
4.	Biocathode-microbial desalination cell	Effluent of a dairy WWTP with COD of 1,000 mg l^{-1} was the substrate at the anode and the *Chlorella vulgaris* was inoculated at the cathode. Two different saline water concentrations i.e., 15 g l^{-1} (MDC-1) and 35 g l^{-1} (MDC-2) were used.	In MDC-2, the maximum power density (PD) and salinity removal rate of 20.25 mW m^{-2} and 0.341 g l^{-1} day^{-1} were observed. The PD and salinity removal rate were over 2 and 1.5 times higher than in MDC-1. Also, high algal growth was observed in MDC-2, being 38% better in performance over MDC-1.	[33]
5.	Hollow fiber microfiltration membrane coupled biocathode MDC	Domestic WW was acquired from downstream of the grid in a WWTP at the Tsinghua campus.	The produced diluate was 86.1 v%, with the final conductivity being 59.2 ± 12.2 μS/cm, COD 35.5 ± 4.0 mg/L, total nitrogen 1.65 ± 0.39 mg/L, and the total phosphorus was 0.14 ± 0.05 mg/L. The corresponding concentrate (13.9 v%] concentrated 6.7 times the initial conductivity, and recovered phosphorous by 98.7% (TP of 40.9 ± 4.9 mg/L) and part of the nitrogen.	[34]

10.3.2 Biocathode MDC (BC-MDC)

The conventional MDCs that have been reported use ferricyanide as the catholyte and platinum catalysts as well as other materials in the AC-MDC. These are expensive, toxic, and require continuous replacement; thus making the conventional MDC less feasible for commercial use [35]. Thus, to overcome these issues, BC-MDC was designed, where the biocathode is the electrode and the microbial population on the surface of the electrode or in catholyte act as a site where the reduction reaction takes place [36]. The self-regeneration and sustainability of the BC-MDC has been gaining attention of the scientific community [37]. The microorganism in the biocathode acts as the catalyst, increasing the oxidation-reduction reaction and enhancing desalination [38]. As the microorganism is responsible for the process in the BC-MDC, the increased growth in the biofilm enhances the potential at the anode [39]. Depending upon the type of the terminal e^- acceptor, the BC can be of two types, i.e., aerobic and anaerobic cathode, and thus, the electrorophs are capable of accepting diverse types of e^- (e.g., oxygen, sulfate, etc.) from the cathode [40]. This has been an efficient system for enhanced generation of current and desalination. Despite the advantage, various limitations are associated with the process. In the case of the aerobic system, maintaining the supply of dissolved oxygen is the issue and in the case of the anaerobic system, the growth of biocathode biofilm is a difficult and time-taking process [41, 42] (Table 10.1).

10.3.3 Stacked structure MDC (SS-MSC)

In the case of the SS-MDC, desalination is enhanced with the incorporation of multiple ion-exchange membranes between the two chambers, i.e., cathode and anode. Here, the ion-exchange membranes enhance the transfer efficiency of the saline water and improve its flow through the MDC, thereby increasing the desalination [27]. The alternate anion and cation exchange membranes create alternate segments of desalination and salt-rich segments, thereby increasing the separation of ion pairs from the high salt region for every e^- passing through the circuit [43]. Compared to other MDCs previously reported, it produces more electricity (in the anodic chamber) with enhanced desalination and is economically more feasible [44]. The ion-exchange membranes and the Dc enhance the current and desalination, with a decrease in resistance, over previously reported MDCs [45]. The limitation of SS-MDC is that the fluctuation in pH, i.e., decrease in pH reduces the microbial activity in the anodic chamber and the increase in pH at the cathodic chamber causes potential loss and reduces the efficiency of the process. The limitation of pH fluctuation is more prominent in the single-cell structure, over the SS-MDC, due to the segregation formed due to stacking [46] (Table 10.2).

Table 10.2: The various types of microbial desalination cell (MDC) reported in literature.

Sl. No	Type of MDC	Application	Removal	Reference
1.	Stacked-MDC, packed with mixed ion-exchange resins of 10 L	Was operated with a salt concentration of 0.5 g/L NaCl, and is a typical level of domestic WW	This is a system with circulation flow rate of 80 mL/min, and desalination efficiency and final effluent concentration of 95.8% and 0.02 g/L, respectively. The system has a stable desalination performance of >93% when the concentrate volume decreased from 2.4 to 0.1 L. It has to be noted that only 0.875 L of nonfresh water was consumed to desalinate 1 L of saline water.	[47]
2.	1-liter upflow-stacked MDC (USMDC)	–	After 24 h, the desalination ratio reached 91.9% when the external resistance was 1.5 Ω, which was 1.18 and 1.48 times higher than in SMDC and UMDC, respectively. While maintaining the desalination efficiency (DE) at 87.2–96.0%, the system was stable for 120 days. The maximum PD of the system was 32.91 W m^{-3}.	[48]
3.	Microbial capacitive desalination cell	Anaerobic sludge from the Englewood-Littleton WWTO	The system removed 72.7 mg total dissolved solids/ gram of activated carbon cloth without any external energy. The DE was 7–25 times higher than traditional capacitive deionization processes.	[49]

Table 10.2 (continued)

Sl. No	Type of MDC	Application	Removal	Reference
4.	Combination of osmotic microbial fuel cell and up-flow microbial desalination cell	Petroleum refinery WW	The combined system removed 93% of COD from the petroleum refinery WW, whilst 48% salts were removed from seawater.	[50]
5.	Osmotic (Os) microbial fuel cell, containing FO membrane, was hydraulically coupled with MDC, having ion exchange membranes	Artificial WW	It was found that the high-power operation mode was more suitable for the Os microbial fuel cell, over open-circuit mode and high-current mode, because of the good desalination performance (95.9% conductivity reduction in the coupled system) and energy production (0.160 kWh/m3 treated saline water). With the FO membrane facing the feed solution, more water flux was attained than the reversal membrane orientation. The coupled system achieved had > 85% reduction in conductivity from the salt solution containing 10–50 g NaCl/L.	

10.3.4 Microbial electrolysis desalination and chemical-production cell (MEDCC)

The MDC, consisting of acid-production chamber and a bipolar membrane in the microbial electrolysis desalination cell, results in the MEDCC. The system desalinates the water and simultaneously generates hydrochloric acid (HCl) and sodium hydroxide (NaOH) [51–54]. The desalination potential of SS-MDC and MEDCC is 1.4 times higher than the conventional MDC [52, 55]. The potential of MEDCC can be enhanced with the stacking of desalination membranes, resulting in an increase in the charge transfer per e^- released [56] (Table 10.2).

10.3.5 Capacitive MDC (C-MDC)

The limitation of conventional MDC is the accumulation of salt in the anolyte and catholyte, preventing its further use in the desalination cell; and thus requires regular replacement. The limitation can be overcome by the use of C-MDC. The system uses the double-layer capacitator on the high surface electrode, which absorbs the ions when it flows over the system. Later, when the potential gradient is removed, the ions are flowed back to the liquid. This technique favors the removal of the salt at the electrodes without contamination at the cathode and anode chambers [57, 58] (Table 10.2).

10.3.6 Up-flow MDC (U-MDC)

The UMDC is a tubular unit, consisting of two compartments separated by ion-exchange membranes. The inner compartment is the anode, consisting of graphite granules, which offers increased surface area for the oxidation to take place effectively. In addition, two graphite rods are immersed in the graphite granules as current collector, which allows the transfer of e^-. The outer chamber consists of saline water for desalination. Similarly, the inner and outer chambers are enclosed with an anion and cation exchange membrane [59]. As it has a tubular structure, the catalyst used is made up of a mixture of platinum and carbon, layered with a carbon cloth around the system/reactor, to make the cathode. As a result, a separate chamber of cathode is not required in the case of U-MDC [59]. In this system, the saline water enters from the bottom of the system and the desalinated water leaves from the top of the anode chamber of the U-MDC. The other advantage associated with the MDC is that the mixing of the liquid component is attained without agitation, and with the help of osmosis, more than 100% of the water can be recovered. The microorganisms will then remain in the suspension of the anode chamber and allow the oxidation of the organic matter [60] (Table 10.2).

10.3.7 Osmotic MDC (O-MDC)

The replacement of the ion-exchange membrane with the forward osmosis membrane leads to the formation of O-MDC. Here, two solutions are present in the membrane, i.e., the feed and the draw solution. The feed solution has higher water potential than the draw solution, resulting in the formation of water flux due to concentration gradient [43, 61–66]. The forward osmosis membrane in the O-MDC possesses certain limitations, such as the fouling of the membrane, which may increase the internal resistance and thus reduce the water flux [43]. The O-MDC has to be studied in detail to understand its working principle for the development of a better and enhanced system [61] (Table 10.2).

10.3.8 Bipolar membrane microbial desalination cell (BP-MDC) and Decoupled microbial desalination cell (D-MDC)

The other type of MDC reported in literature is the BP-MDC, where the anion and the cation layers are laminated using heat-press or glued together into one layer [67]. This single layer allows high permselectivity, i.e., allows a single passage of ions with the same charge, at low electrical resistance, low water splitting voltage, and a longer life duration [67, 68]. In the case of BP-MDC, the bipolar membrane is placed next to the anode, making a system consisting of four chambers, which was previously three in the conventional system, consisting of cathode, anode, and Dc [43]. The main advantage of the BP-MDC is the maintenance of pH due to the efficient segregation, because of the integration of the membrane. Despite the advantage, the major drawback of the system is the requirement of an external power source, which allows the bipolar membrane to split water to hydroxyl and hydrogen ions in the anode and cathode chambers [69]. This increases the cost of the system. If, however, the products, such as HCl and NaOH, are recovered, it can make the system cost-effective [43].

On the other hand, in the D-MDC, the cathode and the anode are placed in the salt solution, rather than being constructed as separate compartments. In this system, the anode has the configuration of a plate, and the anode and the cathode consist of anion and cation exchange membranes on the either side, and the carbon cloth wrapped around stainless steel is used as the electrode and current collector [70]. It also has the advantage that the volume ratio of the liquid can be easily varied and the damaged part can be replaced with ease, compared to a conventional MDC [71].

10.3.9 Separator coupled stacked circulation MDC (SCSC-MDC) and Ion-exchange resin coupled MDC (IERC-MDC)

The SCSC-MDC helps in the reduction of the pH imbalance created across the MDC chambers. The SCSC-MDC consists of a glass fiber that is placed on the water-facing side of the cathode, thereby acting as a separator. The glass separator enhances the columbic efficiency and averts fouling of the cathode. The SCSC-MDC consists of the cathode, anode, (separated by a cation and anion exchange membrane), and Dc, with the presence of the axial cylindrical cavity [72]. In the anode chamber, activated carbon and a graphite rod can be added to allow the growth of bacteria, and to enable the flow of e^- in the external circuit. In this system, the circulation of the electrolyte allows for easy maintenance of pH, and also maintains the flow of current. The separator prevents the growth of biofilm on the cathode and increases the life of the system for a longer functioning [22].

In the case of MDC, the ohmic resistance increases with decrease in the concentration of the salt and conductivity of saline water, and, thus, would eventually limit the desalination and generation of electricity. One way to overcome the limitation would be to reduce the ohmic resistance [17] by the addition of anion- and cation-exchange resins in the Dc, resulting in the development of IERC-MDC. This system allows for the reduction of ohmic resistance and has high conductivity, which can allow efficient desalination in systems with low salt concentration.

10.4 Limitation, prospects, and conclusion

The major limitation faced by the MDC technology is the optimization of the various parameters of the MDC for the efficient and enhanced functioning of the system. The other major issue is the scale-up of the system for commercial use. A detailed understanding of the system, mechanism, and designing will enable developing a system that can be effectively be used for the welfare of the society. The selection of new materials, membranes, and integration of various technologies will pave the way for new and advanced technologies. As can be seen from Figure 10.3, a number of studies have been carried out over the last five years, with a gradual increase in the number of papers published (data was taken from google scholar when the term microbial desalination cell was entered for the individual years). In addition, the limitations, such as pH imbalance, membrane fouling, and internal resistance, also have to be investigated carefully to develop a better and an efficient system. Worldwide, the trend of multidisciplinary approach is gaining importance. Similarly, here as well, diverse

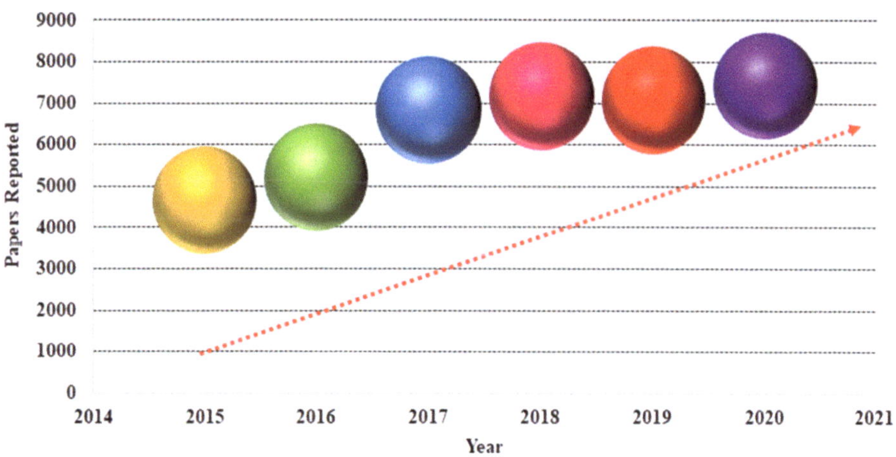

Figure 10.3: The reported papers in google scholar for the search term, microbial desalination cell, from the year 2015 to 2020.

areas, such as biology, chemistry, and engineering sectors can effectively work together for the development of a system with minimal/no limitations. Thus, it can be stated that MDC is a technology with immense scope for future, both at the laboratory and commercial scale, and would simultaneously be beneficial for the ecosystem as well.

Conflict of interest: The authors declare no conflict of interest

References

[1] Boretti, A., Rosa, L. Reassessing the projections of the world water development report. NPJ Clean Water. 2019, 2(1), 1–6.
[2] Ahmed, M., Kumar, R., Garudachari, B., Thomas, J.P. Assessment of pilot scale forward osmosis system for Arabian Gulf seawater desalination using polyelectrolyte draw solution. Desalin. Water Treat. 2019, 157, 342–348.
[3] Olabi, A.G., Mahmoud, M., Soudan, B., Wilberforce, T., Ramadan, M. Geothermal based hybrid energy systems, toward eco-friendly energy approaches. Renew. Energy. 2020, 147, 2003–2012.
[4] Soudan, B. Community-scale baseload generation from marine energy. Energy. 2019, 189, 116134.
[5] Tawalbeh, M., Tezel, F.H., Al-Ismaily, M., Kruczek, B. Highly permeable tubular silicalite-1 membranes for CO2 capture. Sci. Total Environ. 2019, 676, 305–320.
[6] Wilberforce, T., El Hassan, Z., Durrant, A., Thompson, J., Soudan, B., Olabi, A.G. Overview of ocean power technology. Energy. 2019, 175, 165–181.
[7] Al-Othman, A., Darwish, N.N., Qasim, M., Tawalbeh, M., Darwish, N.A., Hilal, N. Nuclear desalination: A state-of-the-art review. Desalination. 2019, 457, 39–61.
[8] Al-Othman, A., Tawalbeh, M., Assad, M.E.H., Alkayyali, T., Eisa, A. Novel multi-stage flash (MSF) desalination plant driven by parabolic trough collectors and a solar pond: A simulation study in UAE. Desalination. 2018, 443, 237–244.
[9] Kiwan, S., Salam, Q.I.A. Solar chimney power-water distillation plant (SCPWDP). Desalination. 2018, 445, 105–114.
[10] Al Bsoul, A., Hailat, M., Abdelhay, A., Tawalbeh, M., Jum'h, I., Bani-Melhem, K. Treatment of olive mill effluent by adsorption on titanium oxide nanoparticles. Sci. Total Environ. 2019, 688, 1327–1334.
[11] Hassan, S.H., Abd El Nasser, A.Z., Kassim, R.M. Electricity generation from sugarcane molasses using microbial fuel cell technologies. Energy. 2019, 178, 538–543.
[12] Mohammed, H., Al-Othman, A., Nancarrow, P., Tawalbeh, M., Assad, M.E.H. Direct hydrocarbon fuel cells: A promising technology for improving energy efficiency. Energy. 2019, 172, 207–219.
[13] Ogugbue, C.J., Ebode, E.E., Leera, S. Electricity generation from swine wastewater using microbial fuel cell. J. Ecol. Eng. 2015, 16, 5.
[14] Rabaey, K., Verstraete, W. Microbial fuel cells: Novel biotechnology for energy generation. Trends Biotechnol. 2005, 23(6), 291–298.

[15] Zhu, Y., Robinson, T., Al-Othman, A., Tremblay, A.Y., Ternan, M. n-Hexadecane fuel for a phosphoric acid direct hydrocarbon fuel cell. J. Fuels. 2015.
[16] Zinadini, S., Zinatizadeh, A.A., Rahimi, M., Vatanpour, V., Bahrami, K. Energy recovery and hygienic water production from wastewater using an innovative integrated microbial fuel cell–membrane separation process. Energy. 2017, 141, 1350–1362.
[17] Cao, X., Huang, X., Liang, P., Xiao, K., Zhou, Y., Zhang, X., Logan, B.E. A new method for water desalination using microbial desalination cells. Environ. Sci. Technol. 2009, 43(18), 7148–7152.
[18] Tawalbeh, M., Al-Othman, A., Singh, K., Douba, I., Kabakebji, D., Alkasrawi, M. Microbial desalination cells for water purification and power generation: A critical review. Energy. 2020, 209, 118493.
[19] Kalleary, S., Abbas, F.M., Ganesan, A., Meenatchisundaram, S., Srinivasan, B., Packirisamy, A.S.B., Krishnan Kesavan, R., Muthusamy, S. Biodegradation and bioelectricity generation by microbial desalination cell. Int. Biodeterior. Biodegrad. 2014, 92, 20–25.
[20] Luo, H., Xu, P., Ren, Z. Long-term performance and characterization of microbial desalination cells in treating domestic wastewater. Bioresour. Technol. 2012, 120, 187–193.
[21] Luo, H., Xu, P., Roane, T.M., Jenkins, P.E., Ren, Z. Microbial desalination cells for improved performance in wastewater treatment, electricity production, and desalination. Bioresour. Technol. 2012, 105, 60–66.
[22] Saeed, H.M., Husseini, G.A., Yousef, S., Saif, J., Al-Asheh, S., Fara, A.A., Azzam, S., Khawaga, R., Aidan, A. Microbial desalination cell technology: A review and a case study. Desalination. 2015, 359, 1–13.
[23] Shinde, O.A., Bansal, A., Banerjee, A., Sarkar, S. Bioremediation of steel plant wastewater and enhanced electricity generation in microbial desalination cell. Water Sci. Technol. 2018, 77(8), 2101–2112.
[24] Utami, T.S., Arbianti, R., Manaf, B.N. Sea water desalination using Debaryomyces hansenii with microbial desalination cell technology. Int. J. Technol. 2015, 6(7), 1094–1100.
[25] Holmes, D.E., Bond, D.R., Lovley, D.R. Electron transfer by Desulfobulbus propionicus to Fe (III) and graphite electrodes. Appl. Environ. Microbiol. 2004, 70(2), 1234–1237.
[26] Rabaey, K., Boon, N., Siciliano, S.D., Verhaege, M., Verstraete, W. Biofuel cells select for microbial consortia that self-mediate electron transfer. Appl. Environ. Microbiol. 2004, 70(9), 5373–5382.
[27] Gude, V.G., Kokabian, B., Gadhamshetty, V. Beneficial bioelectrochemical systems for energy, water, and biomass production. J. Microb. Biochem. Technol. 2013, 6, 2.
[28] Jacobson, K.S., Drew, D.M., He, Z. Efficient salt removal in a continuously operated upflow microbial desalination cell with an air cathode. Bioresour. Technol. 2011, 102(1), 376–380.
[29] Al-Mamun, A., Ahmad, W., Baawain, M.S., Khadem, M., Dhar, B.R. A review of microbial desalination cell technology: Configurations, optimization and applications. J. Clean. Prod. 2018, 183, 458–480.
[30] Zuo, K., Wang, Z., Chen, X., Zhang, X., Zuo, J., Liang, P., Huang, X. Self-driven desalination and advanced treatment of wastewater in a modularized filtration air cathode microbial desalination cell. Environ. Sci. Technol. 2016, 50(13), 7254–7262.
[31] Xu, C., Lu, J., Zhao, Z., Lu, X., Zhang, Y., Cheng, M., Zhang, J. Simultaneous bioelectricity generation, desalination, organics degradation, and nitrogen removal in air–cathode microbial desalination cells. SN Appl. Sci. 2020, 2(2), 1–11.
[32] Ramírez-Moreno, M., Rodenas, P., Aliaguilla, M., Bosch-Jimenez, P., Borràs, E., Zamora, P., Monsalvo, V., Rogalla, F., Ortiz, J.M., Esteve-Núñez, A. Comparative performance of microbial desalination cells using air diffusion and liquid cathode reactions: Study of the salt removal and desalination efficiency. Front. Energy Res. 2019, 7, 135.

[33] Zamanpour, M.K., Kariminia, H.R., Vosoughi, M. Electricity generation, desalination and microalgae cultivation in a biocathode-microbial desalination cell. J. Environ. Chem. Eng. 2017, 5(1), 843–848.
[34] Zuo, K., Chen, M., Liu, F., Xiao, K., Zuo, J., Cao, X., Zhang, X., Liang, P., Huang, X. Coupling microfiltration membrane with biocathode microbial desalination cell enhances advanced purification and long-term stability for treatment of domestic wastewater. J. Memb. Sci. 2018, 547, 34–42.
[35] Clauwaert, P., Van der Ha, D., Boon, N., Verbeken, K., Verhaege, M., Rabaey, K., Verstraete, W. Open air biocathode enables effective electricity generation with microbial fuel cells. Environ. Sci. Technol. 2007, 41(21), 7564–7569.
[36] Croese, E., Pereira, M.A., Euverink, G.J.W., Stams, A.J., Geelhoed, J.S. Analysis of the microbial community of the biocathode of a hydrogen-producing microbial electrolysis cell. Appl. Microbiol. Biotechnol. 2011, 92(5), 1083–1093.
[37] Perazzoli, S., de Santana Neto, J.P., Soares, H.M. Anoxic-biocathode microbial desalination cell as a new approach for wastewater remediation and clean water production. Water Sci. Technol. 2020, 81(3), 550–563.
[38] Wen, Q., Zhang, H., Chen, Z., Li, Y., Nan, J., Feng, Y. Using bacterial catalyst in the cathode of microbial desalination cell to improve wastewater treatment and desalination. Bioresour. Technol. 2012, 125, 108–113.
[39] Kim, B.H., Chang, I.S., Moon, H. Microbial fuel cell-type biochemical oxygen demand sensor. studies. 2006, 3(10), 1–12.
[40] Zaybak, Z., Pisciotta, J.M., Tokash, J.C., Logan, B.E. Enhanced start-up of anaerobic facultatively autotrophic biocathodes in bioelectrochemical systems. J. Biotechnol. 2013, 168 (4), 478–485.
[41] Butler, C.S., Clauwaert, P., Green, S.J., Verstraete, W., Nerenberg, R. Bioelectrochemical perchlorate reduction in a microbial fuel cell. Environ. Sci. Technol. 2010, 44(12), 4685–4691.
[42] Morita, M., Malvankar, N.S., Franks, A.E., Summers, Z.M., Giloteaux, L., Rotaru, A.E., Rotaru, C., Lovley, D.R. Potential for direct interspecies electron transfer in methanogenic wastewater digester aggregates. MBio. 2011, 2, 4.
[43] Kim, Y., Logan, B.E. Microbial desalination cells for energy production and desalination. Desalination. 2013, 308, 122–130.
[44] Shehab, N.A., Logan, B.E., Amy, G.L., Saikaly, P.E. Microbial electrode ionization cell stack for sustainable desalination, wastewater treatment and energy recovery. Proc. Water Environ. Fed. 2013, 2013(19), 222–227.
[45] Kim, Y., Logan, B.E. Series assembly of microbial desalination cells containing stacked electrodialysis cells for partial or complete seawater desalination. Environ. Sci. Technol. 2011, 45(13), 5840–5845.
[46] Qu, Y., Feng, Y., Liu, J., He, W., Shi, X., Yang, Q., Lv, J., Logan, B.E. Salt removal using multiple microbial desalination cells under continuous flow conditions. Desalination. 2013, 317, 17–22.
[47] Zuo, K., Cai, J., Liang, S., Wu, S., Zhang, C., Liang, P., Huang, X. A ten liter stacked microbial desalination cell packed with mixed ion-exchange resins for secondary effluent desalination. Environ. Sci. Technol. 2014, 48(16), 9917–9924.
[48] Forrestal, C., Xu, P., Ren, Z. Sustainable desalination using a microbial capacitive desalination cell. Energy Environ. Sci. 2012, 5(5), 7161–7167.
[49] Sevda, S., Abu-Reesh, I.M. Improved petroleum refinery wastewater treatment and seawater desalination performance by combining osmotic microbial fuel cell and up-flow microbial desalination cell. Environ. Technol. 2019, 40(7), 888–895.

[50] Wang, Y., Xu, A., Cui, T., Zhang, J., Yu, H., Han, W., Shen, J., Li, J., Sun, X., Wang, L. Construction and application of a 1-liter upflow-stacked microbial desalination cell. Chemosphere. 2020, 248, 126028.
[51] Chen, S., Liu, G., Zhang, R., Qin, B., Luo, Y., Hou, Y. Improved performance of the microbial electrolysis desalination and chemical-production cell using the stack structure. Bioresour. Technol. 2012, 116, 507–511.
[52] Chen, S., Luo, H., Liu, G., Zhang, R., Wang, H., Qin, B., Hou, Y. Integrated utilization of seawater using a five-chamber bioelectrochemical system. J. Memb. Sci. 2013, 444, 16–21.
[53] He, Z., Huang, Y., Manohar, A.K., Mansfeld, F. Effect of electrolyte pH on the rate of the anodic and cathodic reactions in an air-cathode microbial fuel cell. Bioelectrochemistry. 2008, 74(1), 78–82.
[54] Mehanna, M., Kiely, P.D., Call, D.F., Logan, B.E. Microbial electrodialysis cell for simultaneous water desalination and hydrogen gas production. Environ. Sci. Technol. 2010, 44(24), 9578–9583.
[55] Chen, X., Xia, X., Liang, P., Cao, X., Sun, H., Huang, X. Stacked microbial desalination cells to enhance water desalination efficiency. Environ. Sci. Technol. 2011, 45(6), 2465–2470.
[56] Qu, Y., Feng, Y., Wang, X., Liu, J., Lv, J., He, W., Logan, B.E. Simultaneous water desalination and electricity generation in a microbial desalination cell with electrolyte recirculation for pH control. Bioresour. Technol. 2012, 106, 89–94.
[57] Yuan, L., Yang, X., Liang, P., Wang, L., Huang, Z.H., Wei, J., Huang, X. Capacitive deionization coupled with microbial fuel cells to desalinate low-concentration salt water. Bioresour. Technol. 2012, 110, 735–738.
[58] Zhang, B., He, Z. Energy production, use and saving in a bioelectrochemical desalination system. RSC Adv. 2012, 2(28), 10673–10679.
[59] Jacobson, K.S., Drew, D.M., He, Z. Use of a liter-scale microbial desalination cell as a platform to study bioelectrochemical desalination with salt solution or artificial seawater. Environ. Sci. Technol. 2011, 45(10), 4652–4657.
[60] He, Z., Wagner, N., Minteer, S.D., Angenent, L.T. An upflow microbial fuel cell with an interior cathode: Assessment of the internal resistance by impedance spectroscopy. Environ. Sci. Technol. 2006, 40(17), 5212–5217.
[61] Cath, T.Y., Childress, A.E., Elimelech, M. Forward osmosis: Principles, applications, and recent developments. J. Memb. Sci. 2006, 281(1–2), 70–87.
[62] Werner, C.M., Logan, B.E., Saikaly, P.E., Amy, G.L. Wastewater treatment, energy recovery and desalination using a forward osmosis membrane in an air-cathode microbial osmotic fuel cell. J. Memb. Sci. 2013, 428, 116–122.
[63] Zhang, B., He, Z. Integrated salinity reduction and water recovery in an osmotic microbial desalination cell. RSC Adv. 2012, 2(8), 3265–3269.
[64] Zhang, B., He, Z. Improving water desalination by hydraulically coupling an osmotic microbial fuel cell with a microbial desalination cell. J. Memb. Sci. 2013, 441, 18–24.
[65] Zhang, F., Brastad, K.S., He, Z. Integrating forward osmosis into microbial fuel cells for wastewater treatment, water extraction and bioelectricity generation. Environ. Sci. Technol. 2011, 45(15), 6690–6696.
[66] Zhao, S., Zou, L., Tang, C.Y., Mulcahy, D. Recent developments in forward osmosis: Opportunities and challenges. J. Memb. Sci. 2012, 396, 1–21.
[67] Buck, R.P. Ion-selective Membranes and Electrodes, Access Science, McGraw-Hill Education, 2014.
[68] Alvarez, F., Alvarez, R., Coca, J., Sandeaux, J., Sandeaux, R., Gavach, C. Salicylic acid production by electrodialysis with bipolar membranes. J. Memb. Sci. 1997, 123(1), 61–69.

[69] Ping, Q., He, Z. Improving the flexibility of microbial desalination cells through spatially decoupling anode and cathode. Bioresour. Technol. 2013, 144, 304–310.
[70] Ping, Q., He, Z. Effects of inter-membrane distance and hydraulic retention time on the desalination performance of microbial desalination cells. Desalin. Water Treat. 2014, 52(7–9), 1324–1331.
[71] Chen, X., Liang, P., Wei, Z., Zhang, X., Huang, X. Sustainable water desalination and electricity generation in a separator coupled stacked microbial desalination cell with buffer free electrolyte circulation. Bioresour. Technol. 2012, 119, 88–93.
[72] Morel, A., Zuo, K., Xia, X., Wei, J., Luo, X., Liang, P., Huang, X. Microbial desalination cells packed with ion-exchange resin to enhance water desalination rate. Bioresour. Technol. 2012, 118, 43–48.

Can Burak Özkal

11 Control of antibiotic resistance by advanced treatment: recent advances, in a nutshell

Abstract: Safe re-use of treated wastewater effluents are globally progressing. Legislations and regulations are being updated concurrently. There are no details or technical procedures yet on how to safeguard human and environmental health in the case of a mixture of pathogens, in particular, antibiotic resistant bacteria and gene developing resistance. Treatment methods aim to control ARGs, while it is hypothesized that wastewater treatment plant effluents may serve as a source of antibiotic resistance genes and mobile elements carrying resistance for downstream aquatic environments. The level of antibiotic resistant bacteria inactivation, antibiotic resistance gene reduction, and bacterial regrowth are directly related to the treatment process, and the interaction with the water matrix and bacteria cell structure. A variety of treatment methods, mechanisms, and their combinations are used and hypothesized to control the spread of antibiotic resistant bacteria and antibiotic resistance gene. In this chapter, we provide a brief summary of the latest research and advances in the field of antibiotic resistance control. In addition, advanced treatment alternatives will be provided.

Keywords: Effective disinfection, ARB, ARG, Advanced treatment, Management of Emerging pollutants

11.1 Antibiotic resistance and mechanism of spread

Misuse and overuse of antibiotics, alongside inefficient removal rates at wastewater treatment plants (WWTPs), contribute to significant proliferation of ARGs. Antibiotics become ineffectual over time and necessitate the development of new antibiotics. Following the release of new antibiotics, bacteria develop new resistance genes, and the formation of multidrug- resistant bacteria becomes unavoidable. Genes resistant to the most widely used antibiotics and the bacteria carrying those genes cause infectious diseases, possess threat to public health, and are listed amongst the emerging concern of contaminants [1, 2]. Antibiotic resistance genes (ARGs) are fragments of deoxyribonucleic acid (DNA), in which, antibiotic resistance is coded. ARGs mostly

Can Burak Özkal, Department of Environmental Engineering, Tekirdağ Namik Kemal University, Çorlu, Tekirdağ 59860, Turkey

https://doi.org/10.1515/9783110727227-011

exist in the forms of plasmid and integron structures, which are highly transmittable. These structures are transmitted between bacteria types and strains by horizontal gene transfer (HGT) [3].

Antibiotic resistance genes are located on chromosomes or mobile genetic elements, and are present as intracellular DNA and extracellular DNA. The extracellular ARGs are persistent and have transformation ability. They are, therefore, able to promote ARG propagation [4]. Microbial resistance can be obtained by the release of active antibiotics from bacterial cells by an efflux pump, alternative metabolic pathways, antibiotic target modification, and inactivation of the antimicrobial by enzymatic pathway. The above listed natural resistance mechanisms are also reported to be observed at WWTPs [5].

By the horizontal gene transfer (HGT) mechanism, microorganisms share genetic information and spread antibiotic resistance to all the neighboring microorganisms. HGT occurs by four different mechanisms –Conjugation, transformation, adaptation, and gene transfer agents. The occurrence of horizontal gene transfer between the different bacteria species was considered to be unlikely, while for the last 20 years, ARG transfer has been widely reported between natural and different species [6]. With the development of new beta-lactam agents and different antibiotic groups, bacteria have become stronger with natural selection mechanisms, as a result of their fight against antibiotics. Multiple-resistant *E.coli* (resistant to more than one antibiotic) is isolated from many water and wastewater sources. The most widely reported understanding is that dissemination of antibiotic resistance is achieved through cloning and horizontal gene transfer. In bacteria that provide multiple resistance to beta-lactam antibiotics, resistance genes can be transferred between different gram-negative bacteria. ESBL-producing bacteria species can be resistant to many different types of antibiotics, such as penicillin, cephalosporin, and carbapenem. Plasmids that contain ESBL genes also contain different antibiotic resistance genes [7]. Treatment mechanisms effective on antibiotics degradation are mostly contributing to multidrug resistance development. Conventional treatment plants mostly consist of ineffective mechanisms for the removal of antibiotics and antibiotic resistance genes (ARG). Based on water quality assessment and determining water treatment requirement, quantity and effect-assessment of micropollutants, and priority pollutants must be taken into consideration, in addition to conventional parameters. Pharmaceuticals and micropollutants, arising from usage in medication, are released into receiving bodies as a result of insufficient treatment efficiencies achieved in most of the present treatment plants. ng/L to ug/L environmental concentration of pharmaceuticals is typical for river, lake, sea, groundwater, and catchments [1].

Apart from the intrinsic properties of certain types of resistances, acquirement of resistance occurs through mutation and selection or through horizontal gene transfer via conjugation, transformation, and transduction mechanisms [6]. The environment takes the role of the source of antibiotic resistance. Resistance genes are transferred from harmless bacteria to pathogenic bacteria, and the environment acts as the transmission medium between certain pathogenic bacteria. The selective pressure caused by antibiotics increases the development and spread of antibiotic resistance [3]. Sub-inhibitory concentrations of antibiotics in the environment and WWTPs promote HGT. At the same time, HGT can be reduced by positive selection – elimination of potential recipients. Some antibiotics are reported to be promoting horizontal transfer of mobile genetic elements. Plasmids can transfer resistance genes to neighboring hosts, even in conditions where selective pressure does not exist. Besides, conjugative plasmids, transposons, and integrons may take action by enabling the capture and expression of exogenous genes. The effect of integrons on ARG spread among bacteria has been mostly observed in gram-negative bacterial species. They also play a role in the generation of phenotypic diversity and environmental adaptation of bacteria to a variety of changes [8].

Even so, after the ARB cells are killed, the genes subsist within the environment for a long period. Neighboring and surrounding bacteria may take those ARGs and gain antibiotic resistance (the transformation mechanism). Even if the resistance genes are damaged, other bacteria cells may assimilate those genes. ARGs are inefficiently removed/separated/damaged by wastewater treatment plant process mechanisms and are mostly released into the environment, generally resulting in selective pressure on microorganisms. Despite the natural and environmental origin of the antibiotic resistance formation mechanism, the way they are released into the environment, their amount, and accumulation accelerate their spread and make it difficult to control (Figure 11.1) Transfer of resistant bacteria to human happens through drinking water, irrigation of plant by treated wastewater, and direct contact with ARG-loaded sludge applied on soil. Through the wastewater reuse (including agriculture and livestock activities), surface water and ground water sources are contaminated by antibiotic resistance determinants (Figure 11.1.) [1]. As a result of the environmental threat caused by the accumulation on surface and groundwater sources, river and sediment layers i) antibiotic-resistant bacteria, and ii) ARG control issues are in the focus of international research. In the last decade, wastewater and wastewater treatment plants have started to be mentioned among the prominent sources for release of antimicrobials into the environment and for the emergence of AMR. The most common usage area of bio-solids, formed as a result of wastewater treatment plant process mechanisms, is agricultural applications. Bio-solids harbor antibiotics and other personal care products, thereby becoming a linear source of pollution and applying selective pressure on microorganisms [9] (Figure 11.1).

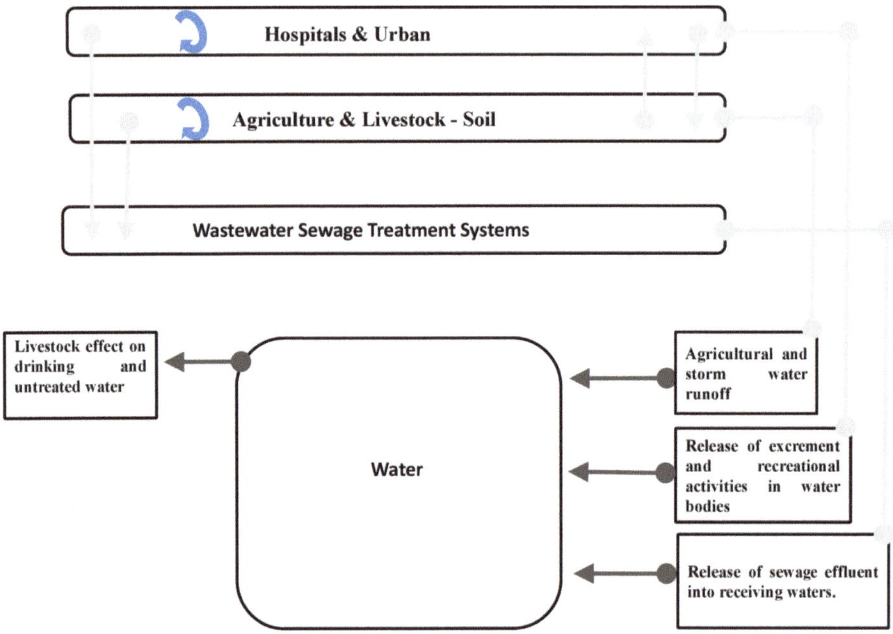

Figure 11.1: Environmental spread of antibiotic resistance.

11.2 Detection and monitoring of antimicrobial resistance

Since the 1950s, antibiotic resistance development has been widely supported by clinical findings and increased antibiotic production. Based on research findings of monitoring studies carried out mostly on soil environment, the idea that the level of antibiotic resistance can be traced through historical soil data and associated with the spread of antibiotic resistance gene (ARG) was supported. The vast majority of the specified antibiotic resistance genes spread over time, and alterations are prevalent between the different genes and regions. This gives information about the factors that will affect the long-term conservation of resistance genes in certain regions. Most findings indicate that wastewater treatment plants contribute to the generation of *E. coli*, producing large-scale beta-lactamase (ESBL) in surface waters [10, 11].

Regional monitoring and bacteria-specific research on antibiotic resistance development should be based on analyzing more isolates in a particular area, sampling, and locating additional pollutant sources in the researched area. The flow rates between the discharge points and the recreational use area, wastewater treatment plants, and other pollutant sources must be considered. In order to determine the main sources of ARG and Antibiotic-Resistant Bacteria (ARB), priority should be set on revealing the relative amounts occurring in wastewater networks. ARB and ARG monitoring in the wastewater treatment plant downstream and the receiving environment is the focus of scientific studies, international collaboration, and project calls in the field of antimicrobial resistance (AMR) control. The transfer of antibiotics, pathogenic bacteria, and antimicrobials resistance genes to the recipient environment, monitoring point, and line sources, mitigating the effects at the source and in the receiving environment, can be listed amongst the principal decisions that are being implemented under the direction of the World Health Organization, European Union, and the USA. In ARG monitoring and measurement studies, historical data of regional and national antibiotic use rates are taken into consideration in determining target-resistance genes [10].

In urban water systems, diversity and abundance of ARB and ARG have so far been investigated by culture-dependent and culture-independent techniques. Culture-dependent methods comprise direct cultivation of ARB, while the culture-independent approaches rely on the genetic material analyses. Metagenomics, quantitative polymerase chain reaction (PCR), and high-throughput qPCR are up-to-date and most widely used molecular methods in the last decade. Culture-based methods are based on determining the level of growth of bacterial isolates in the environment, in response to specific concentrations of antibiotics. qPCR has been widely used for determining the occurrence and abundance of certain ARGs in the environmental matrices. Quantification of multiple ARGs and mobile genetic elements, and the discovery and identification of new ARGs are time-consuming and not feasible with qPCR. To determine the abundance and diversity of ARGs and mobile genetic elements, the high-throughput qPCR (HT-qPCR) method is widely used. Specific smart chips have been developed. They permit the running of multiple PCR reactions (>1000 in an hour) and analyzing multiple primers in parallel. It is an efficient technique and has been frequently used in the field of monitoring the fate of ARGs in varying environmental samples, be it in the form of liquid or soil. It is theoretically not feasible for the discovery and quantification of new ARGs. Besides, there is no technical barrier against the detection of rare ARGs, and can be maintained through the amplification process. While with short-read metagenomics analysis, rare ARGS can be overlapped due to noise of other genes and elements, long-read metagenomics analysis are used to overcome such a situation. Metagenomics analyses are built on the entire gene-pool approach. Therefore, it is not limited to DNA sequences of the total genetic material. With metagenomics sequencing, either the determination of the emergence and relative abundances of certain ARGs or the detection and identification of new ARGs in environmental samples are feasible goals [6]. Despite the ongoing progress, analyzing and monitoring ARG transfer in urban wastewater treatment plants is still an issue to be resolved. Microbiological antibiotic resistance tests have been standardized and widely used to enable distinction between sensitive and resistant bacteria. At a molecular scale, rapid determination of microorganism composition and activities can be achieved by the PCR method (based on the amplification of the polymerase chain reactions), coupled with the DNA microchip methods.

According to research findings, in the last decade, wastewater treatment plants are reported as hotspots for horizontal gene transfer and the growth of multi-resistant bacterial species [12]. Based on an ARG study using different molecular methods for analyzing the effluents of WWTPs with conventional treatment processes, significant ARG reduction in WWTP effluent has been demonstrated [13]. Although conventional wastewater treatment processes are able to reduce the number of bacteria (by >99%), the quantity of ARGs in the effluent are reported to be increasing [14]. Spatial, seasonal and annual monitoring of ARB and ARGs were carried on sediment and water environment at different parts of the ecosystem; downstream-upstream of treatment plants, lakes, marine, and specific reservoir environments. Findings were interpreted with the antimicrobial consumption data, number of treatment steps, and the mechanism of the applied treatment processes. A high correlation was reported between the analyzed ARG concentrations in the WWTP effluents of > 10 European countries and the number of treatment stages applied [8].

According to recent research, rivers, as receiving environment of treated wastewater, have been reported to be hosting an increased abundance of ARB and concentrations of ARG. This clearly demonstrates the significant contribution of wastewater treatment plants on environmental dissemination of ARGs [15] . On the basis of data reported from WWTPs at Europe, America, Asia, and Africa between 1999–20,009 and 2009–2019, the most frequently encountered antibiotics are macrolides, sulfonamide, trimethoprim, quinolones, norfloxacin, and tetracycline. While the most commonly reported ARGs are blaCTXM, blaTEM (against β- lactam antibiotics), sul (against sulphonamide antibiotics), tetQ, tetW, tetA (against tetracycline antibiotics) and ermB genes (against macrolide antibiotics). Except for the case of β-lactam antibiotics and bla genes, a positive correlation was reported between the concentration of the commonly detected antibiotics and the abundance of ARGs in WWTPs [6]. The absolute abundance of ARGs in the effluent (2.9–4.6 logs in median) was much lower than that in the influent, with a reduction of 2 log order of magnitude. Influent and effluent antibiotic concentrations and the corresponding ARGS abundances were found to be in certain positive correlation according to the surveillance study finding of the European WWTPs. The presence of ARGs was reported to be higher in the case of countries with high antibiotic consumption, in comparison to low antibiotic consumption countries (Spain, Portugal, Ireland, Cyprus, Germany, Norway, Finland, etc.) [8].

11.3 Overview of the current legislative regulations

Misuse and redundancy in antibiotics usage diminish the effectiveness of antimicrobials and cause the spread of microorganisms resistant to the last-line of antibiotics.

According to the Organization for Economic Cooperation and Development (OECD), over 700,000 AMR deaths are reported worldwide, 25,000 of which are in the European Union countries. This emphasizes the necessity of cooperation in research, development, and monitoring studies on the subject. Unless effective measures are identified and implemented, deaths due to AMR are expected to exceed 10 million per year by 2050 [2]. According to the European Centre for Disease Prevention and Control (ECDC), the total patient hospitalization cost for multidrug resistant bacteria infection cases have exceeded 1.5 billion EUR, while the costs have been reported to exceed 2.2 billion EUR in United States. Specific actions against the AMR spread are subcategorized as follows; a) Those that avoid the emergence of resistance and b) Those that take the role of preventing AMR transmission [16].

The most recent policy options for antibiotics and ARG reduction comprise a three-layer approach – reducing human and animal antibiotic usage, management of ARB- and ARG-containing wastes, and management of the aquatic environment, soil, and wildlife. Taking control and managing the AMR spread mechanisms as a last barrier approach is becoming widespread, and research in this field is being funded at international scale [16]. The proposed approaches must be integrated with the policy options and compatibility has to be ensured, considering the national and international strategies to tackle/mitigate AMR. These strategies, so far, have focused on rational prescribing, assessment of sensitivities by diagnosis, while literature and research emphasize the necessity of environmental contamination aspects to be considered. Antibiotic resistance was recognized amongst the most important issues related to public health and was the main subject at the G20 summit (where the commitment to combat AMR was affirmed in 2017) and at the General Assembly of the United Nations. There is a growing interest from the scientific community on the global action plan for antibiotic resistance control. In 2015, the 68th World Health Assembly adopted the World Health Organization (WHO) Global Action Plan (GAP) on AMR, which emphasized the need for actions plans to be carried out, at all scales, in line with the GAP [10, 17].

GAP-AMR introduced the **One-Health** approach based on collaborative, multisector, and transdisciplinary viewpoint (including veterinary and human use of medicine, agriculture, environmental, consumer, and financial perspectives). What distinguishes One Health from other approaches is the interconnected perspective of human, animal, and the external environment [3]. According to the EU declaration on the One Health Action Plan, the release of pharmaceuticals, including antimicrobials, into the environment through wastewater and wastewater treatment plays a determining role on the emergence of AMR and has to be addressed by taking appropriate steps.

Global development of action plans and an antimicrobial stewardship framework was suggested by the World Health Assembly in 2015 (early WHO efforts stating the necessity of developing antimicrobial usage-specific measures and control strategies have existed since 1998) [17]. Stewardship programs may include monitoring, guidelines, regulation interventions, education, and campaigns to expand awareness [17]. Policy objectives reported by WHO before the GAP-AMR include policy objectives in line with the context of GAP-AMR. Global action plan of WHO on AMR has put forward the responsible use of medicines and guidance to countries to arrive at a common ground for developing their national AMR actions so as to control infectious diseases by the effective use and supply of safe medicines [9, 16, 17]. Literature and research have highlighted the policy gaps and failure in practice, taking into account the economic, ethical, and sustainability aspects. The disparity between the policy objectives, conditions, and actions in many different aspects (economic, ethics, etc.) is discussed elsewhere [3, 18]. In order to ensure accordance to policies at the national and international levels, GAP has defined a list of objectives: i) increasing awareness by education and training, ii) strengthening knowledge by research and surveillance, iii) sanitation, hygiene for infectious disease control, iv) optimizing antimicrobial use and v) economic aspects, and had also set a deadline for all national action plans to be completed before 2017. By 2017, more than 35% of the WHO member countries have completed their national actions plans. However, more than 60 countries have not yet completed. Thus, GAP-AMR fails to provide satisfactory results [8, 9, 16].

National plans put in action to tackle AMR kicked off in the early 2000's. They became widespread at the European level, between 2011 and 2020. By 2014, United Kingdom, Canada, and USA implemented their antimicrobial resistance action plans [17]. Building effective national surveillance systems, and carrying out an analysis of the antimicrobial resistance, and studying the spread of resistant microorganisms were listed foremost amongst the alternative mitigation plans. Unfortunately, the number of countries that have established an antibiotic resistance microorganism monitoring surveillance system is less than hundred. Southeast Asia is taking the lead by 100% coverage, followed by Western Pacific, Europe, and America with more than 50% coverage. Eastern Mediterranean and African regions are far below with less than 50% coverage. Regulations on antimicrobials usage and preventing ARM spread are two main categories of effective actions to be implemented at the national and international level [17].

As per the various reference documents addressing the issue of AMR, the Group of Seven (G7) countries and EU countries have implemented national policies and priorities under 9 main titles; i) Implementation of antimicrobial stewardship, with an intent to optimize the use of antimicrobial agents ii) Prevention and control of ARMs in a way to reduce the incidence of infection, iii) Building surveillance and monitoring programs, thereby strengthening the knowledge base, iv) Innovation, research & development (strengthening knowledge through research), v) Animal & agricultural sectors: One-Health approach vi) Cross country collaborations and international efforts vii) Increasing awareness and education, viii) Health professionals' training, ix) Sustaining financial stability [17, 19]. The AMR and ARG threat might be mitigated by specific measures, such as improved wastewater treatment processes, reduction of manufacturing emissions, consideration of environmental impacts in procurement and drug approval decisions, and better manure management.

Recent research has highlighted that in order to mitigate the environmental contamination by AMR and to be implemented at different levels of context, the solutions must be scalable, from the point of antibiotics usage up to the ARB- and ARG-containing waste discharge/release into the environment [9]. With an intent to provide sustainable solutions for ARB & ARG control, it is essential to develop and provide dissemination of wastewater treatment alternatives that may provide efficient disinfection (the term which is used to describe treatment processes that considers both bacteria inactivation and antimicrobial resistance control aspects) [5, 9, 20].

11.4 Advanced treatment technologies: Recent advances in the field of antibiotic resistance control

Pathogenic microorganisms threaten the environment and human health as a result of their unrestrained release and spread in the environment, along with the mechanisms of antimicrobial resistance development and resistance gene transfer, within and between the bacteria species. Consequently, studies on effective disinfection methods as well as the control of disinfection by-products have gained momentum. The intent of the studies are set but not limited to achieve an advanced level of bacteria inactivation. It has become essential to examine the factors that have an impact on antimicrobial resistance development. Biological treatment process conditions provide the environment for the transfer of antibiotic resistance genes between and within various bacterial species, thereby proliferating the antimicrobial resistance levels of pathogenic bacteria. Developing disinfection processes that together take into account efficient bacteria inactivation and resistance gene control goals must be paid the utmost attention [8, 21]. As in the case of micropollutants, antibiotics, and antibiotic resistant bacteria, the effect of removal mechanisms are specific to pollutant-treatment process scenarios. For the removal mechanisms of different ARGs, it appears that specific resistance genes may exhibit higher removal efficiencies, as is reported in the case of conventional activated sludge process [22]. Different operating conditions and process technologies in the treatment facilities affect the incidence of ARB and ARG and their spread into the environment through effluents. Therefore, the applicability and the scale-up potential of conventional disinfection and advanced oxidation processes are evaluated based on their effect on ARB and ARG control. The occurrence of antibiotic residues in WWTPs play an important role in the maintenance and spread of ARB and their genes, besides electing resistant phenotypes that may also disturb the indigenous bacterial communities. On the one hand, sub-inhibitory concentrations of antibiotics and metal concentrations in the water matrix, or oxidative stress conditions may promote HGT. In contrast, the resultant positive selection reduces HGT by eliminating the potential recipients in the environment [5, 23].

The operating conditions and the different wastewater treatment technologies influence, for example, the occurrence and the consequent release of ARGs and ARB into the environment. ARB and ARGs possess low abundances in effluent, in comparison to the influent of conventional WWTPs, due to embedding of those components within the treatment sludge, due to the biological treatment mechanism [5, 6, 24]. Treatment technologies, including conventional disinfectants (chlorine and ozone), UV irradiation and advanced oxidation processes (fenton oxidation, solar-driven Fenton oxidation, and photo-catalytic oxidation), filtration-based membrane processes, and membrane bio-reactors are presented and discussed in the context of their application in wastewater treatment and their impact on the fate of ARB and ARGs. Finally, some ideas arising from reviewed literature on how to reduce the spread of antibiotic resistance are presented.

11.5 Conventional disinfection processes mechanisms effective on ARB and ARG reduction

Conventional WWTPs (as a result of biological treatment processes mechanism) have become a source of ARGs for downstream environments. Consequently, the operational efficiency of the WWTP is of utmost importance. Biological treatment processes, such as activated sludge and multiple sludge treatment processes, possess limited capacity on the control of ARG and ARB in plant effluent. Pathogenic microorganisms in bio-solids can also be effectively destroyed in thermophilic sludge digestion process and higher ARG reduction rates are obtained due to the reduced level of microbial diversity, in comparison to mesophilic process conditions [25] For the most detected antibiotics and ARGs/ARB of concern, the removal and reduction rates at WWTPs are 53%-78% for antibiotics and 2 log for ARGs and antibiotic resistant *Escherichia coli*, *Enterococci* [26].

Alongside the determination of bacteria inactivation levels, process mechanism, process duration, and environmental conditions may promote resistance gene transfer, causing an increase of absolute and relative ARG quantities, which have to be determined. Adsorption, sedimentation, physical separation, retention and oxidative stress mechanisms, and their combinations are frequently applied for controlling the propagation of ARB and ARG. Treatment process mechanisms affect horizontal gene transfer, the level of bacterial reactivation, cell structure, and the amount and relative proportions of ARG in the unit bacterial cell or in the process effluent (Table 11.1) [21, 23, 27–29].

Coagulation/flocculation and sedimentation disinfection application have widespread use in the secondary treatment effluent of European WWTPs. On the other hand, chlorination, disinfection with ozone, per-acetic acid oxidation, and UV radiation have a wide field of application. Microfiltration (MF), ultrafiltration (UF), nanofiltration (NF), and reverse osmosis are quaternary treatment alternatives applied for microbial control with an intent to provide drinking water of high purity and/or quality. In WWTPs, suspended solids act as a shield and surround the microorganisms, thereby inhibiting the disinfection mechanism. A large proportion of coliform bacteria group exist in the form of suspended solid matter. Following the chlorination application in secondary treatment effluents, the treated water has the potential to be used in agricultural irrigation. Chlorination and photolysis methods are not effective on *E.coli* bacteria inactivation, and the level of antimicrobial resistance tends to increase the effectiveness of the advanced treatment processes, mostly determined based on the bacterial inactivation rates, but not taking into account resistance transfer by gene transfer and other mechanisms [30, 31].

Even disinfection-based process development and applications have been reported to remain incapable in the control of ARGs and ARB [21]. In comparison to ARB, ARGs were found to be more tolerant to disinfection process mechanisms [10, 29]. Apart from the level of reduction in ARG loads, disinfection processes may end up with regrowth of the survival cells. Bacterial cells with ARGs are reported to have a selective advantage against disinfection. Therefore, increased prevalence of ARG may occur in the bacterial community even after tertiary wastewater treatment processes. Reuse of treated water, following tertiary treatment, must be managed, taking into account the aspects of storage time and conditions, and the distribution system design parameters [32]. Assessment of ARG removal efficiency of the treatment process by using the normalization per volume of wastewater approach has a wide field of application (Table 11.1) [27, 29, 33]. Besides, proportionality to ng of DNA or number of 16S rRNA gene copies has been implemented as the alternative normalization factors. In this manner, it has become possible to interpret the absolute and relative abundancies of target ARGs in different mediums [34]. Normalization factors may provide additional informational, such as frequency of a gene in the total DNA mass or bacterial community represented by 16S rRNA, rather than providing standalone total abundance values. With this approach, it may become possible to observe not only the removal of the total resistance and integrase genes, but also the frequency of genetic determinants occurrence in the post-treated survival microbial cell. Removal of ARGs could be attributed to the efficient removal of micro-organisms in wastewater samples and that the abundance of background bacteria in wastewater may play an important role in the horizontal spread of ARG [22].

There are many examples of counter-view conclusions on antibiotic resistance elimination potential of the chlorination process. Low disinfectant concentrations (in the case of chlorination) act as promoter of selective pressure for antibiotic resistance and proliferate AR. Increases in disinfectant doses bring along higher bacteria inactivation rates and level of ARG damage. Chlorination is studied in different process conditions, but there are no specific regulations or guidelines reported for chlorine usage in ARB inactivation and ARG removal. The WHO and CDC reported chlorine dosages (0.92–112 mg.min/L) are far from providing efficacious ARB and ARG removal in practice, within reasonable process durations. Thus, most of the studied chlorination process conditions are not technically feasible for full-scale application in WWTPs, which is also the matter in the case of ozonation process (the WHO reported dosage for 99% *E.coli* inactivation is 0.02 mg.min/L) [35].

11.6 Advanced treatment processes: Reaction mechanisms effective on ARG control

Chemical and photo-chemical processes end up with reactive species formations that have high oxidation potentials. Reactive oxygen species (ROS) possess a highly toxic effect on the biological molecules. ROS may target bacterial the cell wall and result in bacteria inactivation or end up destroying the total bacterial cell. Up-to-date research has revealed that bacterial cells and photocatalyst surface contact times affect the overall efficiency of the inactivation process. The issue of bacterial re-growth has been reported even after removal efficiencies of over 99% were achieved [23, 32]. Chlorination and almost all of the disinfection and advanced oxidation processes lead to the release of free DNA (harboring ARGs) into the receiving water, and that can be taken up by non-resistant bacteria (described as the ARG transfer via transformation of extracellular ARG). The intracellular ARG have the potential to promote dissemination of antibiotic resistance via transduction and conjugation mechanisms [6]. The mechanism of damage to both intracellular and extracellular DNA and ARG by treatment processes have to be studied using qPCR analysis and gel electrophoresis techniques [36].

As a result of the comparison between ozonation, TiO_2-based heterogeneous photocatalysis, and conventional chlorination, and taking into account the process mechanism effect on the DNA structure. chlorination was found to have no effect, while ozonation/photocatalysis processes may cause degradation on the DNA structure at increased level of oxidant doses [35]. Chlorine and chlorine-based compounds, which are preferred for their easy operation and average cost levels in wastewater disinfection, are also known for their potential to create harmful byproducts during the process. Discussion is underway on the effectiveness of chlorination and UV-chlorination on ARG removal. During chlorination and AOPs on natural effluents of WWTPs, a variety of organic matter, including micropollutants, are subjected to oxidation. As a result, dangerous oxidation intermediates and disinfection byproducts, such as trihalomethanes (THMs) and halo-acetic acids (HAAs), are formed. Those by-products and intermediates are subjected to cytotoxic (in vitro) and genotoxic (in vitro and in vivo) analysis with intent to estimate the potential threat on the environment and living organisms [37].

The main process mechanism of UV-based disinfection is defined to be damaging the bacterial cell DNA. The turbidity of aqueous matrix and the issue of bacterial regrowth are the main drawbacks of the process. The DNA damage, as a result of UV irradiation, can be defined as the damage that is tolerable by bacterial DNA repair mechanisms. This situation also reflects upon the treatment application in which the UV irradiation was used, alone or in combination with different chemicals (UV/H_2O_2, UV/chlorination). UV irradiation is combined with H_2O_2 in order to increase the rate of OH• formation and process efficiency, which were reported to promote bacterial cell structure damage and intracellular DNA degradation mechanisms. UV/H_2O_2 process is effective on ARB removal and reduction of intracellular ARG, while the possible depletion of ROS by intracellular components, insubstantial effect of OH• on DNA structure, and the relatively low dose of oxidant were the underlying reasons for the inadequacy of the process mechanism on extracellular ARGs [38].

Based on the total copy number per unit volume, which is based on the alterations of the blaTEM gene before and after treatment with UV/H_2O_2 process (on *E.coli* culture with multiple antibiotic resistance isolated from the biological treatment effluent of UWWTP), there was not a significant change in the abundance of the blaTEM gene. On the other hand, a number of colony-forming bacterial cell concentrations reduced below the detectable level (5 CFU/mL), based on the increased process duration (>300 min). It was emphasized that the ARG concentration in the aquatic environment is caused by the release of the resistance genes from the cell structure of bacteria whose cellular structure is damaged during inactivation [29]. 2.58–3.79 logs and 2.63–3.48 log removal rates of ARGs (sul1, tetX, tetG, intI1, and 16S rRNA genes) were achieved in WWTPs effluent treatment by fenton oxidation and UV/H_2O_2 processes, respectively. Particularly, fenton oxidation is slightly more efficient in reducing ARGs than the UV/H_2O_2 process, which can be associated to OH• scavenging by the cellular components of the bacteria [22].

Disinfection with membrane filtration is based on the physical separation mechanism, effective on both pathogenic bacteria cells and ARGs, ozone treatment targets DNA, and different cellular structures, causing bacterial inactivation or complete cell death [21]. ARB and ARG removal by membrane filtration provided 6 log inactivation and efficient retention. Ozonation under the effect of water matrix constituent (mostly micro-pollutants) was found almost ineffective for ARG removal (Table 11.1) [24, 34]. Aerobic and anaerobic microorganisms occur in the biological treatment process and their metabolic activities end up with degradation of organic contaminants into biomass, carbon dioxide, methane, and water. A standalone application of biological treatment processes remains incapable of simultaneous elimination of ARB and

ARGs, while sequenced operation of anaerobic and aerobic reactors may provide efficient removal rates for ARB and ARGs.

For the case of membrane bioreactors, a combination of physical separation and biological treatment mechanisms help in the achievement of certain level of ARG removal rates. The anaerobic-anoxic-aerobic (A^2O) MBR process-based ARG removal in wastewater was found interrelated with increased ARG abundances in the treatment sludge matrix. Apart from the obtained removal rates with the wastewater treatment process (Table 11.1), a reasonable rate of ARG transfer from wastewater to sludge is found apparent and directly acting on the fate of AR determinants in plant effluent treatment [39]. The ARG abundances were found higher in sludge rather than in the WWTP effluent [40]. A study examining the fate of ARGs through anoxic/aerobic MBR process has revealed the efficient removal of bacterial intracellular ARGs. Membrane fouling on ARG removal correlated with the content and components of soluble microbial products (SMP) and extracellular polymeric substances (EPS). Significant correlation was noticed between the ARG abundance and SMP and EPS contents and components [41]. Theoretically, the abundance of 16S rRNA genes in WWTP effluent is 3–4 times higher than with other genes. The most effective removal rates, by advanced oxidation processes, have been reported for antibiotic resistance genes, intl1 and sul1. Intl1 gene is directly associated with resistance gene transfer and found in bacterial cells with antibiotics resistance genes harbored in the plasmid structure [22]. Complete inactivation of the bacteria (determined by molecular analysis based on alterations of 16S rRNA abundance in the effluent) and effective ARG degradation were achieved by the TiO_2-modified photocatalytic PVDF membrane process. ARGs located in plasmid were able to be removed to a lesser extent than genome-located ARGs. UV-irradiated photocatalytic membrane process is reported to be inhibiting horizontal gene transfer mechanism throughout the treatment process and scale-up potential of the process is emphasized [42].

Genes, specific to bacteria species and antibiotic resistance genes, exhibit different behavior, post treatment with photocatalysis process. It has been reported that the 23S sRNa gene sequences, specific to sul1, ermB, and enterococcus species, remained unchanged throughout the treatment process, while bacterial species specific genes were degraded and ended with reduced abundances (decreased by number). The presence of bacterial defense mechanisms against cellular damage caused by reactive oxygen species plays a decisive role in the disinfection of total and antibiotic-resistant *E.coli* species [33]. Photocatalytic disinfection studies carried out with graphene-based (in order to increase micro-contaminants adsorption onto composite and visible light adsorption capacities) TiO_2 composite (180 min of process on *E.coli* culture isolated from domestic wastewater MBR treatment plant effluent) was found sufficient for total inactivation of bacteria and no re-activation or re-growth were observed. Restrictions of PC inactivation rate due to UV-irradiation level or wavelength-specific limitations are mostly subdued by the synthesis of the photocatalyst with enhanced photocatalytic materials. With

Table 11.1: Control of ARGs by conventional disinfection and advanced treatment processes.

Treatment process(es) applied	Reduction of ARG	Target genes	Process details / Key findings	Water matrix	Analysis of Resistance gene	Reference
UV irradiation	>2 Log removal,	Sul1, intl1- tetG	>10.000 mj/cm^2	Biological treatment effluent	Quantitative PCR, on treatment process effluent, Relative abundance of selected genes have increased during ozone treatment and at low doses of UV-A, and chlorination concentrations (16S rDNA normalized)	[44]
Chlorination	~ 3 Log removal		NaClO, 120 min Effect of Cl_2: NH_3-N rate,			
Ozonation	1.65–2.55 Log removal		30 min,			
UV irradiation	4–5 Log removal	tetW	5–20 min, 69.8–279 mj/cm^2	Biological treatment effluent	Quantitative PCR on treatment effluents	[30]
Chlorination	0.1–4 Log removal		Cl_2, 10 min			
Chlorination	Drastical decrease over 2 mg/L Cl_2, max at 4 mg/L Over 4 Log removal,	Resistance genes conferring resistance to Vancomycin, tetracycline, sulfonamide, MLSB, FCA, Beta Lactam, aminoglycoside	Cl_2, 30 min.	Sterile distilled water	High capacity Smartchip Quantitative PCR, Relative abundance of 119 of 125 ARGs decreased in chlorination effluent	[31]

(continued)

Table 11.1 (continued)

Treatment process(es) applied	Reduction of ARG	Target genes	Process details / Key findings	Water matrix	Analysis of Resistance gene	Reference
Ozonation	1–2 Log Removal	Ctx, Oxa, Tem,	20 m	Activated sludge process effluent	Quantitative PCR analysis on process effluents	[24]
UV irradiation	1–2 Log Removal	Ctx, Oxa, Tem,	4.5 W/m² 5 min,			
Chlorination	Reduction is not apparent	Ctx, Oxa, Tem,	30 min			
UV/H_2O_2	No removal	blaTEM, qnrS, tetW	60 min.	Sterile, DNA free water	Quantitative PCR analysis on DNA extracted from process effluent and re-inoculated survival bacteria cells	[29]
UV/H_2O_2	Removal average: 2.8–3.5 Log and 1,5–2.32 Log	Sul1, tetx, tetG, intl1 and 16s rRNA	Effect of pH: pH 3.5 and pH: 7 30 min irradiation	Biological treatment effluent	Quantitative PCR on process treatment effluents	[45]
Fenton process	~ 5 Log removal, ~ 4 Log removal and ~ 3 Log removal	Sul1 and intl1, tetX, tetG and 16S rRNA	Most efficient when Fe_2/H_2O_2 ratio > 0.5 2 h	Biological treatment effluent	Quantitative PCR on process treatment effluents	[46]
UV/H_2O_2	2.8–3.5 Log removal		3o min of UV irradiation, pH: 3.5 UV energy: NA			

UV/H_2O_2	1.4–2.7 Log Removal, 2.3–2.9 Log removal	mecA, ampC	12–120 mj/cm^2 UV254	Natural surface water	Quantitative PCR analysis on treatment process effluent	[47]
Ozonation	1.4–2.6 Log removal, 1.6–2.6 Log removal, > 3 Log removal / 3.7–4.7 Log Removal and > 6.7 Log removal	İntl1, 16S rRNA, sul1 (<LOQ) / 16S rRNA and blaTEM	30 min, pH:7	Biological treatment effluent / synthetic effluent	Quantitative PCR analysis on treatment process effluent	[34]
Anoxic-aeration Membrane bioreactor	067–4.73 Log removal	Sul1, intl1, tetX, tetG, tetW		44% domestic wastewater and 56% industrial wastewater mixture	Quantitative PCR analysis on treatment process effluent	[39]
MBR	Overall 2.57 Log removal / 2.37 Log removal	tetW, tetO, sul1, 16S rRNA		MBR plant / conventional treatmen plant	Quantitative PCR analysis on treatment process effluent	[40]

(continued)

Table 11.1 (continued)

Treatment process(es) applied	Reduction of ARG	Target genes	Process details / Key findings	Water matrix	Analysis of Resistance gene	Reference
UV irradiated TiO_2 modified UF membrane / Pristine UF membrane	Plasmid located ARG removal: >97%, 20%, 99%, 23% / > 93%, 15%, 99%, 88% Removal of genome located ARG > plasmid located ARG >98% Removal	Plasmid located ARG: floR, tetC, sul1, intl1	ROS produced by photocatalytic process resulted with enhanced antifouling and stability properties of PVDF UF membrane.	Biological treatment effluent	Quantitative PCR analysis on treatment process effluent	[42]
UF, NF membrane	ARB removal	–	Electrostatic interactions on the membrane contribute to the removal of ARG.	Biological treatment effluent	Quantitative PCR analysis on treatment process effluent	[28]
Ozonation, UF_1, UF_2	4 Log removal, >6-7 Log removal UF2 was not capable to remove intl1 and sul1 genes below the detection limit	Sul1, blaTEM, tetM, CTX-M, blaOXA-48	UF_1: Nominal pore size 20 nm, 70 L/m^2h, UF_2: Minimal pore size 25 nm, 35 L/m^2h ARG concentration in retentate $UF_1 < UF_2$	Biological treatment effluent	Quantitative PCR, Metagenomic sequencing	[21]

MBR	0.6–5.6 Log removal	Sul1, tetC, tetX, int1, ereA	Bacterial intracellular ARGs are efficiently removed. Contribution of membrane fouling to ARG removal have been reported taking into account the contents and components of soluble microbial products, extracellular polymeric substance	Synthetic wastewater	Quantitative PCR analysis	[41]
Photocatalysis	3–4 fold removal of ampC and mecA. Sul1 and ermB were persistent.	sul1, ampC, ermB, mecA,	Level of re-growth varied in the presence of different photocatalyst. Total genomic DNA concentration was not affected by treatment. Importance of the defense mechanism of bacteria against ROS induced cellular damage have been highlighted.	MBR effluent	Quantitative PCR analysis on treatment process effluent. Total genomic DNA content, ARGs and specie specific marker sequences have been analyzed.	[33]

(continued)

Table 11.1 (continued)

Treatment process(es) applied	Reduction of ARG	Target genes	Process details / Key findings	Water matrix	Analysis of Resistance gene	Reference
Solar H_2O_2 **TiO_2/H_2O_2**	1 Log removal of ARGs, except the blaCTX-M removal of 3 Log	16S rRNA, intl1 and sul1, blaCTX-m, blaTEM, qnrS	Loss of ARG with maintenance of cell availability or cell inactivation. the abundance of intl1 and sul1 genes have increased after 3-days storage to values higher than or close to values verified before treatment. Re-growth after 3 days of storage: Part of the ARG removal was due to cell inactivation and not to ARG loss.	Biological treatment effluent	Quantitative PCR analysis on samples collected before and after treatment process.	[48]
Photocatalysis	Observed ARG transfer was lower during PCD of *E.coli* suspended in biological treatment effluent.	Conjugation experiments by synthetic HT-99 and J-53R *E.coli* strains	Effect of oxidative stress conditions on horizontal gene transfer have been examined. Gene pair conjugant numbers showed a four fold increase post PC treatment in distilled water. Organic and inorganic component in secondary effluent matrix lead to ROS scavenging, thus reduce the level of oxidative stress on *E. coli* cells.	Distilled water and biological treatment effluent	Conjugation based microbiological method	[23]

dopant, varying organic/inorganic and metallic/non-metallic additives photocatalysts with advanced properties are developed (visible light active, reduced surface charge recombination) [33].

Bacteria strains, resistant to antibiotics, are more sensitive to photocatalytic disinfection than antibiotic susceptible strains. In the treated domestic wastewater environment, the oxidative stress level and the rate of conjugated gene pair formation on resistant *E.coli* cells (K:12 susceptible, J53: rifampycin resistant strain, chloramphenicol resistant HT99 *E.coli* strain) are reduced (horizontal gene transfer inhibition) due to scavenging of organic and inorganic components of reactive oxygen species [23]. Advanced oxidation processes target bacteria cell DNA and different cellular components [33, 43]. Recent research focuses on monitoring antibiotic resistance genes (sul1, blaTEM, blaCTX-M, tetA, tetM, mcr, etc.), prior and post treatment of the aquatic sample and bacteria cell medium (based on ARG abundance analysis by DNA extracted from the survival bacteria cell cultures) or both (total DNA) by Real-time qPCR analysis, Research demonstrates lower removal rates obtained for eubacteria marker gene 16S rRNA by fenton oxidation, UV/H_2O_2 processes at secondary treated WWTP effluent for ARGS of concern [21].

11.7 Summary

Detailed analysis of antibiotic resistance transfer, post disinfection, would be possible by carrying out qPCR analysis on total DNA, extracted from process effluent, and DNA extracted from survival bacterial culture cells. In virtue of such approach, rather than sole evaluation of bacteria inactivation potential of disinfection process alternatives, effect assessment of the process mechanism on DNA would be possible. During the course of different advanced treatment process mechanisms, ARGs may be removed, survive, or multiply.

The quantity assessment reveal ARG abundances in treatment plant effluents or in cellular environment have the potential to be used in decision-making stages of WWTP design and operation, effluent storage conditions, and retention times within treatment processes [49].

Multi-resistant and sensitive bacteria cultures may possess different resistance behavior against applied water treatment processes; thereby, antibiotic resistance phenotype and bacterial phylogenetic group characteristics are found with a level of resistance development against different inactivation mechanisms. Research based on urban wastewater treatment process effluent mostly lack simultaneous evaluation of disinfection, and ARB and ARG control. Disinfection at urban wastewater treatment effluent and sludge digestion are key processes and have to be improved for ARB control at WWTPs. Research in the field of efficient disinfection development should be carried out not only on bacterial cultures isolated from nature but also with reference bacterial cultures.

The fate of bacterial cells and ARG depend on the ROS level produced by the advanced oxidation processes. The level of ROS-induced damage to intracellular and extracellular ARGs have to be demonstrated[5].

In order to carry out an extensive study dealing with the ROS-induced stress mechanism and response of bacteria, wide range applicability of the proposed process have to be confirmed, both in theory and in practice. Secondly, research has to be carried on bacteria of different types and origins (including wild type, clinically- and environmentally-adapted, and synthetic strains).

Based upon the evaluation of the most recent application of advanced treatment processes in the field of antibiotic resistance control, a combination of different advanced treatment mechanism can be considered as an important research area.

References

[1] Kassinos, D.F., Meric, S., Nikolaou, A. Pharmaceutical residues in environmental waters and wastewater: Current state of knowledge and future research. Anal Bioanal Chem. 2011, 399, 251–275.

[2] Sanderson, H., Fricker, C., Brown, R.S., Majury, A., Liss, S.N. Antibiotic resistance genes as an emerging environmental contaminant. Environ Rev. 2016, 24, 205–218.

[3] Jamrozik, E., Selgelid, M. Ethics and Drug Resistance: Collective Responsibility for Global Public Health, NY, USA, Springer Nature, 2020.

[4] Dong, P., Wang, H., Fang, T., Wang, Y., Ye, Q. Assessment of extracellular antibiotic resistance genes (eARGs) in typical environmental samples and the transforming ability of eARG. Environ Int. 2019, 125, 90–96.

[5] Barancheshme, F., Munir, M. Strategies to combat antibiotic resistance in the wastewater treatment plants. Front Microbiol. 2018, 8.

[6] Ahmed, Y., Jin, M., Guo, J., Wang, Y., Li, J. Control Strategies to Combat Dissemination of Antibiotic Resistance in Urban Water Systems. Handb Environ Chem. 2020, 91, 147–187.

[7] Kamruzzaman, M., Shoma, S., Bari, S.M.N., et al. Genetic diversity and antibiotic resistance in *Escherichia* coli from environmental surface water in Dhaka City, Bangladesh. Diagn Microbiol Infect Dis. 2013, 76(2), 222–226.

[8] Pärnänen, K.M.M., Narciso-Da-Rocha, C., Kneis, D., et al. T.U. Antibiotic resistance in European wastewater treatment plants mirrors the pattern of clinical antibiotic resistance prevalence. Sci Adv. 2019, 5, 1–10.

[9] Goulas, A., Livoreil, B., Grall, N. What are the effective solutions to control the dissemination of antibiotic resistance in the environment? A systematic review protocol. Environ Evid. 2018, 7, 1–9.

[10] Pazda, M., Kumirska, J., Stepnowski, P., Mulkiewicz, E. Antibiotic resistance genes identified in wastewater treatment plant systems – A review. Sci Total Environ. 2019, 697, 134023.

[11] Munir, M., Wong, K., Xagoraraki, I. Release of antibiotic resistant bacteria and genes in the effluent and biosolids of five wastewater utilities in Michigan. Water Res. 2011, 45, 681–693.

[12] Michael, I., Rizzo, L., McArdell, C.S., et al. Urban wastewater treatment plants as hotspots for the release of antibiotics in the environment: A review. Water Res. 2013, 47, 957–995.

[13] Hultman, J., Tamminen, M., Pärnänen, K., Cairns, J., Karkman, A., Virta, M. Host range of antibiotic resistance genes in wastewater treatment plant influent and effluent, FEMS Microbiol. Ecol. 2018, 94, 1–10.

[14] Osińska, A., Korzeniewska, E., Harnisz, M., et al. Small-scale wastewater treatment plants as a source of the dissemination of antibiotic resistance genes in the aquatic environment. J Hazard Mater. 2020, 381, 121221.

[15] Wu, Y., Cui, E., Zuo, Y., Cheng, W., Chen, H. Fate of antibiotic and metal resistance genes during two-phase anaerobic digestion of residue sludge revealed by metagenomic approach. Environ Sci Pollut Res. 2018, 25, 13956–13963.

[16] Bloomer, E., McKee, M. Policy options for reducing antibiotics and antibiotic-resistant genes in the environment. J Public Health Policy. 2018, 39, 389–406.

[17] Cecchini, M., Langer, J., Slawomirski, L. Antimicrobial Resistance in G7 Countries and Beyond, Paris, Organization for Economic Co-operation and Development, 2015, 1–75.

[18] Eurosurveillance Editorial Team. WHO member states adopt global action plan on antimicrobial resistance. Eurosurveillance. 2015, 20(21), 21137.

[19] Mendelson, M., Matsoso, M.P. The World Health Organization global action plan for antimicrobial resistance. SAMJ South African Med J. 2015, 105, 325.

[20] Manaia, C.M., Macedo, G., Fatta-Kassinos, D., Nunes, O.C. Antibiotic resistance in urban aquatic environments: Can it be controlled?. Appl Microbiol Biotechnol. 2016, 100, 1543–1557.

[21] Hembach, N., Alexander, J., Hiller, C., Wieland, A., Schwartz, T. Dissemination prevention of antibiotic resistant and facultative pathogenic bacteria by ultrafiltration and ozone treatment at an urban wastewater treatment plant. Sci Rep. 2019, 9, 1–12.

[22] Chen, H., Zhang, M. Effects of advanced treatment systems on the removal of antibiotic resistance genes in wastewater treatment plants from Hangzhou, China. Environ Sci Technol. 2013, 47, 8157–8163.

[23] Dunlop, P.S.M., Ciavola, M., Rizzo, L., McDowell, D.A., Byrne, J.A. Effect of photocatalysis on the transfer of antibiotic resistance genes in urban wastewater. Catal Today. 2015, 240, 55–60.

[24] Lamba, M., Ahammad, S.Z. Performance comparison of secondary and tertiary treatment systems for treating antibiotic resistance. Water Res. 2017, 127, 172–182.

[25] Ghosh, S., Ramsden, S.J., LaPara, T.M. The role of anaerobic digestion in controlling the release of tetracycline resistance genes and class 1 integrons from municipal wastewater treatment plants. Appl Microbiol Biotechnol. 2009, 729, 791–796.

[26] Wang, J., Chu, L., Wojnárovits, L., Takács, E. Occurrence and fate of antibiotics, antibiotic resistant genes (ARGs) and antibiotic resistant bacteria (ARB) in municipal wastewater treatment plant: An overview. Sci Total Environ. 2020, 744, 140997.

[27] Alaton, İ.A., Karataş, A.Y., Pehlivan, Ö., Hanci, T.Ö. Elimination of antibiotic resistance in treated urban wastewater by iron-based advanced oxidation processes, Desalin. Water Treat. 2019, 172, 235–253.

[28] Schwermer, C.U., Krzeminski, P., Wennberg, A.C., Vogelsang, C., Uhl, W. Removal of antibiotic resistant *E. coli* in two Norwegian wastewater treatment plants and by nano- and ultra-filtration processes. Water Sci Technol. 2018, 77, 1115–1126.

[29] Ferro, G., Guarino, F., Cicatelli, A., Rizzo, L. β-lactams resistance gene quantification in an antibiotic resistant *Escherichia coli* water suspension treated by advanced oxidation with UV/H_2O_2. J Hazard Mater. 2016, 323, 426–433.

[30] Sullivan, B.A., Vance, C.C., Gentry, T.J., Karthikeyan, R. Effects of chlorination and ultraviolet light on environmental tetracycline-resistant bacteria and tet(W) in water. J Environ Chem Eng. 2017, 5, 777–784.
[31] Lin, W., Zhang, M., Zhang, S., Yu, X. Can chlorination co-select antibiotic-resistance genes?. Chemosphere. 2016, 156, 412–419.
[32] Fiorentino, A., Ferro, G., Alferez, M.C., Polo-López, M.I., Fernández-Ibañez, P., Rizzo, L. Inactivation and regrowth of multidrug resistant bacteria in urban wastewater after disinfection by solar-driven and chlorination processes. J Photochem Photobiol B Biol. 2015, 148, 43–50.
[33] Karaolia, P., Michael-Kordatou, I., Hapeshi, E., et al. Removal of antibiotics, antibiotic-resistant bacteria and their associated genes by graphene-based TiO_2 composite photocatalysts under solar radiation in urban wastewaters. Appl Catal B Environ. 2018, 224, 810–824.
[34] Sousa, J.M., Macedo, G., Pedrosa, M. Ozonation and UV_{254} nm radiation for the removal of microorganisms and antibiotic resistance genes from urban wastewater. J Hazard Mater. 2017, 323, 434–441.
[35] Menceloğlu, Y.Z., Balcioğlu, I.A. Comparison of the effectiveness of chlorine, ozone, and photocatalytic disinfection in reducing the risk of antibiotic resistance pollution. J Adv Oxid Technol. 2011, 14, 196–203.
[36] Sharma, V.K., Yu, X., McDonald, T.J., Jinadatha, C., Dionysiou, D.D., Feng, M. Elimination of antibiotic resistance genes and control of horizontal transfer risk by UV-based treatment of drinking water: A mini review. Front Environ Sci Eng. 2019, 13, 1–9.
[37] Costa, A., Lepretti, M., Rizzo, L., et al. Science of the Total Environment Surface water disinfection by chlorination and advanced oxidation processes : Inactivation of an antibiotic resistant *E.coli* strain and cytotoxicity evaluation. Sci Total Environ. 2016, 554–555, 1–6.
[38] Yoon, Y.,., Dodd, M.C., Lee, Y. Elimination of transforming activity and gene degradation during UV and UV/H 2 O 2 treatment of plasmid-encoded antibiotic resistance genes. Environ Sci Water Res Technol. 2018, 4, 1239–1251.
[39] Du, J., Geng, J., Ren, H., Ding, L., Xu, K. Zhang Y Variation of antibiotic resistance genes in municipal wastewater treatment plant with A^2O-MBR system. Environ Sci Pollut Res. 2015, 22, 3715–3726.
[40] Munir, M., Wong, K., Xagoraraki, I. Release of antibiotic resistant bacteria and genes in the effluent and biosolids of five wastewater utilities in Michigan. Water Res. 2011, 45, 681–693.
[41] Zhu, Y., Wang, Y., Zhou, S., Jiang, X., Ma, X., Liu, C. Robust performance of a membrane bioreactor for removing antibiotic resistance genes exposed to antibiotics: Role of membrane foulants. Water Res. 2018, 130, 139–150.
[42] Ren, S., Boo, C., Guo, N., Wang, S., Elimelech, M., Wang, Y. Photocatalytic Reactive Ultrafiltration Membrane for Removal of Antibiotic Resistant Bacteria and Antibiotic Resistance Genes from Wastewater Effluent. Environ Sci Technol. 2018, 52, 8666–8673.
[43] Ferro, G., Fiorentino, A., Alferez, M.A., Polo-López, M.I., Rizzo, L., Fernández-Ibáñez, P. Urban wastewater disinfection for agricultural reuse: Effect of solar driven AOPs in the inactivation of a multidrug resistant E. coli strain. Appl Catal B Environ. 2015, 178, 65–73.
[44] Zhang, Y., Zhuang, Y., Geng, J., et al. Inactivation of antibiotic resistance genes in municipal wastewater effluent by chlorination and sequential UV/chlorination disinfection. Sci Total Environ. 2015, 512–513, 125–132.
[45] Zhang, Y., Zhuang, Y., Geng, J., et al. Reduction of antibiotic resistance genes in municipal wastewater effluent by advanced oxidation processes. Sci Total Environ. 2016, 550, 184–191.

[46] Zhang, Y., Zhuang, Y., Geng, J., et al. Inactivation of antibiotic resistance genes in municipal wastewater by chlorination, ultraviolet, and ozonation disinfection. Environ Sci Pollut Res. 2015, 22, 7037–7044.
[47] Guo, C., Wang, K., Hou, S., et al. H_2O_2 and/or TiO_2 photocatalysis under UV irradiation for the removal of antibiotic resistant bacteria and their antibiotic resistance genes. J Hazard Mater. 2017, 323, 710–718.
[48] Moreira, N.F.F., Narciso-da-Rocha, C., Polo-López, M.I., et al. Solar treatment (H_2O_2, TiO2-P25 and GO-TiO_2 photocatalysis, photo-Fenton) of organic micropollutants, human pathogen indicators, antibiotic resistant bacteria and related genes in urban wastewater. Water Res. 2018, 135, 195–206.
[49] Berendonk, T.U., Manaia, C.M., Merlin, C. Tackling antibiotic resistance: The environmental framework. Nat Rev Microbiol. 2015, 13, 310–317.

Prasann Kumar, E. Lokesh Goud, Priyanka Devi
and Bhupendra Koul

12 Metal pollutants in the environment

Abstract: Heavy metal pollution in the environment is becoming a serious problem and a major concern as it is causing negative effects all over the world. These inorganic pollutants are being discarded in soil, water, and the atmosphere as a result of the rapidly growing agriculture and metal industries, improper waste disposal, pesticides, and fertilizers. This chapter explains how these pollutants enter our environment and cause serious diseases, such as cancer, by interfering with biological functions, causing growth, and accumulating in various organs. This chapter describes the toxicological and pharmacokinetic properties of each metal as well as some biochemical properties; physiological effects of these metals in human as well as in plants are also discussed.

Keyword: agriculture, environment, food science, heavy metal, soil, water

12.1 Introduction

The environment is defined as the surroundings in which humans, plants, animals, and microorganisms live or work. The environment is made up of the Earth's atmosphere, land, and water. Four spheres that work in harmony define the Earth's system: the biosphere, atmosphere, lithosphere, and hydrosphere. Contaminants as well as pollutants are present at higher levels in the environment than in any other section of the environment [1]. For the last century, the rate of industrialization has been very high, resulting in increased demand for exploitation of the Earth's natural resources at an extremely careless rate, which leads to the problem of environmental pollution. The current state of the environment has been severely polluted by various types of pollutants, such as organic, inorganic, organometallic compounds, gaseous pollutants, radioactive isotopes, and nanoparticles [2].

Heavy metals are compounds with densities greater than 4.5 gcm-3 and atomic weights ranging from 65.4 to 20.59 g per mol. Heavy metals, such as Cd (cadmium), Hg (mercury), Pb (lead), Fe (iron), Cr (chromium), and As (arsenic), have a longer biological life than other elements and can easily persist in the environment for a long period of time. All of these heavy metals are thought to be toxic to all living

Prasann Kumar, E. Lokesh Goud, Priyanka Devi, Department of Agronomy, School of Agriculture, Lovely Professional University, Jalandhar, Punjab 144411, India
Bhupendra Koul, Department of Biotechnology, School of Bioengineering and Biosciences, Lovely Professional University, Jalandhar, Punjab 144411, India, e-mail: bhupendra.18673@lpu.co.in

https://doi.org/10.1515/9783110727227-012

organisms in the environment. They are now classified as contaminants or pollutants because of their high concentrations in soil or water, which cause toxicity and a variety of diseases. Heavy metals are required in trace amounts in some essential micronutrients, and their excess concentration causes toxicity. An increase in the growing industry, particularly the mining industry, is one of the primary causes of heavy metal emissions into the atmosphere, which affects both human and animal health, and the plant ecosystem [3]. Both anthropogenic and natural activities emit metals into the atmosphere, which can travel over long distances. Metals released in the air, soil, crop, and water reservoirs from natural and anthropogenic activities can be found in areas near urban areas and industrial units. Heavy metal dust particles in the atmosphere have a long-term and toxic effect on human health. Because the population is concentrated primarily in urban areas, heavy metal dust particles are inhaled, and consumption of polluted crops may pose health risks. Heavy metal accumulation in crops and vegetables is examined only from the root system accumulation. This is due to metal build-up in the soil system and the plants adsorb from the root system. Aside from plant roots, aerial plant parts such as fruits, leaves, and flowers can also absorb heavy metals. The absorbing structure for metal uptake in aerial plant parts is similar to that of the plant's root structure. Following the deposition of atmospheric particles on the surface of the leaf, metal accumulates on plant leaves via foliar transfer. Heavy metal concentrations are increased in plants located near mining areas as well as smelting areas. A study was conducted near that industry area to learn more about heavy metal contamination and deposition of these heavy metals. Less research has been carried out on heavy metal uptake by plant leaves from the atmosphere. The absorption pathway of those metals that are essential for the metabolic activities of plants has been identified and evaluated for the biochemical reaction of plants. Various studies were conducted to learn about foliar absorption of iron, zinc, and copper. These metals are absorbed by plants via cuticle penetration and eventually accumulate in plant leaf tissues. The leaves of plants can absorb non-essential metals such as cadmium, chromium, lead, arsenic, and mercury via foliar transfer. This review discusses the various mechanisms involved in heavy metal uptake by plants, factors that affect heavy metal uptake by foliar, transportation and compartmentation of foliar heavy metal, toxic and detoxification of heavy metal inside the plant after foliar uptake, and a comparison between root and foliar uptake.

Heavy metals are defined as metals with high atomic mass and high density. Heavy metals are now defined as metallic chemical elements and metalloids that have toxic effects on the environment and humans. Some metals, such as arsenic, aluminum, and selenium, are lighter in weight but still toxic. Some metals are classified as heavy metals, but they are not toxic by nature, unlike gold [4]. Titanium, chromium, manganese, iron, cobalt, nickel, copper, zinc, cadmium, tin, platinum, mercury, and lead are some of the heavy metals that have a high density (5 gm/cm3) and are more common in our daily lives (Shi et al., 2014 [64]).

The goal of this chapter is to highlight heavy metal pollution and its entry into our environment, as well as its mechanism and toxicological effects on humans and animals (Figure 12.1).

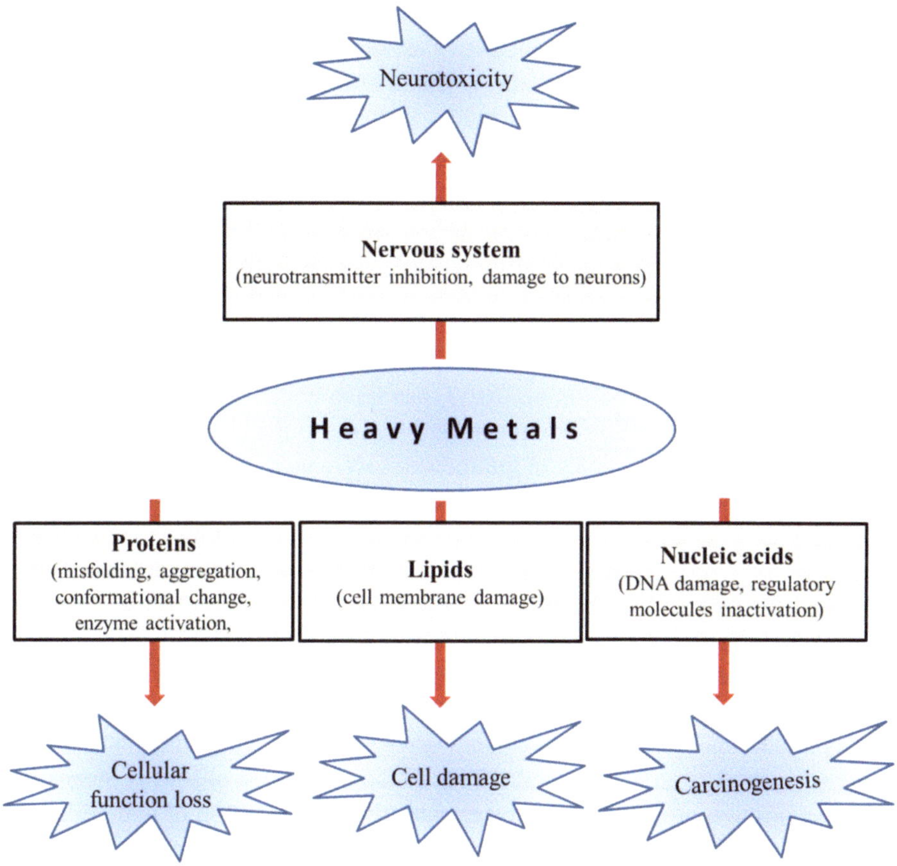

Figure 12.1: Impact of heavy metals on living organisms.

12.2 Different sources of heavy metal pollution

According to the Environmental Protection Agency (EPA), heavy metals were found in drinking water discharged from various industries, such as steel, leather, cement, and from trivalent chromium naturally deposited by erosion in rocks, soil, plants, and volcanic dust found naturally and due to various anthropogenic activities of humans. Heavy metal comes from a variety of sources, including natural processes, such as weathering of rocks, and a variety of anthropogenic activities that cause it to enter the environment. Examples of natural processes include the disintegration of rocks, erosion as well as volcanic eruptions, while human activities include smelting, pesticides use and phosphatic fertilizers, mining, electroplating, industrial effluent and sludge, and biosolids in agriculture [5]. Heavy metals are formed by a variety of processes, including weathering of rocks, which occurs naturally, and a variety of anthropogenic activities, because of which they enter the environment. Weathering, erosion, and volcanic eruptions are examples of natural processes, whereas anthropogenic activities include smelting, the use of pesticides and phosphatic fertilizers, mining, electroplating, industrial effluent a sludge, and biosolids in agriculture [5]. The main cause of heavy metals concentration in the environment is the increased rate of industrialization and urbanization, as well as improper waste disposal. Various anthropogenic activities, such as excessive use of herbicide, fertilizers, and sludge produced are to be blamed for heavy metal pollution in the atmosphere, but the mining process is the most common source of these elements Miransari, 2010. When heavy metal concentrations exceed their permissible limits (Table 12.1), (Zhou et al., 2017) they have negative effects on microbial activities, plants, and the environment. Sharma and Dutta 2017 conducted an experiment in which 240 samples of groundwater were collected in eight districts of Punjab (Malwa region), and the results revealed that the concentrations of various heavy metals were higher than their permissible limits, resulting in groundwater contamination. The primary goal was to investigate the distribution of heavy metals in groundwater as well as the effects of heavy metals on human health. They discovered that the water quality was unsafe for drinking as well as other domestic purposes. Different soil samples were collected from a maize field in China to analyze the heavy metal concentrations, and they discovered that the concentration of different heavy metals in the area where wastewater irrigated was higher in the uppermost layer of the soil. They stated that the concentration of lead is higher and the concentration of chromium is lower, with the concentration of chromium primarily accumulating in maize roots. The weathering and volcanic eruption of chromium-containing rocks, as well as some human activities such as mining and waste disposal, cause chromium accumulation in agriculture, and excessive fertilizer and pesticide use [6]. The holiest place in India, Haridwar, has been found to be affected by modern industrialization. It was discovered that the groundwater around Haridwar was contaminated with five heavy metals, namely Cr, Co, Fe, Zn, and Ni. According to the findings, 216 samples had 99% percentage of Zn, 94% Co, 99% Fe, 98% Cr, and 90% Ni 90, which exceeded the permissible limits. Among the five heavy metals, it has been reported that Cr is the most carcinogenic in nature and has the most severe effects on human health as well as on plants.

Table 12.1: Threshold limits of various metals in air and water.

S. no	Metal	Threshold Limit/ Exposure limit/ Permissible limit			
		Air		Water	
		Occupational	Non- occupational	Drinking water	Packaged water
1	Aluminum (Al)	15 mg/m^3 (total dust); 5 mg/m^3 (respirable fraction)	No data	≤ 0.2 mg/L	0.2 mg/L
2	Vanadium (V)	0.5 mg/m^3 (V_2O_5 dust); 0.1 mg/m^3 (V_2O_5 fumes)	1 $\mu g/m^3$	No data	No data
3	Chromium (Cr)	Cr(VI): 0.005 mg/m^3 Cr(III): 0.5 mg/m^3 Cr (0): 1 mg/m^3	1 $\mu g/m^3$ for lifetime risk of 4×10^{-2}	0.05 mg/L	0.1 mg/L
4	Manganese (Mn)	5 mg/m^3	0.15 $\mu g/m^3$	0.4 mg/L	5 mg/m^3
5	Cobalt (Co)	0.1 mg/m^3	No Data	No Data	No Data
6	Nickel (Ni)	1 mg/m^3	3.8×10^{-4} $(\mu g/m^3)^{-1}$	0.02 mg/L	No Data
7	Copper (Cu)	0.1 mg/m^3(Cu fumes); 1 mg/m^3 (Cu dust)	No data	2 mg/L; EPA: 1.3 mg/ L	No data
8	Zinc (Zn)	1 mg/m^3 ($ZnCl_2$ fumes); 5 mg/m^3 (ZnO fumes & dust)	No data	No data	No data
9	Molybdenum (Mo)	5 mg/m^3 (Soluble Mo dust); 15 mg/m^3 (Insoluble Mo dust)	No Data; NIOSH: 5000 mg/m^3 (Insoluble Mo) and 1000 mg/m^3 (Soluble Mo)	No Data; EPA: 0.08 mg/L	No data
10	Arsenic (As)	10 $\mu g/m^3$	1.5×10^{-3} $\mu g/m^3$; Cancer risk: 1 $\mu g/m^3$	0.01 mg/L	10 ppb
11	Cadmium (Cd)	5 $\mu g/m^3$	5 ng/m^3	0.003 mg/L	0.005 mg/L
12	Selenium (Se)	0.2 mg/m^3	No Data	0.05 mg/L	0.05 mg/L
13	Mercury (Hg)	0.1 mg/m^3 (Organic Hg); 0.05 mg/m^3 (Metallic Hg)	No Data	0.001 mg/L	No Data
14	Lead (Pb)	0.5 mg/m^3	No Data	0.01 mg/L	No Data

12.3 Properties of heavy metal

Toxicological properties reveal that a covalent bond is formed as a result of the appearance of metalloids. When metalloids can covalently bind with organic groups, there are two major consequences. As a result, they establish the compounds and ions that produce lipophilic and toxic effects when cellular-type macromolecules bind with nonmetallic elements because lipophilic behavior is responsible for metalloid distribution within the biosphere, and the response to toxicity is dependent on the simple ionic action formed by similar elements. Methylated and tributyltin oxide are lyophilic compounds with extremely toxic arsenic forms. Heavy metals enter the human body primarily through four routes: inhalation of air from the atmosphere, drinking and ingestion of contaminated water or food, and skin contact from pharmaceutical, agricultural, industrial, residential, and manufacturing areas [1, 2].

Metal is a non-biodegradable and non-degradable substance. Organisms can detoxify metal ions by hiding those active elements inside proteins and intracellular granules, depositing them in a non-soluble form that is excreted in organisms' feces and stored for a long time. When heavy metals are inhaled or swallowed by humans, they bio-accumulate within the human system. As a result, it is classified as hazardous. This bioaccumulation causes physiological and biological complications. Few heavy metals are also called crucial elements that are required for a particular variety of physiological and biochemical functions as well as importance to life. Still, the presence of a large number of heavy metals is toxic in nature for humans and animals. Some of these are widely used by various sectors, industries, medicine and agriculture. The effect is that these have been spreading into the various environments, such as soil, water, and atmosphere [4, 7–9].

The essential elements are classified into three categories: essential elements for the body, trace elements, and macrominerals. For the formation of a block of some living matter, four major important elements are required. According to the atomic number arrangement, hydrogen, nitrogen, oxygen, and carbon are examples. The remaining seven essential elements are known as macrominerals because they contain integral-type elements that balance the ions of structural compounds, nucleic, and amino acids. Sodium, phosphorous, potassium, sulfur, magnesium, calcium, and chlorine are also included in the atomic number sequence. Finally, the trace element group is made up of 13 elements, based on their atomic numbers: vanadium, magnesium, cobalt, copper, arsenic, molybdenum, silicon, chromium, iron, iodine, nickel, zinc, and selenium. These critical elements play a major role in the structural formation of skeleton, the regulation of acid and base equilibrium, and the maintenance of the colloidal system. These are also required for enzymes, hormones, and structural proteins. For example, iron is required for hemoglobin, and selenium is required for the enzyme, glutathione peroxidase [2, 4, 10]. Nonessential metals are not required in the human body and play no significant role. However, these metals have a toxic effect and affect the levels of critical elements in the body [2].

Toxic metals have been shown to have an effect on cell components and organelles, such as nuclei, enzymes, cell membranes, mitochondria, and lysosomes. It has been discovered that toxic metal ions interact with nuclear and DNA proteins, causing DNA damage and, as a result, cell cycle modulation, carcinogenesis, and apoptosis [4]. It has been discovered that toxic metal ions interact with the DNA and nuclear proteins, causing damage to a specific site. Toxic ions can cause both direct and indirect harm. Conformational type changes occur in biomolecules because of the direct harm caused by heavy metals. On the other hand, relative oxygen formation and radicals of superoxide or hydroxyl, which are comprised of nitrogen species, endogenous oxidants, nitric oxide, and hydrogen peroxide (Valko et al., 2005), cause indirect toxic metal harm. It is caused by lipid peroxidation, sulphydryl homeostasis alteration, and DNA damage, caused by free radical formations and toxic metals. Few amounts of change have been observed in the calcium homeostasis, mediated by metal as a result of membrane damage, which causes a type of calcium-dependent system that is activated by endonuclease participation. Cadmium, nickel, iron, chromium, and copper are the most commonly studied metals for free radical formation. Cadmium, chromium, and nickel have carcinogenic properties (Valko et al., 2005 [60]).

Based on the link between oxidative-type damage and carcinogenesis, free radicals mediated by metals initiate the mutagenicity of the DNA. The formation of free radicals causes various types of base DNA modification; the majority of which are pro-mutagenic. As a result, the essential link between carcinogenicity and oxidative damage, caused by toxic metals, is demonstrated. The metals, arsenic, nickel, and cadmium, inhibit the DNA repair mechanism. In DNA, affected by oxidative stress, there was (i) base modification seen by nickel and chromium; (ii) cross-linkage seen by copper, iron, and nickel oxidants; (iii) strand scission seen by chromium, nickel, and cadmium oxidants; and (iv) depuration seen by nickel, chromium, and copper oxidants (Valko et al., 2005).

Antioxidants, both non-enzymatic and enzymatic, can protect against the attack of free radicals, which is mediated with the help of metals. In general, antioxidants protect iron toxicity by (i) reactions of peroxides and preventing molecular O_2 and also by chelating the ions of ferrous, (ii) retaining the state of redox and iron chelation for making, them incapable of iron for decreasing the molecular type of oxygen, and (iii) trapping the radicals produced in this process. Compounds of thiol are the class of major ones that are classified as highly effective and include glutathione, which traps free radicals, maintains the cell's redox state, and decreases peroxide; thus protecting the cells. Damage occurs in the system in vitro due to the presence of metals, which is prevented by non-enzymatic antioxidants, such as vitamin E, and an animal taking an overdose of cadmium, copper, and iron dies if the dose exceeds the permissible limit. The main factors that determine the toxicity effect induced by the metal with carcinogenicity are various reactive type species and radicals in their free form (Valko et al., 2005).

12.4 Effects and transport of pollutants into the ecosystem

Pollutants can enter an ecosystem in a variety of ways, including the lithosphere, hydrosphere, and atmosphere. Aside from that, there are various natural ways for pollutants to enter the ecosystem, such as rock weathering, anthropogenic activity, and volcanic activity. There is a large amount of wastewater released from various industries, sewage disposal, and the overuse of biocides for vector control. Pollutant and heavy metal movement are influenced by a variety of factors, such as temperature, air mass circulation, wind speed, and water surface direction. Aside from that, other factors, such as coefficient partition, vapor pressure, and molecular stability, influence the movement and distribution of pollutants [2].

12.5 Soil pollution

Wastewater irrigation, pesticides, fertilizers, mines, sewage sludge, coal combustion residues, and effluents cause soil pollution from various industries. Untreated wastewater and sewage deposit a large amount of heavy metal on agricultural lands, which then enter the crop grown on that contaminated field and be consumed by humans. Because the atmosphere contains only oxygen, carbon dioxide, and nitrogen, there is widespread concern about the rising heavy metal pollution. Over the last two-three decades, the increasing industrialization as well as urbanization has led to deteriorating the atmosphere by the emissions of different pollutants. Ozone, sulfur dioxide, carbon dioxide, hydrogen fluoride, and nitrogen oxides are examples of common organic and inorganic pollutants. Various heavy metals are emitted into the atmosphere as a result of human anthropogenic activities. Heavy metals are also emitted in large quantities into the atmosphere during the metal processing of ores (heating smelters), contaminating the air. The particles suspended in the air are referred to as particulate matter, which can be liquid or solid, whereas aerosols also pose a threat to environmental health, and some volatile compounds are released from the chimneys of various industries. Different metals are released from various industries, such as chromium, lead, arsenic, nickel, zinc, cadmium, and vanadium, which are carcinogenic and offer potential for a high toxic exposure to living organisms. Because of their ecotoxicity and human bio accessibility, there is potential for oxidative and cytotoxicity metals being emitted into the atmosphere, resulting in haziness in the atmosphere. Waste incineration, domestic oil combustion, power generation plants, industrial units, vehicle traffic, and various remediations of contaminated sites [11] emit heavy metals into the atmosphere. Among all the sources of heavy metal pollution in the atmosphere, industrial and traffic activities are the most significant. Heavy metal emissions into the atmosphere are from industrial processes, such as fusion, crushing, reduction, refining, and processing. Because heavy metal mobility is high in the atmosphere, pollutants can travel for several kilometers due to the wind.

12.6 Water pollution

Water pollution is caused by two major factors: industrialization and urbanization. Pollutants containing heavy metals are transported by runoff from industry, villages, cities, and towns to waterbodies, where they accumulate. This heavy metal is toxic to humans and other ecosystems even at low concentrations. Because of the weathering of Cr-containing rocks, it may enter the water through mechanical operations, soil filtering, and other means. With a focus on the oceanic environment, chromium could reduce the oxidation, disintegration, and precipitation processes [12]. The critical level of chromium (III) in water is 8 micrograms per liter, and the critical level of chromium (VI) is 1 microgram per liter. The level of chromium in effluents from a specific industry or region ranges from 2 to 5 gram per liter [13].

12.7 Air pollution

Pollutants enter the atmosphere in various forms, such as particles, droplets, or gaseous form, and are sometimes associated with particles and droplets. Because particles and droplets cannot travel a long distance, they may fall to the ground after a short distance, whereas particles in gaseous form can travel a long distance, due to air mass. Heavy metals emitted by industrial smelters are transported several kilometers away from their sources via wet and dry deposition. The presence of heavy metals in the targeted organism can pose numerous risks to human health. The World Health Organization established health risk guidelines for heavy metals in the air WHO, 2004. Heavy metals can accumulate in plant leaves, following the deposition of atmospheric particles on the leaf surfaces via foliar transfer Schreck et al., 2014 [60]. The uptake of metal by the roots of plants has been studied extensively, whereas the uptake of metal by the leaves of plants from the atmosphere has lagged. In the context of pollution research, the foliar transfer of metal is overlooked, which is the major pathway of pollution, primarily when fine ultra-particles interact with leaves of plants. The arrangement of the leaves or canopy of plants efficiently filters heavy metals from the atmosphere. Particulate matter is captured on the foliar parts. The plant canopy adsorbs or reduces the particulate matter in the atmosphere. According to various studies, pine forests can retain up to 36.4 tons of airborne particles per hectare per year. The tree's shelterbelt can retain dust particles 38.9 per cent of the time, and it has been reported that plants, as well as growing vegetables, have a high level of foliar-heavy metal. The heavy metal concentration in the foliar plant organ easily defines the risk in an environment, which represents the pollution of the environment. The heavy metal concentration in crop plant tissue was found to be higher than the permissible limits. Vegetables grown in urban areas show enhanced concentrations of heavy metal – lead and cadmium. Heavy metal uptake happens by the foliar surfaces via cuticular cracks, stomata, lenticels, and through the ectodermal, which is the main channel for subsidiary cells and guard cells, or the epidermal cell wall. The cuticle is present above the guard cell and is more permeable than the epidermal cell. According to various studies, the cuticular waxes retain the majority of particulate matter and trichomes on the surfaces of leaves, but some particulate matter enters the plant's leaf tissue. Metal absorption by the foliar is thought to be a surface phenomenon, with the adaxial cuticular playing an important role. According to various studies, copper and nickel transfer rich particles, which may enter the leaves of plants via the stomata [14]. Consumption of heavy metal-polluted plants poses a risk to human health, which requires immediate attention, and there is limited data on the health risk in the kitchen garden near heavy metal contamination areas Mombo et al., 2016 [61].

12.8 Heavy metal pollution in the atmosphere: Need great attention

Because the atmosphere contains only oxygen, carbon dioxide, and nitrogen, there is widespread concern about the rising heavy metal pollution. Over the last two to three decades, increasing industrialization and urbanization have harmed the environment by releasing various pollutants into the atmosphere. Ozone, sulfur dioxide, carbon dioxide, hydrogen fluoride, and nitrogen oxides are examples of common organic and inorganic pollutants [15]. Various heavy metals are emitted into the atmosphere as a result of human anthropogenic activities. Heavy metals are also emitted in large quantities into the atmosphere during the metal processing of ore-heating smelters, contaminating the air. The particles suspended in the air are referred to as particulate matter, which can be liquid or solid, whereas aerosols also pose a threat to environmental health, and some volatile compounds are released from the chimneys of various industries. Different metals are released from various industries, such as chromium, lead, arsenic, nickel, zinc, cadmium, and vanadium, which are carcinogenic and have a high toxic effect on living organisms. Because of their ecotoxicity and human bioaccessibility, there is potential for oxidative and cytotoxicity metals being emitted to the atmosphere, and, consequently, hazard to the atmosphere. Waste incineration, domestic oil combustion, power generation plants, industrial units, vehicle traffic, and various remediations of contaminated sites [11] emit heavy metals into the atmosphere. Industrial and traffic activities are the most significant sources of heavy metal pollution in the atmosphere. Heavy metal emissions into the atmosphere arise from various industrial processes, such as fusion, crushing, reduction, refining, and processing [16] because pollutants can travel for several kilometers aided by wind.

12.9 Heavy metals

12.9.1 Chromium

Heavy metals are formed by a variety of processes, including weathering of rocks, which occurs naturally, and a variety of anthropogenic activities, as a result of which they enter the environment. Weathering, erosion, and volcanic eruptions are examples of natural processes, whereas anthropogenic activities include smelting, pesticide and phosphatic fertilizer use, mining, electroplating, industrial effluent sludge, and biosolids in agriculture [5]. The main cause of heavy metals concentration in the environment is the increased rate of industrialization and urbanization, as well as improper waste disposal Miransari, 2010 [58]. Various anthropogenic activities, such as excessive herbicide use, fertilization, and sludge, are to be blamed for heavy metal pollution in the atmosphere, but the mining process is the most common source of trace elements. When heavy metal concentrations exceed their permissible limits, they have negative effects on microbial activities, plants, and the environment. Sharma and Dutta 2017 [59] conducted an experiment in which 240 samples of groundwater were collected in eight districts of Punjab's Malwa region, and the results revealed that the concentrations of various heavy metals were higher than their permissible limits, resulting in groundwater contamination. The primary goal was to investigate the distribution of heavy metals in groundwater as well as the effects of heavy metals on human health. They discovered that the water was unfit for drinking and other domestic purposes. Different soil samples were collected from a maize field in China to analyze the heavy metal concentrations. They discovered that the concentration of different heavy metals in the area where wastewater irrigated was higher in the uppermost layer of soil. They stated that the concentration of lead is higher and the concentration of chromium is lower, with the concentration of chromium primarily accumulating in maize roots. According to [6], chromium accumulation in agriculture is caused by weathering and volcanic eruption of chromium-containing rocks, as well as some human activities, such as mining, waste disposal, and excessive fertilizer and pesticide use. The holiest place in India, Haridwar, has been found to be affected by modern industrialization. It was discovered that the groundwater around Haridwar was contaminated with five heavy metals, namely Cr, Co, Fe, Zn, and Ni. According to the findings, 216 samples had a percentage of 99% Zn 99, 94% Co, 99% Fe, 98% Cr 98, and 90% Ni 90, which exceeded the permissible limits. Among the five heavy metals, it has been reported that Cr is the most carcinogenic in nature and has the most severe effect on human health and plants.

12.9.2 Toxicology processes

Chromium metal induces oxidative stress, which involves various physiological processes, such as lipid peroxidation in plants, resulting in severe damage to the cell membrane, which further leads to the degradation of photosynthetic pigments and a decrease in plant growth. According to Singh et al., 2011 [62] the purpose of this study was to investigate the different concentrations of chromium and lead in maize seedlings. The results showed that at 300 micromolar concentrations, the growth, photosynthetic pigments, mineral content, and protein content were reduced, whereas metal accumulation increased the content of lipid peroxidation, MDA, and hydrogen peroxide content under chromium and lead toxicity. The shoot and root parts were affected by a 200 micromolar concentration of chromium, whereas lead was not as toxic as chromium. Heavy metals, such as chromium and cadmium, as well as the biofertilizer mycorrhizal fungus, were tested on various parameters, such as growth, proline content, chlorophyll content, and metal accumulation in the flax plant. Within 24 days of the experiment, the chromium concentration ranged from 250 to 500 ppm. As a result, it was discovered that there was a 45% decrease in photosynthetic growth and activity. The use of mycorrhizal fungi can increase plant biomass in metal-stressed environments. Under Cd and Cr stress, the content of chlorophyll decreased by 27.5 and 45.5%, respectively, while the content of proline increased by 84% and 71%. Report the chelating effect on maize-mustard growth grown in cropping system in alluvial soil, with artificial chromium stress in the pot. In addition to EDTA, citric acid, DTPA, oxalic acid, and humic acid, chromium stress was created at five levels: 0, 5, 10, 20, and 30 ppm. Data, such as plant height, leaf number, chlorophyll content, and dry matter accumulation, was collected at various stages – at 30 and 60 DAS, as well as harvesting. As a result of the increased chromium toxicity in mustard and maize, it was discovered that parameters, such as plant height, leaf number, chlorophyll content, and dry matter accumulation reduced. It was discovered that the use of chelating agents significantly increased the growth parameters. The results showed that humic acid at 1 gm/kg produced better results than DTPA at 10 mmole/kg at all chromium levels. Conducted a pot experiment to assess the effects of Al and Cr on various parameters, such as morphophysiological and biochemical parameters, on plant parts. Data was collected at the pre- and post-silking stages when plants were exposed to Al and Cr, with uncontaminated plants serving as the control. It was demonstrated that both the metal and the photosynthetic reduction reduced maize growth and yield. The toxicity of both metals caused damage to the lipid peroxidation and cell membrane, which activated various antioxidative defenses, such as superoxidase (SOD), peroxidase (POD), and catalase (CAT), which mitigated the glutathione content (GSH). In metal toxicity, the proline content increased, while the protein content decreased. The results also revealed that the toxicity of chromium was greater than that of aluminum when applied individually, but when both metals were applied together, the effect was greater than when applied separately. Overall, the toxic effects of Al + Cr were greater than those of Cr > Al > C (control). It was also discovered that the synergistic effects of both metals were more toxic than their individual effects. Islam et al. [17] investigated the effect of chromium, in combination with salicylic acid, on plant growth, heavy metal uptake, and physiological and biochemical responses under Cr contamination.

12.9.3 Fluoride

The atmosphere contains trace amount of airborne fluoride concentration. Fluoride in the air is caused by dust particles, coal combustion, phosphate fertilizer production in an industrial area, and volcanic activity. Fluoride is present in varying degrees in all-natural waters. Drinking water is typically a greater contributor to fluoride consumption on a regular basis. Fluoride concentrations in seawater are typically around one milligram per liter, whereas fluoride concentrations in rivers and lakes are typically less than 0.5 milligrams per liter. Although excess and deficient fluoride concentrations exist in groundwater, they are determined by the fluoride carried by minerals and rocks found in nature. Fluorite is a natural fluoride source in the soil. The fluoride content of the soil ranges between 200 and 300 ppm. Fluoride is a strong soil component that is retained by the soil and does not easily leach down. Fluoride content was found to increase with soil depth, according to scientists WHO, 2004. Water-soluble fluoride accounts for 5 to 10% of the total fluoride in the soil. When the pH of the soil falls below six, fluoride forms complexes with aluminum and iron. Irrigation with fertilized application intensifies the development of Cl^-, SO_4^{2-}, F^-, and NO_3^- in underground water. Fluoride concentrations in irrigated lands with groundwater were high due to alkalization. Fluoride is normally absorbed by the soil surface, with only a trace amount absorbed by the plant. Fluoride accumulation is a direct risk in the food chain caused by fluoride bioaccumulation in grains of various crops, primarily legumes, cereals, and edible parts of various vegetables. A large amount of fluoride accumulates in various popular plants, such as *Chihorium* sp., *Spinacia* sp., *Brassica* sp., and *Apium graveolens*. The primary cause of accumulation is efficient translocation from the contaminated soil, which is rooted in shoots, and direct absorption from the atmosphere. An experiment was conducted and reported that in rice seedlings that were irrigated with contaminated water, with sodium fluoride concentrations of 20 and 30 mg/dm-3 for 15 days, there was an accumulation of 5000 to 4000 mg/kg of fluoride. Plants with a high fluoride concentration are not suitable for animal feeding, and they serve as an indicator for the accumulation of fluoride in grains. When the rice plant was subjected to metal stress by administering sodium fluoride at a concentration of 5 mg/ dm^{-3}, the accumulation of fluoride translocated to the shoot was the greatest. In Tamil Nadu, rice seedlings and black gram were irrigated with groundwater, resulting in fluoride accumulation that was higher than the WHO safe limit of 1.5 mg/kg^{-1}. When *Camellia* sp. was grown in fluoride-contaminated soil, it was discovered that the leaves contain highly accumulated fluoride, that is. 214 mg/kg^{-1}.

12.9.4 Toxicological processes

Stressed seedlings showed significant fluoride bioaccumulation, cell disruption, and growth inhibition. Calcium compounds improve plant efficiency by increasing seed germination percentage, seedling biomass, shoot and root length, preventing chlorophyll deterioration, electrolytes leakage, H_2O_2, and preventing malondialdehyde levels from decreasing. Calcium priming reduces the stress between total phenolic, and carotenoids and flavonoids that are increased by stress. Overall, the results show that a $Ca(OH)_2$ concentration of 0.3 mM is the most beneficial for all three primings. Compounds of calcium are used as a priming agent for revoking the toxicity of fluoride in rice. Ghassemi et al., 2019 [63] reported that they were able to reduce the toxicity of fluoride at various concentrations [free of toxicity, 800, 400, 200, and 100 mg NaF) per kg in safflower seedlings by inducing biochar at 50, 25, and 0 g per kg in soil. When fluoride toxicity exceeds 200 mg NaF kg^{-1}, safflower growth does not slow. Plant growth is reduced when the fluoride concentration in the soil ranges between 800 and 400 mg NaF per kg. Scarcity in plant growth as a result of increased fluoride content in plant tissues, formation of superoxide radicals, H_2O_2, increase in lipid peroxidation, and reduction in photosynthetic pigment synthesis containing carotenoid, chlorophyll, and anthocyanin. A suitable amount of biochar application reduces fluoride solubility, fluoride content in plant tissues, oxidative stress, increases pH, calcium and potassium uptake, and improves photosynthetic pigment synthesis and plant growth. The results show that a 1 kg soil application of 50 g biochar was the best concentration for mitigating fluoride toxicity in the safflower plant. Manoharan et al. [18] investigated whether the amount of fluoride added in acidic soil with phosphatic fertilizer was harmful to the root growth of the barley crop. The increase in fluoride concentration in soil naturally increases the concentration of aluminum in the soil solution with a pH of 4.2 to 5.8. Plant root growth in acidic soils has been observed to be halted due to the addition of an excessive amount of fluoride. According to calculations, increasing fluoride concentration improves the concentration of Al-F complexes in soil. The data showed that the constant application of fluoride into soils resulted in increased acidification of the soil, which could pose a problem for fluoride in the future. Singh et al. [19] reported on fluoride pollution mitigation strategies in plant water and soil. Fluoride is regarded as the 14th essential nutrient physiologically required for human growth and development. FL-1 becomes toxic to humans when the fluoride concentration in drinking water exceeds 1.5 mg. Primarily, geogenic and anthropogenic processes cause fluoride pollution in the ecosystem. Fluoride is widely distributed in all environmental resources, with 0.11–0.61 mg L^{-1} in the air, 151–401 mg kg^{-1} in soil, and 151–401 mg kg^{-1} in water. Humans and animals are exposed to fluoride when the concentrations of fluoride in water and plants range between 0.21–42.1 mg l^{-1} and 0.78–29.6 mg g^{-1}, respectively. The primary goal of this study was to focus on the biological and physiochemical methods for reducing fluoride's impact on the environment.

12.9.5 Manganese

Manganese is abundant in both the aquatic and terrestrial environments, accounting for 0.1% of the Earth's crust. Manganese is always found as compounds with other elements, such as Cl, C, S, Si, and O, all of which have a solid physical state, some of which are water-soluble or are small particles suspended in the air. Despite the fact that Mn can exist in a variety of oxidation states, ranging from −3 to +7; the majority of the compounds are in +4 and +2 oxidation states. This element has a concentration range of 40–900 mg per kg in soil [20]. The concentration of dissolved oxygen in major water bodies on the Earth ranges between 10 and 10,000 g per liter. They have a tendency to settle at the bottom of water resources. Volcanic activity, weathering of rocks on the Earth's crust, forest fires, ocean spray, and vegetation are all-natural sources of Mn. The common oxidation states of Mn are interconvertible via redox reactions caused by either biotic or abiotic processes. Mn^{2+} compounds are soluble, whereas Mn^{4+} compounds are insoluble oxides. The Mn chemical reactions in an oceanic environment are influenced by O_2 concentration, pH, and redox conditions. Manganese oxides have been discovered to settle in oceanic sediments. However, the low oxygen condition, combined with the high temperature, releases free manganese ions into the overlying water, resulting in a 1000-fold increase in Mn concentration in water. Hypoxic (low O_2 concentration) conditions in the marine ecosystem can be caused by either anthropogenic or natural factors.

Despite its widespread distribution in nature, Mn is only required in trace amounts by living organisms – to regulate enzyme activation, metabolic reactions, connective tissue formation, cellular defense, reproductive hormone regulation, bone mineralization, nervous system activities, blood clotting, glycosaminoglycan formation, protein, carbohydrate, lipid, and amino acid metabolism. Mn-containing enzymes include manganese superoxide dismutase, pyruvate carboxylase, and arginase, while manganese-activated enzymes include decarboxylases, hydrolases, DNA and RNA polymerases, kinases, and transferases.

Mn can be released into the environment through industrial processes, such as welding, mining, alloy production, agrochemical production, Fe-Mn operations, and goods processing. Because the metal is bioavailable, there is a greater chance of uptake by living organisms through water. The gastrointestinal tract is the primary absorption route, followed by the inhalation pathway, and the use of narcotics intravenously is a recently reported source [21]. Occupational exposure occurs frequently through inhalation (during welding and smelting), from which it enters the bloodstream and eventually the brain [22]. Toxic effects can occur as a result of contaminated water and food intake. For example, studies show that water in Bangladesh is contaminated with 2 mg L^{-1}, a value four times higher than WHO limits (400 g L^{-1}) and the mean daily uptake of Mn from western diets was 2.3–8.8 mg [23–25]. Excess Mn in soils near a ferroalloy plant in Italy caused serious health problems in schoolchildren, such as impaired odor identification, hand dexterity, and motor coordination [22].

To absorb manganese, the gut can use a simple diffusion or active transport mechanism. Because of the large metal particle size of Mn, which is swallowed after coughing up by the mucociliary action, the diffusion process through the alveoli is difficult. The presence of calcium, iron, and phosphorus in the diet can reduce Mn intake and uptake [26]. The entry of Mn into the brain can be through 3 routes:

(i) Nasal mucosa → Olfactory neuron → Brain olfactory bulb
(ii) Blood → Capillary Endothelial cells → Blood-Brain barrier cells
(iii) Blood → Choroid plexus → Cerebral spinal fluid

The Mn is accumulated in the mitochondria of cells, such as neurons, astrocytes, and oligodendrocytes. It can inhibit NADH dehydrogenase and F1/F0 ATP synthase, causing ATP synthesis to be disrupted. This results in the release of free radicals and, as a result, oxidative stress. Manganism is a neurological disorder caused by Mn exposure that is similar to Parkinson's disease [27].

Manganese is one of the essential micronutrients required by plants, with concentrations ranging from 30 to 500 mg per kg dry weight of different plant species [28]. Plants become toxic when this level is exceeded. A decrease in growth rate, a decrease in root and shoot length, marginal and interveinal chlorosis, necrotic spots on the leaf, dark inclusions, leaf crinkling, a reduction in photosynthesis, a decrease in chlorophyll content, a decrease in carotenoids, and interference in thylakoid stacking due to Mn accumulation in thylakoids are some of the effects of excess Mn concentration in plants [20].

Excess manganese exhibits an antagonistic reaction with similar ions, interfering with their normal functional roles. The proposed theories about Mn phytotoxicity are based on experimental findings. Leaf black spot and callose are formed as a result of apoplastic-mediated phytotoxicity due to the accumulation of oxidized phenols and manganese oxides [29]. Solar radiation causes Mn toxicity by causing

ROS production in the plant. Because the antioxidative enzymatic system lacks the cofactor, the antagonistic activity of Mn has been found to cause oxidative stress.

Chlorosis is a notable symptom of manganese toxicity, caused by photobleaching or the destruction of chlorophyll in the presence of light. As a result, sunlight causes Mn accumulation in leaves. Plants face competition for elements, such as K, Fe, Ca, and Mg, resulting in nutrient stress in plants grown in excess manganese concentrations [30]. Excess Mn in soil resulted in low shoot biomass in the soybean plant due to low stomatal conductance and carbon dioxide assimilation rate, but increased the activity of antioxidant enzymes and hypertrophying symptoms in the adaxial epidermis of the leaves, forming necrotic regions with purple veins. The Mn toxicity causes cell death in plants by interfering with their metabolic activities and causing the degradation of nucleic acids, lipids, proteins, and carbohydrates. Even the photosynthetic apparatus is degraded, resulting in a decrease in the photosynthesis rate [31]. However, some plants, such as *Phytolacca acinosa*, *Polygonum lapathifolium*, *Polygonum hydropiper*, *Phytolacca americana*, *Polygonum pubescent*, *Celosia argentea*, *Schima superba*, and others, can withstand the toxic effects of Mn in high concentrations [32].

12.9.6 Cobalt

Cobalt, an element with the atomic number 27, is required in trace amounts by living organisms and exists as inorganic compounds, most commonly in the +2 and +3 oxidation states. Root nodule development in a leguminous crop can be influenced by Co. Other roles of cobalt include enzymatic activation, being a component of vitamin B12, being a coenzyme in DNA synthesis, being a coenzyme in fatty acid oxidation, and so on. Diseases, such as anemia, appetite loss, bone fragility, and scaly skin, can occur in ruminant animals fed with Co-deficient fodders/feed additives [33, 34]. However, excessive concentrations can have toxic effects on plants such as low shoot biomass, premature leaf fall, vein discoloration, premature leaf closure, and inhibition of greening. It also causes the production of ROS, such as hydrogen peroxide, hydroxyl radicals, proline, and malondialdehyde, which disrupts the electron transport chain, reduces photosynthetic pigments, and impedes N metabolism within the plant body. Grains and grasses have a lower distribution of Co than leguminous crops, indicating that it is species dependent. Leaf necrosis, interveinal chlorosis, chromosome damage, cellular mitosis, inhibition of hypocotyl elongation, root and seed germination inhibit Fe and Ca uptake [35].

Cobalt occurs naturally as carbonates or in association with other metals, such as Ag, Pb, Ni, Fe, Cu, Mn, and Cu, and it is classified as a chalcophile, lithophile, and siderophile element. Mineral forms of Co, such as cobaltite, erythrite, heterogeneity, skutterudite, and spherocobaltite, are abundant in igneous and sedimentary rocks, but their abundance on the Earth's crust is very low in comparison to other heavy metals. Cobalt concentrations of around 100 mg kg^{-1} are found in ultramafic rocks in greater abundance. Hard metal manufacturing units, cement industries, e-waste processing, diamond polishing, paint and pigment industries, incinerators, motor fuel combustion, mining, coal combustion, extraction of metals from Co-containing ores, fertilizer industry working with Co salts, cosmetic products, pharmaceuticals, and so on are examples of anthropogenic Co-release activities. Co dust is a product that is released and suspended in the air, and certain industrial activities cause leaching into the groundwater and soil [35].

Some plant parts, such as grains and fruits, can deposit Co in their tissues, fulfilling Co's essential role in a living body. However, in mining areas, the plant accumulates Co in the range of 111–245 mg kg^{-1}, which is significantly higher than normal values, which are in the range of 0.05–5 mg kg^{-1} dry weight [36]. To ensure secure and safe food production, proper methods for dealing with the remediation of metal-polluted sites must be used, as they do not biodegrade. The uptake of Co and transport across the cell membrane are governed by high affinity binding sites within cells and transport proteins. Toxic metal ion homeostasis is the function of a wide range of proteins such as (i) CPx-type ATPases, that is a class of exporters or transporters protein bound to the cell membrane and facilitate ion export from cells, (ii) NRAMP family, elaborated as natural resistance – associated macrophage protein is found in a broad spectrum of organisms, from bacteria to human, as an integral membrane-bound protein and evolutionarily conserved that transports metal ions, (iii) CDF protein family, known as cation diffusion facilitator, involved in the efflux of divalent heavy metal ions; found in prokaryotes and eukaryotes, while in plants, they are said to be metal-tolerant proteins, (iv) ZIP protein family (zinc – iron permease), which can transport heavy metals, primarily cadmium, cobalt, and zinc, with a structure having 6 TM domains, C-terminal cation binding domain, and N- terminal signature sequence [37, 38]. In the tonoplast of *Arabidopsis thaliana*, Heavy Metal Associated 3 is located (AtHMA3), which is a protein in the family of P-type ATPase and assists in the transport of metals across the membrane. Passive transport is adopted for the translocation of cobalt within the higher plants after the entry via roots, and IRT1 transporters assist in their translocation [39].

Iron deficiency is common in young leaves of plants subjected to excessive Co stress. Catalase activity, chlorophyll concentration, and shoot biomass were reduced, while leaf P and carbohydrate fractions, as well as acid phosphatase, peroxidase, and ribonuclease activities, were increased. Co affects the rate of transpiration and water potential in cauliflower, as well as the translocation of elements, such as sulfur, copper, and phosphorus [40]. At higher Co concentrations, radical growth is hampered in the early stages of the plant, with a reduced rate of germination and the appearance of chlorosis on young leaves. In some cases, the leaf margin is purple. Other visible signs of Co toxicity include leaves with white and dead edges, marginal scorching, rudimentary leaflets with a hook-like appearance at the plant's top, black patches on tomatoes, and diffused chlorosis. The reactive oxygen species produced by metal stress in the bean plant is effectively mitigated by de novo synthesis of phenols and anti-stress protein. As a result, the bean plant is thought to be a metal-resistant species. Cobalt tolerance mechanisms in plants include gene up-regulation, reduced root-to-shoot Co translocation, activation of antioxidant enzymes, amino acid deposition, phytochelatin synthesis, osmoprotectant accumulation, chelation, and organic acid deposition. Exogenous application of epibrassinolide in *Brassica juncea* L., calcium supplementation in a growing medium, and salicylic acid application can all help to mitigate the effects of excess Co stress in plants. Many plants, including *Alyssum murale*, *Lycopersicon esculentum*, and *Thlaspi caerulescens*, are cobalt hyperaccumulator species. Epipremnum and *Pistia stratiotes* are effective for rhizofiltering Co from an aqueous medium [41].

Because of Co's strong affinity for the sulphydryl group, enzymes with critical functions are harmed. This improves mitochondrial respiration, and Co^{2+} creates an antagonistic and inhibitory pathway to Ca^{2+} in binding with the appropriate sites, as well as metal activated enzymes. The Fenton reaction generates ROS, which causes oxidative stress in macromolecules such as DNA, lipids, and proteins. Co can directly cause DNA damage through the harmful effects of DNA-protein cross-linking and sister chromatid exchange. The carcinogenic impact due to cobalt is attributed to the free radicals released in reactions mediated by cobalt. The presence of Co^{2+} in the Fenton reaction increases DNA cleavage at all bases and causes the peptides to act as pro-oxidants, causing cell damage. The lysosome is where Co^{2+}-induced ROS are produced (Valko et al., 2005) [57].

Cobalt can cause cancer in animals, but according to IARC, it is not harmful enough to cause cancer in humans. Cardiomyopathy is a medical condition caused by Co poisoning in humans. Furthermore, the hypoxic condition caused by the Co ions prevents the kidney from receiving oxygen and producing erythropoietin. Furthermore, Co and its ions are genotoxic and induce apoptosis, necrosis, and inflammatory responses [42]. The biological solubility of the metal influences inhalation of Co dust and the subsequent absorption into tissues. Soluble metal ions enter the bloodstream via the alveolar and bronchial walls. The oral pathway for metal ingestion is also affected by Co solubility, fasting, Fe deficiency, and other factors that make Fe and Co compete for absorptive sites in the intestine. In terms of dermal absorption, broken skin is more susceptible to Co uptake than intact skin. The liver absorbs the most Co, followed by the kidneys and lungs, and it is excreted more through urine than feces and bile.

12.9.7 Nickel

Nickel, a group 10 elements in the periodic table, with atomic number 28 and the 24th most abundant element on the Earth's crust, accounts for 3% of the Earth's composition but ranks fifth in terms of abundance in terms of weight. It is a ferromagnetic element that can be found naturally as sulfides and oxides or combined with As and

Sb. This transition metal is highly resistant to corrosive agents such as water, air, and alkali, but dilute oxidizing acids can dissolve Ni. Organic salts and strong acid salts of Ni have high water solubility, whereas weak acid salts and inorganic salts of Ni have low water solubility. Nickel prefers to exist in +2 oxidation states, though a range of −1 to +4 is possible. A large portion of Ni is inaccessible on Earth due to its confinement to the outer core as Fe – Ni alloy, of which 10% is Ni [43].

Nickel is used in armaments, electrical appliances, utensils, stainless steel, clothing fasteners, inexpensive jewelry, coins, keys, alloys, paper clips, catalysts, metallurgical processes, pigments, electroplating, Ni-Cd batteries, orthodontic materials, electroforming, medical prostheses, and so on. This trace metal is widely distributed in soil, air, water, and biological material due to natural and human sources. Natural sources of Ni include volcanic eruptions, forest fires, vegetation, direct leaching of sediments and rocks into water resources, and weathering of rocks and soils. The wind can transport Ni dust from one location to another. Anthropogenic Ni release into the environment is caused by the combustion of fossil fuels, tobacco smoking, food processing in stainless steel utensils, kitchen kettles, sewage, waste incineration, e-waste processing, and various applications of Ni nanoparticles. Nickel carbonyl, a hazardous chemical form, is found in tobacco smoke. The presence of high levels of Ni in nuts, cocoa, chocolate, broccoli, asparagus, tomato, carrots, spinach, and green beans is a serious concern [44].

In industrialized areas, the mean concentration of Ni in the air is higher. In air, Ni was found to have an average concentration of 0.00001–0.003 g m^{-3} in remote areas, while it is 0.07–0.77 g m^{-3} in urban areas with metallurgical or Ni processing industries. This is critical in occupational exposure to Ni dust and fumes via inhalation and dermal contact. 0.2 per cent of the total workforce is at risk of being exposed to Ni fumes or dust. Ni inhalation pathways include insoluble Ni compounds (dust), nickel solutions (aerosols), and nickel carbonyl (gas). Drinking water samples from areas near the ore mining have been reported to contain as much as 200 g of nickel L^{-1} of water. Metal smelters and domestic effluent release are to blame for trace metal contamination in the aquatic ecosystem. Ni deposits phytoplankton, which is known as a sensitive bioindicator of aquatic pollution. Nickel concentrations in soil range between 3 and 1000 mg kg^{-1}. Water-soluble Ni, Ni adsorbed on inorganic cation surfaces, inorganic crystalline minerals, Ni-organic cation exchange complex, chelated metal complexes, or free-ion in the soil solution are the various forms of Ni in soil. Various foods were tested for Ni levels, ranging from 0.1 to 0.5 mg kg^{-1}. Processing and manufacturing can add up to some amounts, either through catalytic hydrogenation of fats/oils by Ni-containing catalysts such as Raney Nickel or flour milling or leaching of Ni from stainless steel food processing equipment. With increased consumption of oatmeal, soy products, nuts, dried beans, dark chocolate, and peas, human Ni uptake can reach nearly 900 g per person per day.

Ni absorption in the gastrointestinal tract is attributed to the metal ion's lipophilicity. Animals absorb trace amounts of Ni via facilitated diffusion and active transport. However, due to carrier saturation, passive diffusion is used for a large amount of Ni. Various ligands and ions can influence these processes. Nickel travels through the bloodstream due to its interactions with albumin and ultra-filterable ligands. Ni competes with Cu to bind to albumin. Ni enters the liver through hepatocytes' Ca channels. Nickel removal or excretion from the body is via urine, sweat, milk, hair, and skin. Because of its solubility in fat, nickel carbonyl passes through the cell membrane, and their reaction in blocking the Ca channel releases free Ca into intracellular compartments, resulting in a negative impact on cell growth, differentiation, and apoptosis. Nickel activates the transcription factor called ATF- 1, which subsequently lessens the activity of the TSP1 regulator, which also stimulates angiogenesis that leads to the growth of tumor. Also, inflammatory response and apoptosis in the body is due to the Ni-mediated activation of another transcriptional factor, NF- κB. The carcinogenic mechanism of nickel is the combination of free radical production, transcription factor stimulation, and gene expression at a controlled level. MEG3 (Maternally Expressed Gene 3) is downregulated as a result of the Ni-induced methylation of the promoter, which consequently upregulates 2 proteins, HIF and 1α, and results in carcinogenesis [45].

A decrease in biomass yield and growth parameters caused by Ni toxicity has a negative impact on a plant's vegetative growth. The dimensions of the vascular bundle, mesophyll cells, epidermal cells on the leaf, and the ultrastructure of the leaf all have negative effects. Cell wall plasticity is compromised, as is the volume of intercellular spaces. Under Ni stress, certain plants may experience disrupted osmosis and diffusion processes, instability in the cytoplasmic ion balance, disruptive membrane functions, decreased rate of transpiration, low rate of stomatal conductance, low water potential in leaf, and reduced leaf area. Other mineral nutrients with radii similar to Ni^{2+} were found to be competitive in plant uptake and binding sites. This results in visual deficiency symptoms caused by those specific metal ions. The lipid peroxidation caused by ROS generation in plants exposed to Ni toxicity causes genetic instability, ionic leakage, membrane instability, loss of turgor pressure, and disruption of membrane-bound enzymes. This is demonstrated by the high level of malondialdehyde (MDA). Any plant species with an active enzymatic or non-enzymatic antioxidative system can recover from stress-induced injuries. Competitive uptake of Ni causes a deficiency of iron (Fe^{2+}), which is an essential component of the antioxidant enzyme cofactor. As a result, enzyme activity is reduced and membrane permeability is increased. Plant photosynthetic activity is hampered by Ni toxicity, as evidenced by structural deformation of chlorophyll, a hindrance to chlorophyll biosynthesis, decreased activity of Calvin cycle enzymes (Aldolase, RuBisCO, NADP dependent phosphoglyceraldehyde dehydrogenase, Fructose – 1,6-bisphosphate, 3- Phosphoglycerate kinase), electron transport reactions, and the number of grana and thylakoids decreased as the lipid membrane's composition changed. The displacement of certain essential metal ions by Ni^{2+}, such as Mg^{2+}, alters the structure of chlorophyll and challenges the activity of RuBisCO. The low content of cytochrome b559, cyt. b6f, plastocyanin, and ferredoxin cause a decrease in the efficiency of the electron transport chain in the photosynthetic process [46].

Some plants in the Brassicaceae family, as well as Arbuscular mycorrhizal fungi, can accumulate Ni in greater quantities. This is a promising technique for the remediation of Ni-contaminated soil that is less expensive and more environmental friendly than other costly and non-environmental friendly techniques. They can also compartmentalize the metal absorbed into vacuoles, which is facilitated by chelating with organic acids. *Vigna radiata, Berkheya coddii, Sebertia acuminate, Pistia stratiotes, Isatis pinnatiloba, Alyssum murale, Alyssum heldreichii, Melastoma malabathricum*, and others are examples.

12.9.8 Copper

Copper (Cu), with atomic number 29, is an essential micronutrient with a high thermal conductivity as well as electrical conductivity, as well as ductile and malleable properties. The most common oxidation states are +1 and +2, which are pronounced cuprous and cupric. They can be involved in a variety of physiological processes such as photosynthetic electron transport, mitochondrial respiration, symbiotic nitrogen fixation, ethylene sensing, photosynthesis, oxidation-reduction reactions, carbohydrate and protein metabolic pathways, and so on. Cu ranks 25th in abundance among all the elements on the Earth's crust. It is the third most commonly used metal in the world. Copper is a structural component of many proteins involved in the regulatory function. Plastocyanins are electron carriers in Cu-based plastids [47].

Fertilizer, metallurgical, refining, fungicides, paints and pigments, printed circuit board production, chemical manufacturing, municipal waste, sewage, agricultural waste, stormwater runoff, mining, traffic emissions, and so on are the major anthropogenic causes for Cu release into the environment. Copper alloys are widely used in pipelines, wire production, power plants, cooling towers, heat exchangers, marine industries, and electronic industries. Copper resists corrosive agents to a large extent by forming a non-conductive layer on its surface. Copper compounds such as Bordeaux mixture, copper hydroxides, copper oxychlorides, and others are used to protect plants from plant diseases, which increase the release of Cu at toxic levels. Cu's antimicrobial properties allow it to be used in clinical or medical settings. The annual increase in copper consumption necessitates more mining. Air, soil, rock, and water are examples of natural pathways. Copper minerals include covellite, malachite, digenite, azurite, tetrahedrite, chalcopyrite, bornite, and chrysocolla.

Copper-based agro-inputs or agrochemicals, such as fungicides, herbicides, miticides, fertilizers, insecticides, and nematicides, are the most common causes of soil contamination. This could amount to more than 5000 tons per year. When copper-rich animal feeds are used in crop production, they cause copper deposition in the soil. Cu is also abundant in some phosphatic fertilizers. Cu has an average abundance of 60 mg per kg on the Earth's crust. They are found attached to the organic matter in the upper soil layers. The majority of copper in the soil cannot be degraded, posing a toxic effect on microbes and disrupting the nutrient cycle. As the degree of contamination in soil increases, so does NH_3 oxidation; so quantification of microbes involved in this process, such as archaea and bacteria, can determine the extent of contamination. Certain micronutrients, such as Mn, Zn, Fe, and macroelement phosphorus, were found to decrease as Cu increased in soils. Cu can be found in water resources and drinking water as particulate matter or complexes. Copper levels in surface water in the United States range from 0.0005 to 1 mg L^{-1}. Aquatic organisms pose a threat to dissolved copper. Particulates released from mine tailings, processing of Cu-containing materials, soil suspension in the air, and combustion sources are all sources of airborne copper [48].

When exposed to copper toxicity, plant growth is slowed, with a bluish appearance that eventually turns brown/yellow. The binding of copper with clay minerals and organic matter creates an imbalance in nutrient availability, reducing productivity. Cu toxicity symptoms include leaf chlorosis with white or cream spots and lesions, necrotic regions on the leaf's edges and tips, and wilting. Cu phyto availability and medium pH have an inverse relationship. Organic matter content, waterlogged soil, N, P, Zn levels, and root growth are other governing factors. The root has a greater proclivity to accumulate Cu and little translocation to upper parts, resulting in a more toxic effect on roots, such as structural deformation, less proliferated root hairs, and a disrupted root cuticle. Free Cu ions can cause ROS to be produced and the signal transduction pathway for cellular damage to be activated, and this reaction is known as the Fenton reaction or the Haber-Weiss reaction. By the Haber-Weiss reaction, Cu^{2+} is reduced to Cu^{+} in the presence of any reducing agent, such as superoxide, which then generates hydroxyl radical from H_2O_2. Unsaturated fatty acids react with OH- in the cell membrane to form lipid radicals and MDA, which is a cytotoxic indicator of tissue damage. Along with the disruption of protein and pigment composition, inefficiency of PS II, reduced CO_2 diffusion, and a negative effect on gas exchange parameters, such as net photosynthetic rate, light saturation point, rate of dark respiration, and light compensation point, it also impedes electron flow in the electron transport chain by inactivating quinone acceptors [49].

ROS production and activity, which is stimulated by Cu, are lethal to almost all cell organelles, including DNA. Cu^{2+} ions have the ability to inhibit enzymes involved in DNA repair, base modification, structural changes of the cell wall, cleaving the double and single strands of genetic material, cross-linking of DNA and protein, lower the mitotic index, chromosomal aberrations, and cell division. The formation of micronuclei in plants is the endpoint of genotoxicity caused by any metal. Copper homeostasis in plants is maintained by limiting the movement of Cu ions to maintain an optimal level in the plant body, which is accomplished through the combined efforts of transporter proteins and a number of genes. Some non-enzymatic antioxidants, such as phytochelatins, salicylic acid, proline, metallothionines, ascorbic acid, and glutathione, are ready for an effective defense response against Cu-induced damage.

Copper becomes toxic to the living body at a concentration of 20 g / g, and 0.9 mg day-1 is recommended for adults to avoid deficiency symptoms. At toxic levels of Cu in the human body, the kidney and liver produce metallothionein (a low molecular weight protein), which binds with copper and produces a water-soluble complex that is easily excreted. The hCtr1 and hCtr2 are two copper transporters that regulate copper intake and absorption in living organisms. Another Cu transporting protein is MNK (P-type ATPase), which allows the passing of Cu via the basolateral membrane of the intestine, delivers the metal to specific enzymes, and ensures copper efflux from the cell. In the bloodstream, translocation of Cu to the liver is possible after binding with histidine and albumin through the portal circulation. In hepatocytes, copper chaperons and copper form complexes that enable the intracellular distribution of this ion. In the liver, another protein, named WND protein, acts during the high Cu concentration by releasing copper ions at the canalicular membrane, which is ultimately excreted through bile. So, a large part of Cu entering the living body is released through bile and small portions are released via sweat, feces, and urine.

12.9.9 Zinc

Zinc enters the aquatic system via artificial pathways such as municipal waste, industrial waste, waste incineration, urban runoff, mine drainage, soil erosion, galvanized materials, and coal-burning power plants. Leachable sources of zinc to groundwater include the use of old and galvanized metal pipes, mineral fertilizers, and well cribbings with Zn coating. Zinc at higher levels in water imparts a metallic and bitter taste, with a 4 mg L^{-1} threshold limit for the particular taste. Above the 5 mg Zn L^{-1}, water has a milky appearance. Boiling drinking water with Zn content greater than 3 mg L^{-1} produces an astringent taste and a greasy layer on top. Because most agricultural fields have zinc-deficient soils, Zn levels below 2 mg L^{-1} are acceptable for irrigation water, but levels above this level are toxic to plants [50].

Excess Zn has a negative impact on plants, causing symptoms such as stunted shoot growth, chlorosis, leaf tip death, and young leaves curling and rolling. Some of the reported effects on plants include root blunting with impaired cell elongation and division, inhibited root growth, damage to cortical cells of roots, late reproductive development, and sometimes no flowers. Any metal's bioavailability is determined by the soil's adsorptive capacity and the release of metal chelators via root exudation. Zinc toxicity affects iron translocation, membrane disruption, cell organelle disintegration, increased nucleus number, distorted mitochondria, and chromatid condensation [51].

More than 300 enzymes, including carboxypeptidase, carbonic anhydrase, alkaline phosphatase, RNA transcriptase, alcohol dehydrogenase, DNA polymerases, and Cu, Zn-superoxide dismutase, rely on Zn to regulate their functioning, catalytic properties, and structural stability. Zn's primary functions are cell proliferation and DNA and RNA synthesis. A diffusible intracellular zinc carrier, CRIP, binds Zn to the intestine's mucosal layer in a saturable process. This carrier protein competes with metallothionein, preventing absorption and allowing transport through the intestinal lumen. When the concentration of zinc increases, metal binding sites in CRIP become rapidly saturated. Zn exhibits non-specific binding on various proteins and ligands, causing brush border cells in the intestinal lumen to be damaged. Citric acid, cysteine, methionine, prostaglandin E2, reduced glutathione, histidine, copper, cadmium, iron, and calcium are all factors that influence Zn uptake in the intestine. Metal complexes with phytate or dietary fibers reduce zinc availability. Carriers in the bloodstream include albumin, 2- macroglobulin, and amino acids. Zinc is stored in various body parts after ingestion, including the liver, kidney, pancreas, bone, and muscles. A transmembrane protein, ZIP (Zinc-Iron related transporter protein), present on the biological membrane, regulates zinc concentration in eukaryotic cells. Zinc's activities include transcription factor activation, dephosphorylation, second messenger metabolism, protein phosphorylation, and extracellular signal recognition. Zinc deficiency can cause apoptosis because zinc is an essential component of cell growth and proliferation. Furthermore, higher and excess zinc concentrations destabilize zinc homeostasis, resulting in cytotoxicity and apoptosis. When Zn levels are high, cells exhibit necrotic symptoms [26].

12.9.10 Mercury

Mercury is a global pollutant that has been identified as a bio accumulative toxin and ranks third on the list of priority-based hazardous substances (Hg). They can survive in the atmosphere for up to two years (ATSDR, 2015). Mercury exists in three different forms in the environment: (a) elemental mercury, denoted as HgO; (b) inorganic mercury, with oxidation states +1 and +2; and (c) organic mercury. Organic mercury is the most toxic of the three, existing as monomethyl or dimethyl – Hg and methylation of inorganic mercury in the absence of oxygen, and the aquatic system can obtain organic mercury [52]. Mercury is used as a vaccine preservative, in the cosmetic industry, for medicinal uses, as disinfectant against microbes, switches, in metal amalgamation, electrolysis chlorine production, barometer, thermometer, manometer, and industries related to dyes, industries, paints, plastic, and pharmaceuticals. Mercury's natural sources include forest fires, volcanoes, fossil fuels, cinnabar ore, and so on. Both human and natural sources are estimated to emit 5000–8000 metric tons of mercury per year. Gold mining, coal combustion, non-Fe metal manufacturing, and cement manufacturing all contribute 38%, 21%, 15%, and 11% of anthropogenic mercury release, respectively. Effluent dumping from mining, paper manufacturing, hydroelectric power projects, and pulp production causes the alarming rate of mercury release. Other human-caused sources include landfills, municipal waste incineration, medical waste incineration, laboratory usage, dental amalgamation, alkali manufacturing, power generation via coal combustion, agricultural activities, and so on [53].

The global mean concentration of mercury in soil is estimated to be 1.1 mg kg^{-1}. Contamination, on the other hand, adds a several-fold increase to this value; sometimes it is 2456 mg kg^{-1} near industrial sites. The determinant of mercury toxicity is chemical speciation, which is determined by soil chemical properties such as pH, cation exchange capacity, electrical conductivity, and organic matter. The elemental form is gaseous because of its high vapor pressure, which allows it to vaporize easily. The Hg reductase enzyme converts Hg^{2+} to Hg^0 in soil microbes and is found in bacteria with the Mer gene, including *Cryptococcus* sp., *Thiobacillus ferrooxidans*, *Pseudomonas* sp., *Streptomyces* sp., and *Staphylococcus aureus*. The increase in temperature boosts microbial activity in the soil, which in turn boosts Hg^{2+} methylation.

Mercury is primarily absorbed through inhalation, followed by oral ingestion. Incineration of waste containing mercury-related compounds or substances can produce Hg vapors that are inhaled and have a half-life of 60 days. With a half-life of 20 days, these vapors are lipid-soluble and encourage bioaccumulation in body parts. Hg exposure can have a negative impact on the cell, ranging from changes in cell membrane permeability to DNA damage. Because of its affinity for the sulphydryl group, mercury forms bonds with S-containing amino acids such as cysteine and methionine, which disrupts the cell's normal functioning and metabolism. Hg binds to RBC, allowing it to pass easily through the placenta and the blood-brain barrier. The pancreas, lungs, breast, thyroid, adrenals, liver, myocardium, salivary glands, testes, skin, kidney, sweat glands, prostate, enterocytes, and breast milk are all sites of Hg deposition. Elemental mercury is excreted from the body as mercuric Hg. Mercurous chloride, also known as calomel, has a low water solubility and intestinal absorption. However, a small amount is oxidized and absorbed in the body, which can result in acrodynia or pink disease. Inorganic mercury is excreted primarily through urine and feces, whereas organic mercury is excreted primarily through feces and breast milk. Cardiomyopathies, pulmonary fibrosis, bronchitis, and Young's syndrome are all heart-related impairments and disorders. Chronic renal disease, nephrotic syndrome, glomerulonephritis, renal cancer, and acute tubular necrosis are the kidney disorders caused by Hg exposure. Because the nervous system is the primary target of mercury toxicity, a chronic malfunction state results from the blockage of the enzymatic process, P-450, which inhibits energy generation in neurons [54].

Transmembrane proteins in plants can mediate mercury in and out of translocation. In higher eukaryotes, ABC transporters (ATP Binding Cassette) are in charge of the influx and efflux of Hg ions. Mercury, whether present as Hg^{2+} in organic or inorganic form, is primarily toxic. Low productivity, genotoxicity, lipid peroxidation, nutrient imbalance, and impaired photosynthetic process all impede plant growth. Mercury, a genotoxin, can cause genetic mutations by interfering with the proteins that control DNA repair, and they can break the DNA strands. The cell cycle is disrupted by low mitotic and cell proliferation indices. Excessive mercury reduces the total protein content of the plant and also causes the proteins to precipitate. The normal function of some components in the photosynthetic apparatus in the chloroplast is altered in the presence of Hg, such as the protein involved in oxygen evolution, the donor site of PS – II, the – subunit of ATP synthase, substitutive inhibition of Hg ions in the place of other essential metallic ions, inhibition of chlorophyll biosynthesis enzymatic activity, chlorophyll degradation, and the lipid bilayer of the cell membrane is damaged by the hydrogen peroxides or other free radicals by a process called lipid peroxidation, which mostly yields two cytotoxic substances, namely, Thiobarbituric acid relative substances (TBARS) and malondialdehyde. These compounds are the indicators of cells under stress. In *Cucumis sativus*, $HgCl_2$ applied exogenously was observed to have an increase in TBARS by 250%. Consumption of Hg-contaminated food items and its presence in edible parts pose potential health risk among the public. Around 22 to 42% of mercury exposure is through vegetable consumption. A severe mercury poisoning, first reported to be in Japan was named Minamata disease, which is a neurological syndrome suffered by the people who consumed the fish and shellfish with methyl mercuric contamination discharged from a nearby chemical plant [53].

12.9.11 Lead

Lead, the second most toxic metal and defined as a non-disintegrative and noxious heavy metal, accounts for 0.002% of the Earth's crust. Alloys can be formed due to their ductility and malleability. Inorganic lead can be found in old paint, soil, and dust, whereas organic lead, Tetraethyl lead, can be found in leaded gasoline. Both types of lead are toxic, but organic lead complexes have a greater toxic impact on the biological system. Physiological body functions require 29 ng per gram of dietary intake for reproductive functions, cell development, and hematopoiesis to occur. Lead enters the body via contaminated water and food, the canning industry, atmospheric dust, paint, and automobile exhaust, among other routes. The human body can absorb a large portion of the ingested lead, approximately 70%. Natural causes include volcanic activity, weathering, remobilization of water, soil, or sediment from mining areas, and sea spray emissions, while anthropogenic causes include the use of paint, stained glass, batteries, cosmetics, agrochemicals, toys, jewelry, pottery, water from old pipes, smoking, radiators of cars and trucks, door frames, smelting, peeling windows, oil-processing, and plumbing. Lead can be found in soils as a free metal ion, in association with sulfates, carbonates, chlorides, and bicarbonates, or as organic ligands such as humic acid, fulvic acid, and amino acid. Because plants have a high affinity for organic matter, their uptake of Pb is more pronounced [55].

The devastation caused by Pb in the human body is determined by age, duration of exposure, amount of intake, route of intake, and other factors. The Center for Disease Control and Prevention has established the standard level of lead in the blood of adults as 10 g/dL and in children as 5 g/dL [56]. From the total absorbed Pb, only a small fraction (5%) is moved to the upper ground portion of the plant. Lead from the soil solution was found to be bound to polysaccharide or uronic acid in the rhizome of some plant species, including *Festuca rubra*, *Lactuca sativa*, *Brassica juncea*, *Vigna unguiculata*, and *Funaria hygrometrica*, after adsorption on the roots. Passively entered Pb uses the apoplastic water stream for translocation up to the endodermis, but due to the impermeable casparian strips, symplastic movement of Pb is observed after that. Pectins immobilize Pb in the root cell wall, while insoluble salts of Pb precipitate the intercellular spaces. Pb sequestration in the vacuole or accumulation in the cell membrane has also been reported in root cells. A large portion of the Pb is excreted from the endodermis region as part of the plant's detoxification mechanisms. These may or may not be cited as reasons for the low Pb translocation to the upper ground portion.

Pb toxicity causes a decrease in growth, disrupted cell division, shortened roots, impaired chlorophyll structure and chloroplast lamella, reduced seed germination, and seedling development in plants. Some plants' internal detoxification mechanisms include compartmentalization (vacuolar metal sequestration), maintaining selectivity in heavy metal uptake, allowing specific ligands to bind with the metal for complexation, excretion, and so on. Plants have a critical limit of 2 mg Pb kg^{-1}, above which there may be biochemical, physiological, and morphological effects. Excess lead inhibits radical emergence from seeds by increasing protein and carbohydrates via negative effects on the enzyme peroxidase, acid phosphatases, – amylase, acid invertases, – amylase, and polyphenol oxidase. Roots lose their ability to oxidize. The Pb toxicity reduced the turgidity of a cell, decreased the plasticity of the cell wall, decreased the rate of transpiration, disrupted stomatal movements, reduced water use efficiency in plants and relative water content, and caused the development of a waxy layer on a leaf, which reduced the respiration rate and transpiration ratio. It competes with the plants for the uptake of divalent cations as well as monovalent K^+.

It can be inhaled as Pb-laden dust, ingested in soils, or ingested orally through polluted water and food, resulting in adverse health conditions. People who live in close proximity to industrially polluted areas have varying lead levels in their blood during different seasons, such as summer and early fall. Lead primarily binds to hemoglobin and is transported throughout the body in the bloodstream. Large Pb particles are prevented from entering the respiratory tract by mucociliary cells, but they are swallowed after passing through the oropharynx. The distribution of lead in the body is independent of the route of exposure and is primarily confined to the bones. Body conditions, such as lactation, pregnancy, osteoporosis, and menopause, increase bone resorption. This raises the blood lead level once more. Breastfeeding can be a way for the newborn to be exposed to the mother. Pb forms complexes with ligands such as albumin, ALAD, sulphydryls, and others. Human excretion pathways include feces, urine, saliva, breast milk, hair, sweat, nails, and seminal fluids (ATSDR, 2019). Some clinical symptoms of lead poisoning in humans include hearing loss, hallucinations, a blue line in the gums, jaundice, pallor, depression, decreased libido, memory loss, coma, encephalopathy, insomnia, nausea, diarrhea, reproductive impairments, anemia, fatigue, seizers, convulsions, malaise, and decreased bone density.

12.10 Conclusion

As already mentioned, metal toxicity can hamper the biochemistry and physiology of both animals and plants. It is indeed true that anthropogenic activities such as mining, industrialization (smelting, foundries etc.), automobiles exhausts, landfills (metal leaching), etc. have brought in heavy metal pollution. In order to cope with the heavy metal pollution and environment clean-up, various soil and water in-situ and ex-situ remediation strategies should be deployed. Various approaches such as phytoremediation, phycoremediation, mycoremediation, microbe-mediated remediation, biochar-mediated remediation, and nanoremediation shall be crucial in the near future for environment clean-up. However, their use depends on the type of metal pollutant, its spread, bioavailability, and severity (toxicity).

References

[1] Masindi, V., Muedi K.L. Environmental contamination by heavy metals. In: Heavy Metals, InTech, 2018, 2018.
[2] Walker, C.H., Sibly R.M., Hopkin S.P., Peakall, D.B. Principles of Ecotoxicology, 4th edition, CRC Press, 2012.
[3] Harguinteguy, C.A., Cofré, M.N., Fernández-Cirelli, A., Pignata, M.L. The macrophytes Potamogeton pusillus L. and Myriophyllum aquaticum (Vell.) Verdc. as potential bioindicators of a river contaminated by heavy metals. Microchem J. 2016, 124, 228–234.
[4] Tchounwou, P.B., Yedjou, C.G., Patlolla, A.K., Sutton, D.J. Heavy metal toxicity and the environment. Mol Clin Environ Toxicol. 2012, 133–164.

[5] Wuana, R.A., Okieimen, F.E. Heavy metals in contaminated soils: A review of sources, chemistry, risks and best available strategies for remediation. Isrn Ecology. 2011.
[6] Khan, M.U., Shahbaz, N., Waheed, S., Mahmood, A., Shinwari, Z.K., Malik, R.N. Comparative health risk surveillance of heavy metals via dietary foodstuff consumption in different land-use types of Pakistan. Human Ecol Risk Assess: Int J. 2016, 22(1), 168–186.
[7] Lenntech, B.V. Heavy Metals, Available online: https://www.lenntech.com/processes/heavy/heavy-metals/heavy-metals.htm. (Accessed 29 June 2018).
[8] Duffus, J.H. " Heavy metals" a meaningless term? (IUPAC technical report). Pure Appl Chem. 2002, 74(5), 793–807.
[9] Wang, L.K. Heavy Metals in the Environment, CRC Press, 2009.
[10] Villanueva, R., Bustamante, P. Composition in essential and non-essential elements of early stages of cephalopods and dietary effects on the elemental profiles of Octopus vulgaris paralarvae. Aquaculture. 2006, 261(1), 225–240.
[11] Manno, E., Varrica, D., Dongarrà, G. Metal distribution in road dust samples collected in an urban area close to a petrochemical plant at Gela, Sicily. Atmos Environ. 2006, 40(30), 5929–5941.
[12] Kimbrough, D.E., Cohen, Y., Winer, A.M., Creelman, L., Mabuni, C. A critical assessment of chromium in the environment. Crit Rev Environ Sci Technol. 1999, 29(1), 1–46.
[13] Chandra, P., Sinha, S., Rai, U.N. Bioremediation of chromium from water and soil by vascular aquatic plants, 1997.
[14] Kozlov, M.V., Haukioja, E., Bakhtiarov, A.V., Stroganov, D.N., Zimina, S.N. Root versus canopy uptake of heavy metals by birch in an industrially polluted area: Contrasting behaviour of nickel and copper. Environ Pollut. 2000, 107(3), 413–420.
[15] Cruz, A.M.J., Sarmento, S., Almeida, S.M., Silva, A.V., Alves, C., Freitas, M.C., Wolterbeek, H. Association between atmospheric pollutants and hospital admissions in Lisbon. Environ Sci Pollut Res. 2015, 22(7), 5500–5510.
[16] Zhuang, P., McBride, M.B., Xia, H., Li, N., Li, Z. Health risk from heavy metals via consumption of food crops in the vicinity of Dabaoshan mine, South China. Sci Total Environ. 2009, 407(5), 1551–1561.
[17] Islam, F., Yasmeen, T., Arif, M.S., Riaz, M., Shahzad, S.M., Imran, Q., Ali, I. Combined ability of chromium (Cr) tolerant plant growth promoting bacteria (PGPB) and salicylic acid (SA) in attenuation of chromium stress in maize plants. Plant Physiol Biochem. 2016, 108, 456–467.
[18] Manoharan, V., Loganathan, P., Tillman, R.W., Parfitt, R.L. Interactive effects of soil acidity and fluoride on soil solution aluminium chemistry and barley (Hordeum vulgare L.) root growth. Environ Pollut. 2007, 145(3), 778–786.
[19] Singh, G., Kumari, B., Sinam, G., Kumar, N., Mallick, S. Fluoride distribution and contamination in the water, soil and plants continuum and its remedial technologies, an Indian perspective–a review. Environ Pollut. 2018, 239, 95–108.
[20] Millaleo, R., Reyez-Diaz, M., Ivanov, A.G., Mora, M.L., Alberdi, M. Manganese as essential and toxic element for plants: Transport, accumulation and resistance mechanisms. J Soil Sci Plant Nutr. 2010, 10(4), 476–494.
[21] O'Neal., S.L., Zheng, W. Managnese toxicity upon overexposure: A decade in review. Curr Envir Health Rpt. 2015, 2, 315–328.
[22] Lucchini, R.G., Guazzetti, S., Zoni, S., Donna, F., Peter, S., Zacco, A. et al. Tremor, olfactory and motor changes in Italian adolescents exposed to historical ferromanganese emission. NeuroToxicology. 2012, 33, 687–696.
[23] Frisbie, S.H., Ortega, R., Maynard, D.M., Sarkar, B. The concentrations of arsenic and other toxic elements in Bangladesh's drinking water. Environ Health Perspect. 2002, 110(11), 1147–1153.

[24] Khan, K., Wasserman, G.A., Liu, X., Ahmed, E., Parvez, F., Slavkovich, V. et al. Manganese exposure from drinking water and children's academic achievement. Neurotoxicology. 2012, 33, 91–97.
[25] U.S. EPA. Manganese. EPA Integrated Risk Information System 7439- 96-5, Washington, D.C., Environmental Protection Agency, 1995.
[26] Briffa, J., Sinagra, E., Blundell, R. Heavy metal pollution in the environment and their toxicological effects on humans. Heliyon. 2020, 6, e04691.
[27] Harischandra, D.S., Ghaisas, S., Zenitsky, G., Jin, H., Kanthasamy, A., Anantharam, V., Kanthasamy, A.G. Manganese-induced neurotoxicity: New insights into the triad of protein misfolding, mitochondrial impairment, and neuroinflammation. Front Neurosci. 2019, 13, 654. doi: 10.3389/fnins.2019.00654.
[28] Clarkson, D.T. The uptake and translocation of manganese by plant roots. In: R.D. Graham, R.J. Hannam, N.J. Uren, eds. Manganese in soil and plants, Dordrecht, The Netherlands, Kluwer Academic Publishers, 1988, 101–111.
[29] Wissemeier, A.H. and Horst, W.J. Effect of light intensity on manganese toxicity symptoms and callose formation in cowpea *Vigna unguiculata* L. walp. Plant Soil. 1992, 143, 299–309.
[30] Fernando, D.R., Lynch, J.P. Manganese phytotoxicity: New light on an old problem. Ann Bot. 2015, 116, 313–319.
[31] Santos, E.F., Santini, J.M.K., Paixao, A.P., Junior, E.F., Lavres, J., Campos, M., Reis, A.R. Physiological highlights of manganese toxicity symptoms in soybean plants: Mn toxicity responses. Plant Physiol Biochem. 2017, 113, 6–19.
[32] Liu, K., Yu, F., Chen, M., Zhou, Z., Chen, C., Li, M.S., Zhu, J. A newly found manganese hyperaccumulator – *Polygonum lapathifolium* Linn. Int J Phytorem. 2016, 18(4), 348–353. doi: 10.1080/15226514.2015.1109589.
[33] Lee, H.J. Trace elements in soil–plant–animal systems animal production. In: Nicholas, D.J.D., Egan, A.R., eds. Trace elements in soil–plant–animal systems, New York, Academic Press, 1975, 39–54.
[34] Smart, M.E., Gudmundson, J., Christensen, D.A. Trace mineral deficiencies in cattle: A review. Can Vet J. 1981, 22, 372–376.
[35] Mahey, S., Kumar, R., Sharma, M., Kumar, V., Bhardwaj, R. A critical review on toxicity of cobalt and its bioremediation strategies. SN Appl Sci. 2020, 2(1279). https://doi.org/10.1007/s42452-020-3020-9.
[36] Yaman, M. Teucrium (*Teucrium polium* L.) as novel discovered hyperaccumulator for phytoextraction of Ni contaminated soils. Ekol. 2014, 90, 81–88.
[37] Hall, J.L., Williams, L.E. Transition metal transporters in plants. J Exp Bot. 2003, 54(393), 2601–2613.
[38] Kolaj- Robin, O., Russell, D., Haye, K.A., Pembroke, J.T., Soulimane, T. Cation diffusion facilitator family: Structure and function. FEBS Lett. 2015, 589, 1283–1295.
[39] Sarma, H. Metal hyperaccumulation in plants: A review focusing on phytoremediation technology. J Environ Sci Technol. 2011, 2, 118–138.
[40] Nagajyoti, P.C., Lee, K.D., Sreekanth, T.V.M. Heavy metals, occurrence and toxicity for plants: A review. Environ Chem Lett. 2010, 8, 199–216.
[41] Akeel, A., Jahan, A. Role of cobalt in plants: Its stress and alleviation. In: Naeem, M., Ansari, A.A., Gill, S.S., eds. Contaminants in agriculture: Sources, impacts and management, Springer, 2020.
[42] Simonsen, L.O., Harbak, H., Bennekou, P. Cobalt metabolism and toxicology-A brief update. Sci Total Environ. 2012, 432, 210–215.
[43] Cempel, M., Nikel, G. Nickel: A review of its sources and environmental toxicology. Polish J of Environ Stud. 2006, 15(3), 375–382.

[44] Genchi, G., Carocci, A., Lauria, G., Sinicropi, M.S., Catalano, A. Nickel: Human health and environmental toxicology. Int J Environ Res Public Health. 2020, 17, 679.
[45] Engwa, G.A., Ferdinand, P.U., Nweke Nwalo, F., Unachukwu, N.M. Mechanism and health effects of heavy metal toxicity in humans. In: Poisoning in the modern world – new tricks for an old dog, IntechOpen, 2019.
[46] Shahzad, B., Tanveer, M., Rehman, A., Cheema, S,A Fahad, S., Rehman, S., Sharma, A. Nickel; whether toxic or essential for plants and environment – A review. Plant Physiol Biochem. 2018, 132, 641–651.
[47] WHO. Copper in Drinking Water. Background document for development of WHO Guidelines for Drinking-water Quality, 2004.
[48] Rehman, M., Liu, L., Wang, Q., Saleem, M.H., Bashir, S., Ullah, S., Peng, D. Copper environmental toxicology, recent advances, and future outlook: A review. Environ Sci Pollut Res. 2019. https://doi.org/10.1007/s11356-019-05073-6.
[49] Shabbir, Z., Sardar, A., Shabbir, A., Abbas, G., Shamshad, S., Khalid, S. et al. Copper uptake, essentiality, toxicity, detoxification and risk assessment in soil-plant environment. Chemosphere. 2020, https://doi.org/10.1016/j.chemosphere.2020.127436.
[50] Noulas, C., Tziouvalekas, M., Karyotis, T. Zinc in soils, water and food crops. J Trace Elem Med Biol. 2018, 49, 252–260.
[51] Rout, G. and Das, P. Effect of metal toxicity on plant growth and metabolism: I. Zinc. Agronomie. EDP Sci. 2003, 23(1), 3–11.
[52] Kumari, S., Amit., Jamwal, R., Mishra, N., Singh, D.K. Recent developments in environmental mercury bioremediation and its toxicity: A review. Environ Nanotechnol Monit Manage. 2020, 13, 100283.
[53] Natasha Shahid, M., Khalid, S., Bibi, I., Bundschuh, J., Niazi, N.K.N., Dumat, C. A critical review of mercury speciation, bioavailability, toxicity and detoxification in soil-plant environment: Ecotoxicology and health risk assessment. Sci Total Environ. 2019. https://doi.org/10.1016/j.scitotenv.2019.134749.
[54] Bernhoft, R.A. Mercury Toxicity and Treatment: A review of the literature. J Environ Public Health. 2012, Article ID 460508. https://doi.org/10.1155/2012/460508.
[55] Kumar, A., Kumar, A., Cabral- Pinto, M.M.S., Chaturvedi, A.K. et al. Lead toxicity: Health hazards, influence on food chain, and sustainable remediation approaches. Int J Environ Res Public Health. 2020, 17, 2179.
[56] CDC. Advisory Committee on Childhood Lead Poisoning Prevention (ACCLPP), London, UK, Jessica Kingsley Publishers, 2012.
[57] Valko, M.M.H.C.M., Morris, H., Cronin, M.T.D. Metals, toxicity and oxidative stress. Curr Med Chem. 2005, 12(10), 1161–1208.
[58] Miransari, M. Contribution of arbuscular mycorrhizal symbiosis to plant growth under different types of soil stress. Plant Biol. 2010, 12(4), 563–569.
[59] Sharma, R., Dutta, A. A study of heavy metal pollution in groundwater of Malwa Region of Punjab, India: Current status, pollution and its potential health risk. Int J Eng Res Appl. 2017, 7(3), 81–91.
[60] Schreck, E., Dappe, V., Sarret, G., Sobanska, S., Nowak, D., Nowak, J., . . . Dumat, C. Foliar or root exposures to smelter particles: Consequences for lead compartmentalization and speciation in plant leaves. Sci Total Environ. 2014, 476, 667–676.
[61] Mombo, S., Foucault, Y., Deola, F., Gaillard, I., Goix, S., Shahid, M., . . . Dumat, C. Management of human health risk in the context of kitchen gardens polluted by lead and cadmium near a lead recycling company. J Soils Sediments. 2016, 16(4), 1214–1224.
[62] Singh, A., Prasad, S.M. Reduction of heavy metal load in food chain: Technology assessment. Rev Environ Sci Bio/Technol. 2011, 10(3), 199.

[63] Ghassemi-Golezani, K., Farhangi-Abriz, S. Biochar alleviates fluoride toxicity and oxidative stress in safflower (*Carthamus tinctorius* L.) seedlings. Chemosphere. 2019, 223, 406–415.
[64] Shi, Peng, Jun Xiao, Yafeng Wang, Liding Chen. Assessment of ecological and human health risks of heavy metal contamination in agriculture soils disturbed by pipeline construction. Int J Environ Res Public Health. 2014, 11(3), 2504–2520.
[65] Zhou, Q., Yang, N., Li, Y., Ren, B., Ding, X., Bian, H., Yao, X. Total concentrations and sources of heavy metal pollution in global river and lake water bodies from 1972 to 2017. Glob Ecol Conserv. 2020, 22, e00925.

Atif Aziz Chowdhury, Nilendu Basak, Taniya Roy and Ekramul Islam

13 The language of gram-positive and gram-negative bacteria

Abstract: To survive, communication is essential. Bactria is no exception. Communications help them to survive in diverse areas, from hot springs to dental surfaces. Significant results have brought this field into a new worldview to examine the unpredictable occurrence of biofilm networks and far-off cells. Bacteria use different modes to communicate with each other, which includes chemical signaling, sound signaling, and the latest discovered, electrical signaling. The most studied mechanism is quorum sensing, in which, microbes secrete molecules, known as autoinducer or quorum sensing molecule (QSM), utilized for their correspondence with other encompassing microbes (interspecies and intraspecies). This little-emitted diffusible particle is a key regulator of the communication mechanism. Studies have been carried out to understand the complexities of bacterial communication and how the underlying mechanisms could be exploited in the management of bacterial diseases. This field of research is extended to integrate computational modeling, simulation, and to some extent, application in synthetic biology. In this chapter, new signaling components have been briefly portrayed and how they have the potential to ascend to a long-reach correspondence capacity in bacterial networks.

Keywords: Microbial communication, chemical signaling, sound signaling, electrical signaling, quorum sensing, biofilm communication

13.1 Introduction

Being a prokaryotic unicellular organism, bacteria is thought to be the first organism that originated on the Earth. They started living this planet approximately 2 billion years prior to eukaryotes. They have coevolved with the planet and are thought to be responsible for shaping the environment and other organisms on the planet. Both bacteria and higher organisms have evolved sensory mechanisms for sensing environmental factors as well as to communicate with other organisms. Multi-cellular eukaryotes

Acknowledgment: This work was not financially supported.

Atif Aziz Chowdhury, Nilendu Basak, Taniya Roy, Department of Microbiology, University of Kalyani, Kalyani, Nadia, West Bengal 741235, India
Ekramul Islam, Department of Microbiology, University of Kalyani, Kalyani, Nadia, West Bengal 741235, India, e-mail: ekramul.rs@gmail.com

https://doi.org/10.1515/9783110727227-013

have developed sensory organs for environmental response and communication, but bacteria do this job by regulating the intracellular signaling system and secreting extracellular molecules. The communication between them is regulated by cell-to-cell signaling. Bacteria exhibit both intra and interspecies communication for growth and survival.

The mode of communication in microbes differs across cells, depending upon the condition they reside in. They can secrete extracellular molecules, known as quorum sensing molecule (QSM) or autoinducer. These molecules interact with a specific sensory system of another bacterium and start a signaling cascade, known as “quorum sensing” (QS). The quorum sensing mechanism is the most explored communication mechanism among the microbes. However, the signaling molecules involved differ in gram-positive and gram-negative bacteria [1]. These molecules impact the gene expression of the cell and allow it to synthesize the required peptide. It is essential for communication as well as their growth, virulence, biofilm formation, resistance, and environmental adaptation. By using modern molecular techniques, researchers have tried to figure out the underlying mechanism. New antimicrobial approaches are being developed, exploiting the factors involved in bacterial cooperation and thus, bacterial community behavior [2]. Newly discovered mechanisms, such as electrical signaling and sound signaling, are also being researched currently. Electrical signaling can help in communication from a distance. Exposure to noise can also affect the bacterial behavior [1].

The microbial communication provides the bacteria with the ability of synchronized behavior. After reaching a certain signaling molecule concentration, it provides them pathogenicity and resistance to some antibiotics. Thus, by exploring the underlying mechanism, new strategies can be developed to limit microbial infection. Communication gives an incoming external bacterium an idea of the metabolic state of a particular community, thus allowing it to start producing the necessary molecules before jumping in.

This chapter gives an overview of the signaling molecules and the mechanisms used by bacteria to communicate with one another or with their corresponding host, and focuses on the well-studied QS mechanism as well as electrical and sound signaling. Synthetic biological approaches, using communication, have been briefly discussed.

13.2 Signaling molecules

The first idea of a signaling molecule came from the study of Dr. Alexander Tomasz, who identified this molecule as a bacterially-produced mediator that influences their phenotypic behaviors; and they are heat-labile, peptidase-sensitive, and are able to synchronize *Streptococcus* population [3]. Since then, several types of molecules that are essential for microbial communication have been discovered that can also help in shaping the microbial community structure. Depending upon the type and the species of bacteria, these can change widely.

13.2.1 N-Acyl Homoserine Lactones (AHLs)

For their behavior and communication, known as QS, microbes possess a wide range of molecules (diffusible signal molecules). N-acyl homoserine lactones (AHLs) are the prime signaling molecules for QS of gram-negative bacteria. It has been a topic of interest among researchers for the last few decades and to date, extensive studies are being carried out on bacterial intracellular signaling, focusing on AHLs and their role in signaling [4].

Rhizospheric microbes also synthesize AHLs to control gene expression in a population. AHL signaling provides them some beneficial properties, which include symbiosis with legumes for nitrogen fixation. It also triggers plant response (interkingdom signaling) and helps in maintaining balance in plant-microbe interaction. The AHL synthesis process and traits under quorum sensing differ with species in the rhizosphere. The AHL synthesis mechanism is dependent on bacterial genes, environmental features, and transcriptional and/or post-transcriptional mechanism in the bacteria [5].

13.2.2 Peptides

The diffusible chemical messenger, known as pheromones, regulates the expression of genes and activity of a bacterial community (Quorum sensing). It is the main signaling molecule of gram-positive bacteria. Bacteria, such as *Bacillus* and *Streptococcus*, produce pheromones that are small, linear peptides that are detected by the neighboring cells of the RRNPP family cytoplasmic sensory pheromone receptors. The pheromone-binding site works allosterically and controls receptor activity [6].

The gram-positive version of quorum sensing combines a regulatory system made of two components to form a three-component combined system. This is made up of cell membrane-localized histidine kinase (HSK) sensor protein, cytoplasmic response regulator protein (RR), and autoinducing peptide (AIP) [7]. Following the synthesis by cellular ribosomes, they undergo modification to form a stable signaling molecule. Membrane-bound ABC transporters transport autoinducer peptides out of the cell, and after attaining a certain concentration, activate a signaling cascade by phosphorylating the receptor kinase. The quorum sensing mechanism is named depending upon the species, for example, it is agr- quorum sensing for Staphylococci, ComX for Streptococci, and for Bacilli, it is Rap-quorum sensing mechanism [8].

13.2.3 Volatile organic compounds

Volatile organic compounds (VOCs), low molecular weight, and lipophilic compounds are also important groups of compounds that can function in bacterial signaling, synthesized by multiple genotypes or species-specific metabolic pathways [9]. For long distance communication with other microbes and plants, microbial VOCs are a key component [10]. VOCs that are produced by fungi and bacteria include terpenoids, alkene, alkanes, alcohol, ketones, and sulfur compounds, which can also function in antimicrobial activity as well as in the QS system. Moreover, they influence gene expression, hormonal signaling, defense mechanism, stress tolerance, virulence, and biofilm formation [11].

13.2.4 γ -Butyrolactones

As gram-positive bacteria lack a porous outer membrane, like gram-negative bacteria, they do not use the AHL-mediated signaling system. Some of them use γ-butyrolactones as a signaling molecule, which offers some structural remembrance to AHL. The majority of gram-positive cells use peptide signaling. γ-butyrolactones are the first discovered bacterial signaling molecule in the 1960s (from *Streptomyces* sp). From different *Streptomyces* species, a range of γ-butyrolactones were isolated. In spite of being homologs to AHLs, they differ in fatty acid side chain length, the number of branches in them, and also in stereochemistry. AHLs and γ-butyrolactones do not respond to one another's sensory system. Both of them are biologically active at very low concentrations. γ-butyrolactones cause the repressor protein to detach from the promoter sequence of the gene of interest, thus resulting in gene expression.

13.2.5 Other types of signaling

13.2.5.1 Autoinducer-2 and Autoinducer-3

A new class of novel universal signaling, called Autoinducer 2 (AI-2), other than AHL (gram-negative) or peptide (gram-positive), was reported in 1997. AI-2 is a family of connected furanosyl-borate, produced by more than 55 species of gram-positive and gram-negative bacteria. LuxS, a synthase gene, is present in all of these bacteria, whose inactivation causes loss of AI- 2 productions. AI-2 function in signaling was reported in *Salmonella enterica* serovar Typhimurium and *Vibrio harveyi*. Autoinducer 3 (AI-3), chemically different from AI-2, and a potential second universal signal was reported in 2003 [12].

13.2.5.2 Muropeptides

Muropeptide is the product of enzymatic degradation of peptidoglycan, and studies suggest that these can be recognized as signals by bacteria and their host. Bacterially-detected signal of muropeptide signals the presence of antibiotics that can degrade the cell wall in the surrounding. Spore-forming bacteria for vegetative growth utilize this as germinant, whereas eukaryote-sensed muropeptide indicates the presence of bacteria within the host tissue [12].

13.2.5.3 Eavesdropping

Some bacteria don't produce signals as such but are able to sense the signals. This eavesdropping allows the foreign bacteria to understand the metabolic condition of the community and allows them to express the specific trait. *Pseudomonas aeruginosa*, an opportunistic pathogen, primarily colonizes on the lungs cell of cystic fibrosis patients and worsens the formation of mucus [12].

Researches suggest SdiA, the AHL-binding site of bacteria, such as *Escherichia coli* and *Salmonella enterica*. It can attach AHL from other bacteria, and thus they can control gene expression. SdiA gene of *Salmonella* controls its adhesion, invasion of the epithelial cell, and offer resistance to complement by regulating the rck gene. *E. coli*. SdiA regulates acid resistance, motility, biofilm formation, virulence, and AI-2 transport [13].

13.2.5.4 Quorum quenching and quorum sensing inhibition

The phenomenon of degrading AHL signal or interfering with the signaling pathway of other microbes is called "quorum quenching" [14]. Over 20 bacterial genera are reported to degrade AHL signals. The enzymes, AHL lactonase and AHL acylase, break down AHL by opening homoserine lactone ring or by cleaving fatty acid chains from homoserine lactone, respectively.

Nitrogen fixers, such as *Sinorhizobium spp.*, colonize the legume roots, producing six enzymes that are involved in AHL degradation [15]. *Bacillus cereus*, a gram-positive bacterium, reduces AHL signal activity (up to 1000-fold) by producing aiiA-encoded lactonase. When cloned into potato plant, it showed less virulence to soft rot pathogen *Erwinia caratovora*, an AHL-dependent organism [14].

13.3 Signaling system in bacteria

13.3.1 Quorum signaling

Quorum sensing is the chemical-mediated communication system for bacteria. QS is the mode of bacterial cell-to-cell communication, where an extracellular gathering of autoinducer coordinates specific gene expression and senses population density. Bacteria secrete chemical molecules only when they are present at a particular concentration. A variety of bacterial populations use this mechanism to control diverse biological functions, such as virulence factor production, bioluminescence, secondary metabolite production, biofilm formation, motility, etc. [16]. QS-based mode of communication is also associated with PGPR colonization in plant roots. The plant receives a cognate signal and thereby alters gene expression, depending on the bacterial community density [11].

Signaling molecules that mediate between cells are called autoinducers. These autoinducers can be of various types, depending on the bacterial cells and the purpose of the communication. There is a distinct difference between gram-positive and gram-negative bacterial autoinducers.

13.3.1.1 Mechanism

The main mechanism of QS is based on autoinducers (AI). AI is synthesized by the bacterial cells, which are acylated homoserine lactone (AHL) in gram-negative bacteria and peptides in gram-positive bacteria, respectively. Once these AI cross the threshold level, cell receptors recognize these molecules and a signaling cascade begins. This ultimately triggers the essential genes to express and help the bacteria to maintain their social behavior through chemical communication [17].

Agrobacterium tumefaciens uses TraI/TraR regulatory protein system, which helps Ti-plasmid conjugation. CepI/R regulatory protein system is employed by *Burkholderia cepacia* for siderophore and protease activity. *Erwinia stewartii* has the EsaI/EsaR system; thus helping it in exopolysaccharide production and developing virulence property [18].

Nisin is a well-studied signaling molecule that helps *Lactobacillus lactis* to synthesize antibiotics. *Enterococcus faecalis* has GBAP, FsrB, CylL AMP-like peptide pheromone (EntF) as a signaling molecule to synthesize virulence class II bacteriocin [19].

13.3.2 Electrical communication

Quorum sensing, the most extensively studied micro-mechanism, is the best-characterized cell-to-cell signaling mechanism, through which microbes coordinate for their collective behavior. A new approach for cellular communication is based on ion channel-mediated electric signaling. Detailed study has provided insights into molecular foundations of neuronal signaling of ion channels, in spite of having unknown native functions. The transmission of potassium waves in bacterial biofilm through ion channels produces long-range signals. Metabolic stimuli cause the release of intracellular potassium in a positive feedback loop manner, and the released potassium depolarizes the adjacent cells. Biofilm microbial activities are directed by this depolarizing wave. This spread of signaling can be stalled by a genetic alteration in potassium ion channel gating [20].

A microfluidic approach was used to study whether this signaling can propagate to the distant non-biofilm planktonic cells, and it revealed that *Bacillus subtilis*'s biofilm generated potassium ion channel-mediated signals that can attract cells that are not part of the biofilm. As revealed by modeling and experiments, these signals can guide distant bacterial motility by altering their membrane potential. This can be cross species signaling, where *Pseudomonas aeruginosa* cell are guided by *B. subtilis*'s biofilm signaling. Furthermore, the affection relies on the membrane potential mediated sensitivity of the cells to generate biofilm potassium signaling. Thus, biofilm signaling not only communicates within itself but also with distant cells, and can affect the activity of diverse cells through electrical signaling [21].

13.3.3 Sound signaling

Microbes can also effectively communicate over the sound wave. Research suggests that under stress conditions, growth of *Bacillus carboniphilus* is controlled by sound. Sound waves can also be detected by sensitive microphones of *Bacillus subtilis* [18].

In spite of not being able to grow in high salinity and high temperature conditions, the neighboring cell of the same or different species on a different plate stack up, stimulating their growth. The sound wave from external speakers can also stimulate growth under non-permissive conditions. Carbon materials (graphite or activated charcoal) in the medium convert external electromagnetic radiation into sound. A dense lawn of *Bacillus subtilis* produces reproducible sound spectra whereas *B. carboniphilus* and other bacteria or yeast does not produce such sound. The average spectrum with three broad frequency peaks was found to be similar to the frequency that was responsible for *B. carboniphilus* growth promotion. Therefore, it is evident that some bacterial species can produce sounds for communication [22].

13.4 Signaling in gram-positive and gram-negative bacteria

13.4.1 Signaling via quorum sensing in gram-negative bacteria

The main signaling molecule of gram-negative bacteria is AHL. Research suggests that rhizospheric proteobacteria, such as *Burkholderia, Pseudomonas syringae, Pseudomonas putida*, respond to and produce AHLs [11]. The gram-negative bacterial quorum sensing system is compromised with two protein molecules, LuxI-type protein and LuxR-type protein. LuxI protein synthesizes Acylated homoserine lactone (AHL), a type of autoinducer. The concentration of AHL secreted by gram-negative bacteria gets higher with increasing cell density. Once it reaches the threshold level, AHL binds with the LuxR-type protein and forms a LuxR-AHL complex, which upregulates the quorum sensing-regulated genes by binding with the promoters.

13.4.1.1 Quorum sensing in *Vibrio fischeri*

The marine organism, *Vibrio fischeri*, is a symbiont and resides in the light organ of a eukaryotic marine organism. *V. fischeri* produces bioluminescent light, employing a quorum sensing mechanism. In this QS mechanism, two main regulatory proteins, LuxI and LuxR, participate, and LuxI initiates the production of the autoinducer molecule called N-(3-oxohexanoyl)- homoserine lactone. N-(3-oxohexanoyl)- homoserine lactone diffuses through the cell membrane and increases both intra and extracellularly with the growing concentration cells. Once it reaches the required level, this autoinducer binds with LuxR, which promotes LuxR to bind with the luxICDABE promoter region. Thus, the luxICDABE operon gets activated and starts expressing. Upon expressing this operon, bioluminescent light is produced, which is utilized by the Hawaiian squid, *Euprymna scolopes*, for antipredator purposes [23].

13.4.1.2 Quorum sensing in *Pseudomonas aeruginosa*

Pseudomonas aeruginosa is a well-known human pathogen often associated with cystic fibrosis disease. The QS system in *P. aeruginosa* plays a major role in bacterial virulence, pathogenicity, and biofilm formation, which are directly linked with disease occurrence. In *P. aeruginosa*, two signaling cascades work simultaneously, LasI/LasR and Rh1I/Rh1R, homologous to LuxI/LuxR. LasI, and Rh1I, produce the autoinducers n-(3-oxododecanoyl)-homoserine lactone and N-(butryl)-homoserine lactone, respectively [24]. LasR binds with n-(3-oxododecanoyl)-homoserine lactone and the binding complex attaches to the gene's promoter to initiate the expression of genes such as aprA, toxA, lasA, and lasB, which encode for alkaline phosphatase, exotoxinA, proteases, and elastase, respectively. These virulence factors take part in disease progression. LasR-autoinducer complex activates the expression of Rh1R. N-(butryl)-homoserine lactone, produced by Rh1I, diffuses into the cell to bind with Rh1R, and this newly formed complex turns on the expression of rpoS, rh1AB, lecA genes. This complex triggers aprA and also lasB. *Pseudomonas* also have *Pseudomonas* quinolone signal (PQS) using 2-heptyl-3-hydroxy-4-quinolone, which controls lasB gene expression [18].

13.4.2 Signaling in gram-positive bacteria

Signaling peptides or pheromones control gene expression at a population level in gram-positive bacteria. They are also involved in various cellular functions, such as competence, virulence, etc. Bacteria produce a range of extracellular peptides, some of which are involved in communication. Despite the recent development, this complex topic is far from well-studied. Thus, more insight is needed to formulate new tactics to control gram-positive bacteria, which in turn can serve as an alternative to antibiotics [25].

Cyclic Oligopeptides molecules serve as autoinducers in gram-positive bacterial quorum sensing pathways [26]. Oligopeptides are modified from precursor molecules, which are synthesized from peptide signal precursor locus. ABC transporter system is responsible for the intracellular transport of peptide molecules [18]. After attaining a certain concentration, peptides interact with the outer membrane, histidine kinase receptor molecules, of other cells. After a series of phosphorylation events, autoinducer-mediated upregulation of targeted genes starts to initiate.

13.4.2.1 Quorum sensing in *Streptococcus pneumoniae*

The main aim of quorum sensing in *Streptococcus pneumoniae* is to make the cell competent, so that it can take up exogenous DNA molecules. Signaling peptides, called competent stimulating peptides (CSP), create competency in the cell. ComAB produces processed CSP, which is transported through ABC transporters. At high cell density, CSP is detected by ComD, a sensor kinase protein that auto-phosphorylates itself. ComE is phosphorylates. Later, it activates the targeted gene [18].

13.4.2.2 Quorum sensing in *Bacillus subtilis*

Bacillus subtilis exhibits a unique property –sporulation, to overcome the stress environment. Again, quorum sensing plays a major role to achieve it. ComX and CSF (competence and sporulating factor) are two major peptides that participate in this bacterial QS mechanism [18].

13.4.2.3 Quorum sensing in *Bacillus cereus*

Bacillus cereus is a human pathogen causing food-related intestinal diseases, such as diarrhea. QS controls the expression and secretion of hemolysis, phospholipase, toxin, and degradative enzyme-related genes. PlcR/PapR signaling system accomplishes this target [17].

13.4.3 Quorum sensing in biofilm

Biofilm is a complex environment consisting of clusters of bacteria and deep-seated by extracellular polymeric substances (EPS), secreted by the bacterial cells. Biofilm is distinctively different from free-living bacterial cells in many aspects. Free-living bacterial cells, swimming with their flagella, get attached to a substratum and start dividing the cell. When cell density increases and autoinducers reach the required level, quorum sensing guides the bacterial cells to form biofilm via secretion of EPS. EPS helps to retain water in the biofilm, ceasing it from dehydration. Quorum sensing is an integral part of biofilm formation and this biofilm confers protection to the bacterial cells against antibiotics, predators, host immune system, and other environmental stresses [27]. Biofilm formation by quorum sensing has an in-depth role in bacterial pathogenesis and disease occurrence.

13.5 Other communication mechanisms

Communication in bacteria is mainly with extracellular factors that are released from the cell. Along with this, several latest researches have shown a diverse way through which intra and interspecies communication can take place.

13.5.1 Bacterial communication through membrane vesicles

Bacteria that live in an aqueous system produce extracellular signal molecules, and upon reaching a certain concentration, trigger targeted gene expression. For hydrophobic molecules that do not diffuse in the aqueous system and rather excreted in vesicles, research suggests that these molecules, such as long-chain N-acyl homoserine lactone from *Paracoccus denitrificans,* are carried by membrane vesicles

(MVs). Other than blebbing of the cell membrane, in various modes, MVs can be formed, as discovered in *Pseudomonas aeruginosa* and *Bacillus subtilis* [28].

MVs are 40–400 nm in size, produced by most bacteria, and are involved in various roles, such as cellular communication, horizontal gene transfer, and increased cell survivability (by serving as a decoy to threats). This can be found in the natural environment and plays a role in element cycling. It can transport virulence factors and is thus used in vaccine delivery [28].

Removal of MVs from *Pseudomonas aeruginosa* causes seizing of 2-heptyl-3-hydroxy-4-quinolone signal, that is, *Pseudomonas* quinolone signal (PQS), to the population and thereby prevents PQS-controlled actions of the population [29].

13.5.2 Intercellular nanotubes mediate bacterial communication

Dubey et al. proposed nanotubes, which arise from the plasma membranes and cross the cell wall, as suggested by the cryo-EM analysis. It is the essential feature for the bacteria to sense, respond to the environment, and communicate with fellow bacteria. In the absence of cells in close proximity, they produce highly-extended nanotubes to form the nanotube network with the cells that are used for the distribution of nutrients and cellular communication. Further extension of nanotubes can give rise to MVs [30].

Using *Bacillus subtilis* as a model organism, Dubey et al. found that this complex system can exist as intercellular or as extended tubes. Real-time study further suggested that they can form within minutes. Calcineurin, like protein, and YmdB, in nanotubes, are essential for their production and molecular exchange [30].

Acinetobacter baylyi and *Escherichia coli* can form nanotubes, within and between, to cross-feed essential amino acids to these two distant bacteria [31]. Nanotubes can form a bridge among *Bacillus subtilis* and *Staphylococcus aureus* (interspecies) or with distantly-related bacteria, such as *B. subtilis* and *Escherichia coli*. These nanotubes play a key role in transferring nonhereditary resistance [32]. Intracellular nanotubes minimize the loss of factors (due to diffusion) and benefit the cell. Thus, this mode is more appropriate in an aqueous system or in nutrient-limiting condition [33].

13.6 Communication and synthetic biology

Bacterial physiological functions are mediated at a cellular level by many interlinking signal transduction pathways. These characteristics are a combination of multiple "parts" of a system and to understand the properties of an organism as a whole, the idea of the functionality of these parts is important, on which the theme of synthetic biology relies.

Synthetic biology is an emerging research area, where researchers design a biosystem, with known phenotypic behavior. It allows researchers to alter quorum sensing. Alternatively, synthetic cultures can be made by exploiting bacterial quorum sensing as a tool [34]. Synthetic biological application is vast and covers from green chemistry to therapeutical approaches. As bacterial multipurpose QS mechanism works as a regulatory body, "eavesdropping" has a fabulous possibility in synthetic biological use, by regulating the gene expression [35].

In natural conditions, the symbiotic microbial consortia are difficult to study for their complexity. But, they share biomolecules for their communication. QS-mediated designing of synthetic community allows researchers to identify the underlying mechanism and draw hypotheses from it. The metabolic modification allows getting molecular products from a coculture and multi-population community [34].

13.7 Concluding remarks

In spite of being considered unicellular, bacteria are often found in a crowd of organisms of the same species as well as with other bacterial species, forming a community of bacteria. They are also found in association with organisms from other kingdoms. Residing in this condition, they can sense and respond to other organisms and environmental parameters. This sensing or communication is generally mediated by different types of chemical molecules, as well as some physical modes, some of which are known, and some are still unknown. This communication provides a lot of advantages to the microbes, including virulence, biofilm formation, resistance, etc. by the expression of certain genes that are crucial for that function.

Multiple signaling molecules have been revealed that are essential for bacterial chatting. They differ in chemical nature and species or family specific. For example, AHLs are fatty acid derivatives that are used as a messenger in gram-negative bacteria, whereas gram-positive bacteria use peptide signaling. There are also some molecules, namely, AI-2, AI-3, γ-butyrolactone, muropeptides, etc. that are also associated with their communication.

By commencing communication, quorum sensing to be specific, the sensing molecule can influence the expression of certain genes bi-directionally, that is, host produce signals that can influence bacterial gene expression or vice versa. Thus, communication can affect the ecology by stating themselves as the key player. A vivid knowledge of communication can therefore provide us an advantage in multiple areas, including host-pathogen interaction, virulence suppression, waste treatment, and bioremediation on a large scale.

Recent development in the communication of microbes has given rise to two more (mainly) and several other modes of bacterial talking. Electrical signaling, often considered as ion channel-mediated signaling, found in several species and biofilms, can portray the message to a distant cell, thereby reducing the need for proximity. On the other hand, sound signaling in bacteria, produced artificially or by other bacteria, has an astonishing effect on their growth. New modes, such as nanotube- or nanowire-mediated signaling, and membrane vesicle-mediated signaling are coming up as new candidates.

The understanding of communication mechanisms has given rise to a new area of research hotspot called synthetic biology. It includes the manipulation of the bacterial signaling (QS) system. This can positively impact human health, environmental restoration, and sustainable agriculture and also to construct synthetic consortia, study the hidden mechanism, and produce some valued products from the synthetic consortia. Though it is in its early stages, there is potential for promising applications in multiple areas.

In this chapter, we have tried to touch upon different modes associated with bacterial communication. To explore the current status and to present the current understanding, separate sections were created where the current stand has been discussed briefly. Though there are still researches that are being carried out to fill the void, a proper knowledge of communication networks, as a whole, can help manipulate the tool for the greater good. Synthetic biology can be a stepping stone for the future development of humankind. Therefore, we would like to suggest the future researchers to explore the knowledge of sound and electrical signaling, as both are comparatively new, and also to exploit the synthetic biological tool for multiple areas.

References

[1] Majumdar, S., Pal, S. Cross- Species Communication in Bacterial World. J Cell Commun Signal. 2017, 11(2), 187–190. https://doi.org/10.1007/s12079-017-0383-9.
[2] Diggle, S.P. Microbial Communication and Virulence: Lessons from Evolutionary Theory. Microbiology. 2010, 156(12), 3503–3512. https://doi.org/10.1099/mic.0.045179-0.
[3] Tomasz, A. Control of the Competent State in Pneumococcus by a Hormone-Like Cell Product: An Example for a New Type of Regulatory Mechanism in Bacteria. Nature. 1965, 208(5006), 155–159. https://doi.org/10.1038/208155a0.
[4] Wang, J., Quan, C., Wang, X., Zhao, P., Fan, S. Extraction, Purification and Identification of Bacterial Signal Molecules Based on *N*-Acyl Homoserine Lactones: Extraction, Purification and Identification of HSLs. Microb Biotechnol. 2011, 4(4), 479–490. https://doi.org/10.1111/j.1751-7915.2010.00197.x.
[5] Calatrava-Morales, N., McIntosh, M., Soto, M. Regulation Mediated by N-Acyl Homoserine Lactone Quorum Sensing Signals in the Rhizobium-Legume Symbiosis. Genes. 2018, 9(5), 263. doi: https://doi.org/10.3390/genes9050263.
[6] Neiditch, M.B., Capodagli, G.C., Prehna, G., Federle, M.J. Genetic and Structural Analyses of RRNPP Intercellular Peptide Signaling of Gram-Positive Bacteria. Annu Rev Genet. 2017, 51(1), 311–333. https://doi.org/10.1146/annurev-genet-120116-023507.

[7] Lyon, G.J., Novick, R.P. Peptide Signaling in Staphylococcus Aureus and Other Gram-Positive Bacteria. Peptides. 2004, 25(9), 1389–1403. https://doi.org/10.1016/j.peptides.2003.11.026.
[8] Verbeke, F., De Craemer, S., Debunne, N., Janssens, Y., Wynendaele, E., Van de Wiele, C., De Spiegeleer, B. Peptides as Quorum Sensing Molecules: Measurement Techniques and Obtained Levels In Vitro and In Vivo. Front Neurosci. 2017, 11(April). https://doi.org/10.3389/fnins.2017.00183.
[9] Kanchiswamy, C.N., Malnoy, M., Maffei, M.E. Chemical Diversity of Microbial Volatiles and Their Potential for Plant Growth and Productivity. Front Plant Sci. 2015, 6(March). https://doi.org/10.3389/fpls.2015.00151.
[10] Bitas, V., Kim, H.-S., Bennett, J.W., Kang, S. Sniffing on Microbes: Diverse Roles of Microbial Volatile Organic Compounds in Plant Health. Mol Plant-Microbe Interact®. 2013, 26(8), 835–843. https://doi.org/10.1094/MPMI-10-12-0249-CR.
[11] Bukhat, S., Imran, A., Javaid, S., Shahid, M., Majeed, A., Naqqash, T. Communication of Plants with Microbial World: Exploring the Regulatory Networks for PGPR Mediated Defense Signaling. Microbiol Res., 2020, 238(September), 126486. https://doi.org/10.1016/j.micres.2020.126486.
[12] Pierson, L.S., Maier, R.M., Pepper, I.L. Microbial Communication. In: Environmental Microbiology. Elsevier, 2015, 461–481. https://doi.org/10.1016/B978-0-12-394626-3.00020-X.
[13] Smith, J.L., Fratamico, P.M., Yan, X. Eavesdropping by Bacteria: The Role of SdiA in Escherichia Coli and Salmonella Enterica Serovar Typhimurium Quorum Sensing. Foodborne Pathog Dis. 2011, 8(2), 169–178. https://doi.org/10.1089/fpd.2010.0651.
[14] Dong, Y.-H., Wang, L.-H., Xu, J.-L., Zhang, H.-B., Zhang, X.-F., Zhang, L.-H., Wang, L.-H., et al. Quenching Quorum-Sensing-Dependent Bacterial Infection by an N -Acyl Homoserine Lactonase. Nature. 2001, 411(6839), 813–817. https://doi.org/10.1038/35081101.
[15] Krysciak, D., Schmeisser, C., Preuß, S., Riethausen, J., Quitschau, M., Grond, S., Streit, W.R. Involvement of Multiple Loci in Quorum Quenching of Autoinducer I Molecules in the Nitrogen-Fixing Symbiont Rhizobium (Sinorhizobium) Sp. Strain NGR234. Appl Environ Microbiol. 2011, 77(15), 5089–5099. https://doi.org/10.1128/AEM.00112-11.
[16] Bassler, B.L. Small Talk. Cell. 2002, 109(4), 421–424. https://doi.org/10.1016/S0092-8674(02)00749-3.
[17] Bramhachari, P.V. ed. Implication of Quorum Sensing and Biofilm Formation in Medicine, Agriculture and Food Industry, Singapore, Springer Singapore, 2019, https://doi.org/10.1007/978-981-32-9409-7.
[18] Majumdar, S., Roy, S. Microbial Communication: Mathematical Modeling, Synthetic Biology and the Role of Noise, Singapore, Springer, 2020. https://doi.org/10.1007/978-981-15-7417-7.
[19] Gobbetti, M., De Angelis, M., Di Cagno, R., Minervini, F., Limitone, A. Cell–Cell Communication in Food Related Bacteria. Int J Food Microbiol. 2007, 120(1–2), 34–45. https://doi.org/10.1016/j.ijfoodmicro.2007.06.012.
[20] Prindle, A., Liu, J., Asally, M., Ly, S., Garcia-Ojalvo, J., Süel, G.M. Ion Channels Enable Electrical Communication in Bacterial Communities. Nature. 2015, 527(7576), 59–63. doi: https://doi.org/10.1038/nature15709.
[21] Humphries, J., Xiong, L., Liu, J., Prindle, A., Yuan, F., Arjes, H.A., Tsimring, L., Süel, G.M. Species-Independent Attraction to Biofilms through Electrical Signaling. Cell. 2017, 168(1–2), 200–209. e12. https://doi.org/10.1016/j.cell.2016.12.014.
[22] Reguera, G. When Microbial Conversations Get Physical. Trends Microbiol. 2011, 19(3), 105–113. https://doi.org/10.1016/j.tim.2010.12.007.
[23] Hammer, B.K., Bassler, B.L. Quorum Sensing Controls Biofilm Formation in Vibrio Cholerae: Biofilms in V. Cholerae. Mol Microbiol. 2003, 50(1), 101–104. https://doi.org/10.1046/j.1365-2958.2003.03688.x.

[24] Kievit, T.R.D. Quorum Sensing in Pseudomonas Aeruginosa Biofilms. Environ Microbiol. 2009, 11(2), 279–288. https://doi.org/10.1111/j.1462-2920.2008.01792.x.
[25] Monnet, V., Juillard, V., Gardan, R. Peptide Conversations in Gram-Positive Bacteria. Crit Rev Microbiol. 2014, (September), 1–13. https://doi.org/10.3109/1040841X.2014.948804.
[26] Saeki, E.K., Kobayashi, R.K.T., Nakazato, G. Quorum Sensing System: Target to Control the Spread of Bacterial Infections. Microb Pathog., 2020, 142(February), 104068. https://doi.org/10.1016/j.micpath.2020.104068.
[27] Nadell, C.D., Xavier, J.B., Levin, S.A., Foster, K.R. The Evolution of Quorum Sensing in Bacterial Biofilms. Edited by Nancy A Moran. PLoS Biol. 2008, 6(1), e14. https://doi.org/10.1371/journal.pbio.0060014.
[28] Toyofuku, M. Bacterial Communication through Membrane Vesicles. Biosci Biotechnol Biochem. 2019, 83(9), 1599–1605. https://doi.org/10.1080/09168451.2019.1608809.
[29] Mashburn, L.M., Whiteley, M. Membrane Vesicles Traffic Signals and Facilitate Group Activities in a Prokaryote. Nature. 2005, 437(7057), 422–425. doi: https://doi.org/10.1038/nature03925.
[30] Dubey, G.P., Mohan, G.B.M., Dubrovsky, A., Amen, T., Tsipshtein, S., Rouvinski, A., Rosenberg, A., et al. Architecture and Characteristics of Bacterial Nanotubes. Dev Cell. 2016, 36(4), 453–461. https://doi.org/10.1016/j.devcel.2016.01.013.
[31] Pande, S., Shitut, S., Freund, L., Westermann, M., Bertels, F., Colesie, C., Bischofs, I.B., Kost, C. Metabolic Cross-Feeding via Intercellular Nanotubes among Bacteria. Nat Commun. 2015, 6 (1), 6238. https://doi.org/10.1038/ncomms7238.
[32] Dubey, G.P., Ben-Yehuda, S. Intercellular Nanotubes Mediate Bacterial Communication. Cell. 2011, 144(4), 590–600. https://doi.org/10.1016/j.cell.2011.01.015.
[33] D'Souza, G., Kost, C. Experimental Evolution of Metabolic Dependency in Bacteria. Edited by Diarmaid Hughes. PLoS Genet. 2016, 12(11), e1006364. https://doi.org/10.1371/journal.pgen.1006364.
[34] Stephens, K., Bentley, W.E. Synthetic Biology for Manipulating Quorum Sensing in Microbial Consortia. Trends Microbiol. 2020, 28(8), 633–643. https://doi.org/10.1016/j.tim.2020.03.009.
[35] Mitchell, R.J., Lee, S.-K., Kim, T.-S., Ghim, C.-M. Microbial Linguistics: Perspectives and Applications of Microbial Cell-to-Cell Communication. BMB Rep. 2011, 44(1), 1–10. https://doi.org/10.5483/BMBRep.2011.44.1.1.

Index

advanced oxidation processes 113
aerobic granular sludge 59
aerosol 41
air cathode 252
antibiotic resistance
– conjugation 266
– horizontal gene transfer 266
– multidrug resistance development 266
– multiple resistance 266
antioxidants 297
aquatic 41
autoinducer 330
ABC transporters 317
absorbance 122
ACC-deaminase 239
acidophilic chemolithoautotrophs 174
active transport 306
adaptive strategies of microbiota 172
advanced treatment 275, 286
alkaliphilic bacteria 167
Alpine 37
alumina industry 162
alumina manufacturing 162
aluminium oxide 163
anaemia 307
anthropogenic activities 293
antibiotic consumption
– abundance of ARG 270
apoptosis 297
AR control
– advanced treatment 265
– resistance gene 265
– water matrix 265
arbuscular fungal mycorrhiza 227
arbuscular mycorrhizal fungi 179
arcinogenesis 297
ARG monitoring
– abundance 269
– ARG transfer 269
– metagenomics 269
– mobile genetic elements 269
– qPCR 269
– quantitative polymerase chain reaction 269
– receiving environment 269
– wastewater treatment plant downstream 269
atrazine 132

Bayer process 163
bauxite ore 161
bauxite residue 162
biocathode 254
biodegradation 80, 91–92, 94–95, 105, 138–139, 145–146, 148, 154, 156–158
biofilm 80, 82, 84, 334
biofilm reactors 81, 85, 88, 105
biofilms 33
biohydrometallurgy 174
biome 2
biomineralization 80, 100–101, 103–105
biomining 240
bioremediation 138, 145–148, 150–152, 154–155, 157–158, 167
biosorption 80, 96–101, 104–105, 239
bipolar 258
butyrolactone 328
bioaccessibility 300
bioaccumulate 97
bioaccumulation 80, 96, 242, 303
bioaugentation 243
biofilm carriers 81
bioindicators 242
bioinoculants 244
biominerals 100, 104
biopolymer 67
bisphenol A 117

carcinogenic 300
catalase 303
catalyst 113
capacitive 257
cardiomyopathy 309
control of AR
– global action plan 272
– national actions plan 272
– process mechanisms 267
– Regulations 272
coral 46
chemical oxygen demand 116
chlorosis 306
chromatid condensation 315
chromosomal aberrations 314
concentration 113
conduction band 115

https://doi.org/10.1515/9783110727227-014

contaminants 115
critical elements 296
culture dependent approaches 168
culture-independent approach 170
cytotoxicity 298

decoupled 258
desert 39
diffusible chemical messenger 327
DDT 132
degradation percentage 125
desalination 249
detoxification 318
dilution and plating method 170
dioxins 132
disintegration 71
dispersants 185–186, 193–198, 201–202, 205, 207–213
disposal of red mud 165

ecotoxicity 298
ectomycorrhiza 234
eavesdropping 329
ectomycorrhizal 34
estuarine 43
effluents 116
electric signaling 331
electrical conductivity 312
electroplating 301
endophytic bacteria 234
enrichment media 169
environmental pollution 291
Environmental Protection Agency 293
essential elements 296
exchangeable sodium percentage 180
exopolysaccharides 237
expolysaccharides 242
ex-situ remediation 319
external factors 138
extracellular polymeric substances 64, 179

fertilizers 298
fluxomics 20
freshwater 42
flocculation 113
framework 271
free radical 297
functional gene 8

gastrointestinal tract 306
genetic mutations 317
genotoxicity 317
geochip 9
gut microbiota 45
glomerulonephritis 316
glutathione 303
graphitic carbon nitride 127

HCB 132
heavy metal leaching 175
heavy metals 292
horizontal gene transfer
– wastewater treatment plants 270
hot springs 44
hydrogen peroxide 128
Horikoshi medium 173
hydroperoxyl radical 121
hydroxyl radical 120
hyperaccumulators 222

ion-exchange 258
industrial effluent 301
industrialization 294

lignin 118
lindane 132
lipid peroxidation 302
LuxI 332
LuxR 332

matrix 274, 277–278, 284
mercury toxicity 316
metallothionein 315
metal-microbe interactions 237
metals values 162
microbe mediated leaching 174
microbial 249
microbial activity 165
microbial communities 164
microbiome 223
microbiota 167
microorganisms 138, 145, 147–149, 151–152, 154, 156–158
mineralization 305
mining 319
Mirex 132
mangrove 37

marine 42
mechanism of spread
– Bio-solids 267
– Horizontal gene transfer 266
– integron 266
– mobile genetic elements 266
– number of treatment steps 270
– phenotypic diversity 267
– plasmid 266
– proliferation of ARG 265
– transmission medium 267
– wastewater reuse 267
membrane vesicle 334
metabolomics 17–20
metagenome 2
metaproteome 2
metatranscriptomic 2
muropeptide 329
molecular 269–270, 278
multi-resistant bacteria
– wastewater treatment plant 270
mutagenicity 297
mycoremediation 319

N-acyl homoserine lactone (AHL) 327
nanoremediation 319
necrosis 307
nephrotic syndrome 316
Ni-hyperaccumulators 232
nanotube 335
next-generation sequencing 9

oil 185–195, 197–198, 204–205, 208–213
oligotrophic bacteria 170
omics 2
one health plan
– emergence of AMR 271
organic pollutants 137, 139–140, 144–145, 148–149, 153, 155, 159
oxidant 127
oxidation 114
oxone 128

Pb toxicity 318
peroxidase 303
persistent organic pollutants 141, 144, 158
pesticides 298
phosphatic fertilizers 294
phycoremediation 319
phyllosphere 40, 235
phytochelatins 314
phytoextraction 223, 238, 241
phytomining 228, 241
phytoremediation 319
phytotechnologies 242
phytotoxicity 306
plant growth promoting bacteria 239
pollutants 291
polyhydroxyalkanoates 237
polymetallic raw material 175
permafrost 45
phylochip array 7–8
process mechanism 274, 276–277, 285
process mechanisms 274–275
proteobacteria 230
public health
– antibiotic resistance 271

quorum quenching 329
quorum sensing 330

radioactivity of the red mud 176
rare earth elements 175
reaction kinetics 115
red mud 162
reduction 114
rehabilitation of bauxite residue 176
renal cancer 316
rhizobacteria 229, 239
rhizobiome 229
red mud spillage 166
resistance genes
– conjugative plasmids 267
– integrons 267
– transposons 267
rhizosphere 34

semiconductor 115
serpentine soil 222
serpentinomorphoses 223
serpentinophytes 222
siderophores 237
simple diffusion 306
sintering process 163
sludge 113
slurry 123
spills 185–189, 192–195, 197, 208–213
saprotrophic 34

sequencing batch reactor 63
shotgun metagenomic 6
signaling molecule 326
signaling peptide 333
single-cell genomics 11–12
site factors 146, 155
sound wave 331
stacked 254
synthetic biology 336
storage of the bauxite residue 165
superoxidase 303
superoxide radical ion 115

thermal conductivity 312
total organic carbon 116
Tropical 35
temperate 36
treatment
– water matrix 274

ultramafic soils 237
urbanization 294
up-flow 257

valence band 115
visible light 123
volatile organic compound (VOC) 328

wastewater 113, 249, 298
water splitting reaction 115
weathering 293
WHO 271, 276
World Health Organization 269, 271
wastewater treatment plant
– abundance of ARB 270

Zn-hyperaccumulator 236

Bei Fragen zur Produktsicherheit wenden Sie sich bitte an:
If you have any questions regarding product safety,
please contact:

Walter de Gruyter GmbH
Genthiner Straße 13
10785 Berlin
productsafety@degruyterbrill.com